# On Genetic Interests

# On Genetic Interests

## Family, Ethnicity, and Humanity in an Age of Mass Migration

Frank Salter

With a new introduction by the author

Transaction Publishers
New Brunswick (U.S.A.) and London (U.K.)

This book is printed on acid-free paper that meets the American National Standard for Permanence of Paper for Printed Library Materials.

Library of Congress Catalog Number: 2006044532
ISBN: 1-4128-0596-1
Printed in the United States of America

Library of Congress Cataloging-in-Publication Data

Salter, Frank K.
On genetic interests : family, ethnicity, and humanity in an age of mass migration / Frank Salter ; with a new introduction by the author.
p. cm.
Originally published: Frankfurt am Main ; New York : Peter Lang, 2003.
Includes bibliographical references and index.
ISBN 1-4128-0596-1 (pbk. : alk. paper)
1. Human genetics. 2. Sociobiology. 3. Ethnicity. I. Title.

QH431.S274 2006
599.93'5—dc22 2006044532

This book is dedicated to four scientists whose contributions to the understanding of genetic interests inspired this book—the late William Hamilton (1936–2000), Irenäus Eibl-Eibesfeldt, Pierre van den Berghe, and Edward O. Wilson.

# Contents

# List of Tables

# List of Figures

# List of Maps

# Introduction to the Transaction Edition

This new printing of *On Genetic Interests* comes three years after the first. The reviews and comments received in that period have made some useful points but have not challenged the core argument.

One criticism is that the book commits the naturalistic fallacy, that it derives values from facts alone. The naturalistic fallacy is an error in logic made when propositions about values are purportedly deduced from propositions about facts alone. A common form of the error is to go from a description of nature, such as 'humans evolved to survive and reproduce', to an assertion of value, such as 'humans ought to survive and reproduce.' I was well aware of the fallacy when writing the book, and took care to avoid it (see pp. 83, 285-6). However, preconceptions sometimes infiltrate arguments. Perhaps my personal values snuck into the conclusions disguised as deduction. The question is whether the book overall, or at any strategic point, makes an argument that generates values from facts alone. I do not believe it does.

*On Genetic Interests* is an attempt to answer the empirical question: How would an individual behave in order to be adaptive in the modern world? I adopt the neo-Darwinian meaning of adaptive, which is to maximize the survival chances of one's genes. I begin by describing humans as an evolved species and thus as creatures for whom genetic continuity consists of personal reproduction or reproduction of kin. Chapter 2 describes concentrations of kinship from self to family to ethny and finally to the species as a whole. I discuss some ways in which genetic interests inform human values (Chapters 4 and 6) and conclude that, for many, neo-Darwinian adaptiveness is indirectly valued through personal striving and nepotism directed to kin and fellow ethnics. I ask whether growing knowledge about genetic interests might influence these values, and consider arguments to the contrary (Chapter 4). Because genetic continuity is greatly affected by the fate of one's ethnic group, the analysis necessarily goes beyond the reproductive success of individuals and their close kin to encompass ethnic behaviour (Chapters 2, 3, and 5). In Chapter 7 I argue that the nation state is, in effect, an ethnic strategy for defending extended genetic interests, suggesting that 'universal nationalism' is the best way to globally optimize adaptiveness. Only in Chapter 9 do I switch to normative issues. The various sections discuss in some detail whether it is moral to advance one's genetic interests at the cost of other interests, whether genetic or not. I tentatively suggest answers by developing an ethic of 'adaptive utilitarianism.'

Evolutionary theory does play an important role in *On Genetic Interests*, in the same way that economic theory is used to offer prognoses and recommend policy solutions. In the case of economics, the utilities being pursued are not generated by theory (at least they need not be), but from observation of the common aspiration for resources. If people did not care for resources, an economic theory would have no policy implications, even if its internal logic were impeccable. Similarly, it can be rational to use evolutionary theory to understand the distribution and course of reproductive interests. Should those interests be widely valued, then evolutionary theory will be of great practical application. It is at least logically possible that millions of years of evolution have produced an organism that does not care about genetic survival, directly or indirectly. But that seems unlikely.

To quickly summarize, the role of evolutionary theory in *On Genetic Interests* is shown in the following syllogism:

1. *Evolutionary Theory premise:* Neo-Darwinian theory explains how nepotism evolved as the means for spreading the genes coding for that trait. In their evolutionary environments, species that evolved kin altruism obeyed Hamilton's Rule for adaptive altruism, directing helping behaviour preferentially towards genetic kin, who share substantial proportions of each other's genes (Chapters 2, 3, 5).
2. *Motivational premise:* Assume a fitness preserver who cares about his or her genetic interests. This caring might take the practical form of being nurturent towards genetic kin (in family or ethny). With growing knowledge it might even become partly abstracted as a desire for adaptiveness in the neo-Darwinian sense of genetic continuity (Chapter 4, pp. 77-85).
3. *Strategy conclusion:* Then the individual from premise 2 should pursue certain strategies, such as those described in premise 1 and others, depending on circumstances (Chapters 6, 7, 8).
4. *Moral premises:* Morality is one set of human values that constrains behaviour. Ethical issues arise when interests conflict, as they inevitably do in fitness competition. Universal features of morality include duty to aid kin as well as reciprocity (Chapter 9).
5. *Ethical conclusion:* Moral actors will defend their own genetic continuity, not maximize fitness at other's expense, according to the principle of 'adaptive utilitarianism' (Chapter 9). One practical implication is universal nationalism (Chapter 7). An evolutionary ethic for a crowded world is: Go forth and perpetuate (Chapters 6, p. 150).

It is an empirical matter whether these premises are true. But the argument does not attempt to deduce values or policy from evolutionary theory alone. Moral sentiment enters in premise 4, but as an assertion, not as a deduction from theory. Readers are free to disagree with reciprocity or any other asserted value. For some, the motivational second premise will elicit moral feelings, unsurprisingly so since many will feel that it is right for their families or ethnic lineages to survive. Those moral feelings, while giving the argument greater practical force, are not necessary for its validity. Furthermore, the argument remains valid whether or not the 'care' in premise 2 consists of moral or non-moral sentiment, or both. If moral, then logically the 'should' in clause 3 will also be moral. Scientific knowledge does play a role by informing about the genetic significance of kinship. Instead of logically entailing any value such as nepotism, that knowledge can only release or channel it—a function long performed by myth and kin metaphors such as that of shared blood. For example a man might develop nepotistic feelings towards a stranger after discovering through a genetic test that that stranger is his daughter. Those feelings are not carried by the information about their shared genes. Instead, they are released and directed by it.

The argument remains valid for those who are not fitness conservers—such as those who are not interested in sex or nurturing kin or ethny. One need only substitute for premise 2: 'Assume a fitness neutralist...' The argument is even valid for those intent on suicide or genetic extinction. One need only substitute for premise 2: 'Assume a fitness minimizer...' Of course the conclusions would be rather different. Hopefully such an analysis would not be taken to have policy implications except as a negative model. *On Genetic Interests* seeks to develop 'social and political theory about what individuals should do *if* they want to behave adaptively' (p. 325; italics altered). Criticisms that ignore the 'if' in that statement miss the point. Contemplating the conscious pursuit of adaptiveness is a reasonable part of the project of integrating the social and biological sciences.

I now turn to two technical points. One commentator has identified a terminological error that needs to be corrected (David B. 2005c, point 1). Throughout *On Genetic Interests* I define an individual's genetic interests as consisting of his or her distinctive genes, some of which are also found in kin and fellow ethnics. Such genes exist, but kinship generally consists of shared frequencies of genes, even when the genes involved are not unique to any one individual. 'Distinctive genes' should be read as 'distinctive gene frequencies' or 'distinctive allele frequencies.' The same commentator thinks that making such a change weakens the emotive force of genetic interests (evidence that even some critics are impressed by it). 'Why anyone should consider this a "vital", "fundamental", or "ultimate" interest is beyond me. Why so much fuss about shifting a gene frequency from, say, 50 percent to 60 percent?' But kinship found within ethnies is homologous with that found within families. Are we to believe that parents do not have a vital, fundamental, or ultimate interest in their children? The point is that Hamilton's theory of inclusive fitness still applies whether genetic interests consist of unique genes or gene frequencies. If any branch of genetics is likely to have emotive force it is surely that which explains the evolutionary and social significance of kinship, whether in family or tribe.

The same author makes many other criticisms (2005a; 2005b), including some thoughtful philosophical comments, but none are as decisive as that concerning distinctive genes. An example is his criticism of the argument I make in Chapter 8 (section 8.1) that exogamy, marrying outside the ethny, can reduce fitness (David B. 2005a). This is an interesting argument, but one made previously by Grafen, which I discuss on pp. 262-5.

The second technical matter concerns the recent measurement of the variability of kinship among siblings (Visscher et al. in press). While parent-child kinship is 0.25, sibling kinship is 0.25 ± 0.02, or a coefficient of variation of 8 percent. (Parent-child and sib-sib pairs sharing half their genes have a relatedness of 0.5; the same relationship expressed as kinship is 0.25, for reasons explained on p. 45). The magnitude of this variation is approximately equivalent to the kinship between a child and its great great grandparent. Ethnic kinship is likely to be more variable due to assortative mating, chance, and population mixing. This will not reduce the fitness of contributing to collective goods such as group defence, because the large numbers involved will reduce the likelihood of benefiting mainly non-kin. But such variability could affect the fitness of ethnic nepotism directed towards one or a few random fellow ethnics. This might explain Rushton's (1989) finding that in a sample of Englishmen, friends showed elevated genetic similarity. If genetic kinship within an ethny is highly variable, and is detectable by phenotypic similarity, Rushton might be correct in his view that assortative friendship and marriage is a fitness strategy even within ethnically homogeneous populations. This would be contrary to my original position stated on p. 260.

While I have done my best to write a book that is free of error, constructing a political theory from basic evolutionary premises is probably too large a task to be achieved flawlessly in one fell swoop. Nevertheless, the effort will have been worthwhile if the book contributes to the integration of behavioural biology and the policy disciplines.

*Frank Salter*

*References*

David B. (2005a). Interracial marriage: Salter's fallacy. http://www.gnxp.com/MT2/archives/003447.html?entry=3447. Posted 15 January.

David B. (2005b). Ethnic genetic interests. http://www.gnxp.com/MT2/archives/003501.html. Posted 23 January.

David B. (2005c). Ethnic genetic interests—part 2. http://www.gnxp.com/MT2/archives/003504.html. Posted 24 January.

Rushton, J. P. (1989). Genetic similarity in male friends. Ethology and Sociobiology 10: 361-73.

Visscher, P., S. E. Medland, M.A.R. Ferreira, K. I. Morley, G. Zhu, B. K. Cornes, G. W. Montgomery, and N. G. Martin (in press 2006). Assumption-free estimation of heritability from genome-wide identity-by-descent sharing between full sibs. Public Library of Science.

# Preface

This book deals with important and sensitive subjects on which passionately-held convictions are the norm. Readers should therefore take heed of the caveats in the Afterword. To summarize, the analysis deals with aspects of genetics, yet I am not a geneticist. I have done my best to verify formulas and interpretations with geneticists, but the technical parts of the book would, no doubt, have been better written by someone trained in that discipline.

Since 1997, when the idea first occurred to me that ethnic groups might constitute large aggregations of kinship, I have been asking experts for advice on how to quantify the concept. The going was not easy, geneticists being busy professionals with little time to consider ideas expressed in unfamiliar terminology or educate novices outside the lecture theatre. As late as my September 2000 presentation at the American Political Science Association in Washington DC, I was despairing of ever being able to translate measures of variance into kinship coefficients and was relying instead on Iceland's ancient marriage records. This genealogical information indicates the tip of the iceberg of ethnic kinship, which, in its many-fold concentric layers and migratory overlaps goes back tens of millennia to humanity's emergence from Africa. This tip indicated a kinship coefficient one two thousandth of that computed from the global assay data in Chapter 3. Yet I could only mine that rich vein of information with expert advice. The mathematics of Henry Harpending and others forced me to revise sections dealing with distribution of genetic interests and strategies for defending them. All advisers would agree, I am sure, that our confidence in these formulas and interpretations will rise if they weather the broader peer review that this volume will hopefully attract.

# Acknowledgements

I am grateful for help received from many quarters. In the earlier stages Luigi Luca Cavalli-Sforza offered advice on converting his genetic distance measures to kinship coefficients. Steve Sailer provided stimulating ideas and leads on Icelandic genetic data. I first presented the concept of ethnic genetic interests at the September 2000 meeting of the American Political Science Association in Washington, DC.[1] The ensuing discussion brought useful remarks from Kevin MacDonald, Jim Schubert, and the symposium's chairwoman Laurette Liesen. The manuscript at that stage was read by several scholars, and I thank them for their remarks: Hiram Caton, Johan van der Dennen, Irenäus Eibl-Eibesfeldt, and Stephen Sanderson. Subsequently, Henry Harpending and Vince Sarich coached this novice in genetics, and I thank Henry especially for patiently explaining the logic of his method for converting $F_{ST}$ measures into kinship coefficients. Rick Michod and Vince Sarich reviewed Henry's method, and Alan Rogers offered an alternate derivation. Towards the end of writing, new chapters were read and suggestions made by Irenäus Eibl-Eibesfeldt, Antony Flew, and Wulf Schiefenhövel, and Hiram Caton took the trouble to subject the near-final version to close criticism, testing the thesis at several points and prompting amendments, though not as many as he would have liked. Alexandra von Mutius provided research assistance and made several useful criticisms. I received further useful comments following a second presentation, this time at the August 2002 meeting of the International Society for Human Ethology, and wish to thank Nick Thompson for alerting me to criticisms from a group selectionist perspective. Of course I remain responsible for any errors, whether of fact or reasoning.

Luke Burdon helped design the cover. The children's photographs were provided by Irenäus Eibl-Eibesfeldt, Mathias Michel, and Wulf Schiefenhövel. Permission to republish Appendix 1 from *Population and Environment* (Vol. 24, No. 2, Nov. 2002) was provided by the author Henry Harpending and by Kluwer Publishers. Map 3.1 was provided by Rodger Doyle.

*Note*

1 Panel on Biobehavioral Approaches to the Study of Politics, Annual Meeting of the American Political Science Association, Washington DC., 31 August to 3 September 2000.

Methinks I see in my mind a noble and puissant nation rousing herself like a strong man after sleep, and shaking her invincible locks. Methinks I see her as an eagle mewing her mighty youth, and kindling her undazzled eyes at the full mid-day beam.

John Milton, *Areopagitica,* 1644

# Part I: Concepts

# 1. Introduction: Genetic Continuity as the Ultimate Interest

*Summary*
Since life is evolved, reproduction is the ultimate interest, one of overriding importance. Organisms are evolved to reproduce at the cost of all other interests, even personal survival. Reproduction is achieved by passing on genes down the generations, both in one's children, collateral kin, and fellow ethnics. Reproductive interests are therefore genetic interests. An individual's genetic interest is the sum of all his or her distinctive genes in the species. A distinction can be made between the ultimate interest of reproduction and proximate interests such as life, liberty and resources. The latter are subsidiary goals we are evolved to value because they serve, or once served, our genetic interests. Despite their overriding importance, genetic interests have not been explicitly incorporated into political theory.

The sight of a bald eagle reintroduced to the wild, soaring over the Grand Canyon and tending its hatchlings, is reward enough for conservationist efforts. But on deeper reflection, what is it about staving-off extinction that is worth celebrating? Surely not the mere sight of an object that resembles a beautiful living creature. Replenishing the Grand Canyon with robotic eagles would somehow be less satisfying than saving the real thing, even if the outer resemblance were perfect, even if the robots hunted and laid eggs. So what is it about the bald eagle that we treasure? One might work through a checklist of characteristics including appearance, behaviour, contribution to the ecosystem, and biodiversity, all of which we value. One might even appreciate authentic eagles because of the possibility that some of their genes will someday prove of commercial or medicinal value to humans. Still something is missing. Real eagles are descended genetically from other real eagles.

The same applies to ersatz humans. If robots could be made that imitated our children perfectly in outward appearance and behaviour I doubt that many parents would be willing to make the substitution, even if their real children were to go to loving foster homes and robot children were cute and healthy, were toilet trained and winners at school. I have never come across a parent willing to substitute a child even for another human child. Why? The objection is emotional, based on the bond that has developed between parent and child. And this bond serves genetic interests, the preservation of the parent's distinctive genes.

In this essay I argue for the importance of genetic continuity as an end in itself, for humans as well as for other species. Conserving any species or one of its races entails preserving its genes, in addition to a conducive environment; not only because genes code for the properties that we value, but because we affiliate with life for its own sake. And we know that life is not only dependent on ecology but on phylogeny, the evolutionary experience of a species impressed on its genes. If eagles could speak they would probably demand the right—or at least the chance—to survive and flourish, as do we. That is life's overriding goal, its ultimate interest.

Even the embryo strives to live, as do its precursors, the gametes of ovum and sperm and fertilized ovum, or zygote. The sperm is most spectacular in its striving to reach and fertilize the egg, and the prize is large—all the sperm's (and ovum's) genetic material is copied into the new organism should the embryo develop into a viable foetus, child, and adult. Half the genes of that new individual are passed on to the next generation should it reproduce. When born into an environment for which they are adapted, all forms of life, including human beings, strive to survive and to multiply.

Life is the ultimate interest, though we are all destined to die. Phenotypes—organisms put together from information supplied by genes plus environment—are mortal. The causes of life are in the transgenerational evolutionary process stretching back three billion years to the first self-replicating entity. It follows that ultimate interests do not reside in individual survival but in the reproduction of the information used by the organism to construct itself. The basic bits of information are 'germ-line replicators', reproduced with great accuracy. They are the 'units of selection' on which ultimate processes of selection and mutation operate. They are genes, the digitally coded bits of information coded in deoxyribonucleic acid, known as DNA.

Genes decompose with the organism but the information they carry can have great longevity compared to individual lifespans. This is because genes reproduce by cloning, making perfect copies and, rarely, not-so-perfect copies of themselves. But without the aid of modern science, sexually reproducing organisms cannot clone themselves. Children are made by blending a random sample of the genes of two individuals. Individuals are assemblages of tens of thousands of genes in combinations that are never repeated (except in identical twins), no matter how many offspring the individual produces. But genes reproduce with digital fidelity, by copying genetic 'words' spelt from just four molecular 'letters', the nucleic acids: adenine, cytosine, guanine, and thymine.

Except for rare mutations, genes are descended from long chains of clones often originating in pre-human ancestors who lived millions of years ago. All genes, including mutants, are potentially the parents of similarly long chains of descendants. According to neo-Darwinian theory humans, like all other species, are constructed using information carried in genes interacting with environments conducive to life. Like other species, humans have evolved a set of behaviours for propagating their genes. Indeed, in the tradition of Darwinian evolutionary theory, propagating one's genes is life's *raison d'être*. Darwin realized that a major philosophical import of his 1859 book, *On the Origin of Species by Means of Natural Selection, or The Preservation of Favoured Races in the Struggle for Life*,[1] was that it represented nature as governed by physical laws and placed humans in the continuum of life, with the same origin in the struggle for existence.

Individuals might choose any purpose in life, including ones that prevent their genes from being passed on to the next generation. However, maladaptive choices tend to eliminate genes that contribute to those choices within prevailing environments. Genes will not survive the organism in which they reside unless they launch the organism on an adaptive life course—avoiding predators, metabolising food, learning the local language, resisting parasites, finding mates and, in social species, nurturing offspring and defending the kin group. The individual phenotype is a survival vehicle constructed by a parliament of genes, each cooperating to perpetuate itself.[2] This modern evolutionary view of humans (and all other species) has been dramatically expressed by Richard Dawkins[3]: 'They are in you and in me; they created us, body and mind; and their preservation is the ultimate rationale for our existence. . . . [T]hey go by the name of genes, and we are their survival machines.' Genes are not the ultimate rationale for anything, of course, since only a proposition can perform that function. But the process of genetic evolution is certainly the ultimate *cause* of our existence. Individual humans are links in a chain of life stretching back millions of generations of human and prehuman species that managed to perpetuate their genes. A developmental program coded in the genes and enabled by the environment, guides individual behaviour over the lifespan, for example by producing a distinctive physiology and psychology. These 'proximate mechanisms' evolved due to the ultimate cause of biological evolution—the process of differential survival and reproduction of genomes within populations.

The needs and wants of the phenotype are interests, and conventional social theory adopts this perspective. All interests have some importance, even if only subjectively. But some interests are more important than others. Abraham Maslow[4] pointed this out quite explicitly in his 'hierarchy of needs'. We consider some interests vital. Most of us give priority to the survival and well being of ourselves as individuals. For emotional reasons we also give priority to the welfare of kith and kin and other individuals and groups for whom we have sympathy. Nutrition and freedom from disease also rate highly on our hierarchy of needs. Resources and rank are two interests usually considered vital, once survival and good health have been secured. Yet as compelling as these may seem, and despite contributing to reproductive fitness, they are actually of secondary importance to the ultimate interest. Biologist Richard Alexander[5] notes that when 'interests are seen as reproductive, not as individual survival, . . . pleasure and comfort are postulated to have evolved as vehicles of reproductive success'. These secondary goals can be referred to as proximate interests. They are the short-term goals served by adaptations, including motivations and appetites. Yet in relation to the ultimate interest of genetic continuity, the means distilled in proximate interests are expendable. For example, the avoidance of injury is a readily understood proximate interest served by the adaptation of the pain reflex. Still, adaptations such as the pain response are not ends in themselves. As evolutionary theorist George C. Williams points out, these adaptations have the 'ultimate purpose' of promoting 'genetic success'—transmitting one's genes into the next generation.[6] From the biological perspective even personal survival—usually taken to be an end in itself—is a means to genetic success. Since it is an important means, individuals are adapted to defend life and limb. Nevertheless, self-sacrificial behaviour can be observed in all societies. Examples exist in heroic self sacrifice by parents trying to save children and warriors defending their homelands.

The most comprehensive concept of genetic success, 'inclusive fitness', was introduced by the ethologist William D. Hamilton in a famous 1964 paper. Hamilton showed that an individual's genetic interests are advanced not only through personal reproduction, but by aiding the reproduction of other individuals who share some of its genes, typically kin. Richard Alexander expresses this well:

> [H]umans like other organism[s] are so evolved that their "interests" are reproductive. Said differently, the interests of an individual human (i.e., the directions of its striving) are expected to be toward ensuring the indefinite survival of its genes and their copies, whether these are resident in the individual, its descendants, or its collateral relatives.[7]

From an evolutionary perspective, based on observation of many species, genetic continuity is the ultimate interest of all life, since it has priority over other interests. People are prone to risk their lives for close relatives in emergencies, testament to the power that inclusive fitness has to shape human action. Valuing of proximate interests such as self preservation evolved to the extent that they enhanced the ultimate reproductive interest. Edward O. Wilson puts proximate interests in neo Darwinian perspective thus:

> In a Darwinist sense the organism does not live for itself. Its primary function is not even to reproduce other organisms; it reproduces genes, and it serves as their temporary

> carrier. . . . [T]he individual organism is only [the genes'] vehicle, part of an elaborate device to preserve and spread them with the least possible biochemical perturbation. . . . [T]he hypothalamus and limbic system are engineered to perpetuate DNA.[8]

For most of their existence humans have been unaware of the evolutionary process, and so have not been aware of what underlay kinship and other values, even while their physiology and behaviour were shaped by natural selection. Alexander believes that genetic interests exist independently of whether they are perceived.

> We need not be concerned with the possible argument that interests are only definable in terms of what people consciously believe are their interests or intentions. Biologists continually investigate the life interests of nonhuman organisms while lacking knowledge on this point and nonhuman organisms live out their lives serving their interest[s] without knowing in the human sense what those interests are.[9]

Since genes and modern biology are recent discoveries, proximate interests have stood in for ultimate interests. But can phenotypes have ultimate interests at all? It might be argued that an entity can only have ultimate interests if it produces perfect copies of itself. But outside the laboratory sexually reproducing phenotypes such as humans never duplicate themselves. Only genes and other replicators truly reproduce in the form of clones. The answer to this objection lies in the mutuality of interests between phenotypes and genes. The organism is the means by which genes replicate themselves. If that were not the case one would expect more competition between genes within the genome. In general genes cooperate with each other to build viable organisms. According to Richard Dawkins[10] it is the gregariousness of genes within the genome that is remarkable, not their fractiousness. Genes cooperate to build organisms that strive to replicate those same genes by reproducing and by helping kin reproduce. Dawkins again: 'If fitness is correctly defined in Hamilton's way as "inclusive fitness" it ceases to matter whether we speak of individuals maximising their inclusive fitness or of genes maximising their survival. The two formulations are mutually intertranslatable.'[11] Subsequent development of evolutionary theory has been implicitly or explicitly based on the notion that phenotypes have genetic interests[12]. Further challenges to the idea that genetic fitness is the ultimate interest are discussed in chapter 3.

Through evolutionary time proximate interests have served their functions well enough—inevitably so because the adaptations serving those interests were *selected* to be effective over evolutionary time. A proximate interest is a goal sought by an organism. When a proximate interest is adaptive, it serves to boost reproductive fitness. Adaptations mirror not only the environmental forces pressing on a population—such as climate and predators—but the proximate interests most exposed to those forces. The need to avoid predators selected for alertness and the ability to deter, defend, or flee. Need for nutrition selected for food-acquiring behaviours that differ according to ecological conditions and phylogenetic constraints. Earth's millions of species show a great variety of adaptations. Proximate interests are less variable, being closer to the vital functions such as acquiring energy and nutrients and avoiding predators. By observing the 'direction of . . . striving' in many species, as Alexander put it, we can infer that advancing genetic interests is the prime function for which all life is adapted in myriad ways and to various degrees of completeness.[13]

*Humans can no longer rely on their instincts*

There is nothing immutable or necessarily perfect about adaptations or the understanding, appetites and preferences they organize. Natural selection is constrained by evolutionary history and environment. It shapes bodies and behaviours in small increments by modifying existing species. Much in nature is badly designed, if one examines it from an engineer's viewpoint. For example, G. C. Williams[14] points out that the connection between testes and penis in humans is ludicrously indirect. Adaptations as basic as our visual and hearing systems show 'design flaws', though they have been good enough to bring us to the present.

Like adaptations that advance them, proximate interests can be imperfect in promoting genetic interests. The main problem is the slowness of natural selection compared to the rapidity of technological and social change since the Neolithic. The inertia of adaptations can cause them to continue to promote proximate interests that no longer serve fitness. For most of humans' evolutionary history, adaptations tracked slow-moving environmental change, including technological advances. In the species' distant hominid and pre-hominid past, proximate interests that reduced an actor's fitness were valued less and less as the genes that coded for such valuation failed to reproduce. For this reason, at most moments in time proximate interests have correlated with ultimate interests because the environment has changed so slowly that physiology and behaviour could keep track with it. Proximate and ultimate interests have been in equilibrium except where rapid changes in environment occurred. The equilibrium applying to humans has been upset in recent generations, so that we can no longer rely on subjectively designated proximate interests to serve our ultimate interest. We must rely more on science to perceive the causal links between the things we value and formulate synthetic goals based on that rational appraisal.

Proximate interests, often reflected in consciously held values, have become increasingly fallible guides to ultimate interests because modern humans live in a rapidly changing world. Humans evolved in small bands consisting of a few families, sometimes grouped into tribes numbering in the hundreds. For most of their evolutionary history humans made a living by hunting and gathering in largely natural environments. They lacked formal organization and hierarchy. Adults coordinated activities by negotiating simple demographic role specializations—by age and sex—on an egalitarian basis with familiar band members. Most information was common. Humans now live in societies numbering in the millions where the great majority of interactants are strangers or acquaintances. They make their living through a great diversity of occupations resulting in radical asymmetries in information. They live and work in largely man-made urban environments. They are formally organized into states administered by extended hierarchies of rank and resources actuated by authoritative commands, impersonal contracts enforced by the state authority, and powerful forms of indoctrination performed by universal education, centralized media and entertainment. The great complexity, power, and interdependence of modern societies call for either totalitarian control or the voluntary cooperation that comes from a climate of trust. But the human psychological outfit makes it difficult to trust our fellow citizens, a mistrust that is often justified according to ethologist Irenäus Eibl-Eibesfeldt.[15] Trusting, intimate societies no longer happen by themselves, as they did for much of the previous 100,000 years. Similarly, pollution can no longer be solved by vacating a campsite for a year. Industrial society's despoliation of the biosphere can only be prevented or mitigated using long-range planning, planning that constrains our own behaviour. Human nature resists such constraint, if only because man is evolved for short-term thinking, for satisfying wants and winning competitions now. What was for millennia unambiguously adaptive is now alloyed with risk. Eibl-Eibesfeldt[16] argues that there is no alternative but to use conscious strategies to avoid the traps of social fragmentation and short-term thinking.

Modern life is thus evolutionarily novel in several respects. These changes have increased the yield of land and of human labour beyond the imagination of our hunter-gatherer forebears. As a result the *absolute* fitness of the species as a whole has risen dramatically, increasing the world-wide human population from millions before the Neolithic into teeming billions today. Those changes have reduced the effectiveness of some evolved mechanisms for recognizing and defending *relative* fitness, meaning particular individuals' or bands' or tribes' genetic representation in the population. For most of the species' existence the human phenotype was in equilibrium with its environment. But the rapid rates of demographic, cultural and economic change experienced since the Neolithic

Revolution 12,000 years ago—when technological innovation made the crucial leap of domesticating plants and animals—have created a large disequilibrium between some phenotypic characteristics and our man-made environment. In particular, the set of mechanisms for recognizing and investing in ethnies has become inadequate and often downright maladaptive.

By ethny I mean a population sharing common descent. 'Ethny' is a preferable term to 'ethnic group' because members of such a category usually do not form a group. Ethnies are usually concentric clusters of encompassing populations, such as tribe, regional population, and geographic race. An ethny is typically 'a named human population with myths of common ancestry, shared historical memories, one or more elements of common culture, a link with a homeland and a sense of solidarity among at least some of its members'.[17] During much of our species' evolutionary past ethnies corresponded to bands of related families and tribes numbering in the hundreds or low thousands, which were demarcated from other groups by markers of territory, language, distinctive styles of material culture, and ritual. Individual competition continued within these proto-ethnies, for example competition for mates. Directed altruism was one competitive strategy, with most intense investment going to the family, which is the highest concentration of distinctive genes apart from self. But the band and the encompassing tribe also received some investment in the form of food sharing, cooperative child care, and mutual exploitation and defence of territory. Much of this was not altruistic, paying off due to reciprocity and the resulting synergies that benefited the group as a whole.[18] But altruism played its part, especially in group defence and in contributing to the trust that facilitates delayed reciprocity.[19] Conflict within bands and tribes, though endemic, was mitigated by familiarity and by rituals that extended familiarity across cohorts of males.[20]

(By competition I mean the ultimate form—between genes. This technical meaning needs to be distinguished from the everyday use of 'competition', which implies hostility or aggression or status consciousness. Genetic competition need not involve competitive motivations. All that is necessary is that one individual's or group's distinctive genes are able, even accidentally, to replace those of another individual or group in the next generation. Hunting on someone else's land might be motivated by nothing more than hunger; so can migrating to a far off land. Yet both actions can reduce others' fitness by reducing their share of resources, thus reducing the number of surviving offspring, and thus their representation in gene pool. A larger-than-average family can constitute an assault on others' fitness even though it is motivated by nothing more hostile than love of children.)

Tribes numbering in the hundreds or low thousands have now grown to ethnies numbering in the hundreds of thousands and often many millions of individuals, distributed across vast territories. Uniting rituals have broken down due to geographic spread and cultural change. At the same time states correspond less and less to ethnic boundaries as they have absorbed other ethnies due to conquest or immigration. States and markets have imposed modern indoctrination techniques, most notably universal education and the mass media, that tend to break down ethnic solidarity, causing altruism to be directed towards genetically distant individuals. Due to modern economic pressures the extended family is often dispersed and plays a reduced role as an economic unit, further reducing investment in kin outside the nuclear family. Aggregation of populations and atomisation of social organization have contributed to rising absolute fitness of the species overall and to higher standards of living. But growth has not been uniform, causing some ethnies to become marginalized and the remnants to be assimilated into larger expanding populations. Even members of powerful ethnies can unknowingly squander ethnic genetic interests.

Certainly we can no longer rely on our instincts to guide us through the labyrinths of modern technological society. But there is one innate capacity we possess that, combined with one or more motivations, is capable of solving this problem. Humans are uniquely equipped with analytic intelligence, the ability to tackle novel challenges. This 'domain general' problem-solving capacity evolved because it allowed our ancestors to find solutions to novel threats that arose in the environments in which they lived.[21] General intelligence is distinguished from 'domain specific' capacities, such as face recognition and speech, specialized mental modules for solving recurring problems in the environments in which we evolved. We are flexible strategizers *par excellence*, able to construct our own micro environments across a great diversity of climates and ecosystems. Abstract intelligence is physiologically costly because it requires a large brain, difficult childbirth and extended childhood. Nevertheless it has been so adaptive that it distinguishes our species. It allows us to consciously assess dangers and opportunities and to devise novel solutions, or to choose a well-rehearsed routine from our extensive repertoire to apply in a given situation. Now changed environments have effectively blinded us to large stores of our genetic interests, or to put it more accurately, for the first time situated us where we need to perceive those interests and be motivated to pursue them. This blindness is not cured by a people's economic and political power, as documented in Chapter 3, regarding the decline of Western populations. We must rely on our intelligence to adapt, not only using science to perceive our fundamental interests in the abstract but devising ways to realize these interests through proximate interests, the short-term goals of which we are aware and towards which we are motivated to act.

There is reason for hope. We are the only species that can understand its place in nature and thus the ultimate causes of its existence and conditions of survival.

At the beginning of the twenty-first century, with the human genome mapped and evolutionary theory mathematically articulated, human interests are still largely formulated through intuition. Why the delay? The cause is partly scientific and partly political and ideological.[22]

The pace of scientific progress is uneven, relying as it usually does on the inspiration and perspiration of a handful of exceptional individuals. It is true that Darwin's theory of natural selection was formally accepted in the scientific mainstream from the latter part of the nineteenth century, and that the actions of units of inheritance—genes—were discovered by Gregor Mendel as early as 1866. However, Mendel's work only received widespread recognition in 1900. Genetic theory was finally incorporated into Darwinian theory by R. A. Fisher in the 1930s, but genes themselves were described by Watson and Crick as late as 1953. Until Fisher's neo-Darwinian theory, genetic interests were represented metaphorically by concepts such as shared blood. The post-Second World War resurgence of popular evolutionary books by R. Ardrey, K. Lorenz, D. Morris, and others, while contributing to our understanding of behaviour, were written before Hamilton formulated inclusive fitness theory or before it had been incorporated into mainstream theory. Furthermore, most of these authors were zoologists rather than sociologists and political theorists, and thus distant from the social sciences in terminology and often in collegial relations. Neo-Darwinism finally entered popular discourse in the 1970s and 1980s with books such as E. O. Wilson's monumental *Sociobiology* (1975), Richard Dawkins' essay *The Selfish Gene* (1976), and Richard Alexander's groundbreaking yet accessible *Darwinism and Human Affairs* (1979). In the German-speaking world kin selection theory was incorporated into the study of humans by Irenäus Eibl-Eibesfeldt in *Human Ethology* (original German edition 1984), Heiner Flohr and W. Tönnesmann in *Politik und Biologie* (1983) and others. Eibl-Eibesfeldt and Alexander actually identified fitness as an interest.[23]

The uneven acceptance of neo-Darwinism was also due to exciting breakthroughs in research on learning. Attention was diverted from the study of evolutionary influences by the discovery of the conditioned reflex by Pavlov around 1900, the discovery of instrumental conditioning by Edward L. Thorndike around the same time, the formulation of behaviourist theory by J. B. Watson in 1913, and the refinement and popularisation of behaviourism by J. B. S. Skinner and others over the middle part of the century. Behaviourism appeared capable of discovering universally-applicable laws of behaviour that could help resolve social problems.[24] Attention was diverted from innate and evolutionary causes, though conditioning theory is consistent with both.[25]

Social theorists have also been impeded in embracing evolutionary concepts by political, ideological, and ethnic concerns. I suspect that the subject of ethnic kinship would long ago have become a standard concept in the social sciences if not for the role of non-scientific motives. After all, the evolutionary biology of the late nineteenth century informed the sociology of Herbert Spencer, Edward Ross, who coined the term 'social control', and William Sumner, who coined the term 'ethnocentrism'. It inspired Edward Westermarck's 1921 anthropologically-based theory of ethical behaviour, and Sir Arthur Keith's 1931 study of prejudice. But by the end of the 1920s politicised social science in the United States and Europe was putting pressure on evolutionary thinking, especially concerning race and ethnicity.[26] For much of the century it was politically incorrect in the West or in communist societies to report group differences or any biological influences on behaviour. The trend was for race to be relegated as a causally unimportant or even unscientific category, and ethnicity was conceived as a purely cultural substitute. Group feeling was portrayed as the manifestation of irrational prejudices and psychopathologies induced by capitalism, religion, or simply sick authoritarian minds.

Anti-biological sentiment can be partly attributed to a reaction against the fascist variant of social Darwinism as enacted by extreme nationalists before and during the Second World War. In Nazi Germany politicized social science greatly overemphasized race as a causal factor in social processes. However, politically motivated critics tended to lump together all biological approaches, as with Montagu's bitter assault on the distinguished American anthropologist Carleton Coon for advancing the multi-regional theory of race,[27] S. J. Gould's *ad hominem* attacks on biologists who study race differences,[28] or the hounding of E. O. Wilson for alleged 'conservatism' in his 1975 tome *Sociobiology*.[29] It made no difference that Hamilton's theory was built on a refutation of the vague and often mystical group selectionism dear to racial nationalists of the inter-War period. None of these evolutionary thinkers was politically extreme or even politically oriented. Nor did it matter that social Darwinism had not led to ethnic cleansing or genocide in most Western societies where it had become popular, or that the opposed ideology of communism that explicitly repudiated Darwinian theory (and Mendelian genetics) had been responsible for atrocities on a mass scale, directed against ethnic minorities and majorities alike. Ideological resistance kept evolutionary biology separated from Western social science for much of the twentieth century, perhaps inevitably so. Evolutionary psychologist J. P. Rushton has observed that opposition to biological research on race and ethnicity, as well as its support, is often itself ethnically motivated:

> The propensity to defend one's own group, to see it as special, and not to be susceptible to the laws of evolutionary biology makes the scientific study of ethnicity and race differences problematic. Theories and facts generated in race research may be used by ethnic nationalists to propagate political positions. Antiracists may also engage in rhetoric to deny differences and suppress discoveries. Findings based on the study of race can be threatening. Ideological mine fields abound in ways that do not pertain in other areas of inquiry.[30]

Neo-Darwinism is still largely shunned by the social sciences, though the theoretical dispute was decided by the 1980s.[31] Remaining resistance is largely political and institutional. Kin selection theory is part of the standard theoretical tool kit for a growing number of anthropologists and psychologists.[32] There is also a three-decades old tradition of biopolitics and biosociology, mainly in the United States, that deploys sociobiology. These schools apply a range of ethological concepts and research methods, such as inclusive fitness theory, proximate and ultimate causation, and naturalistic observation.[33] The subdiscipline has already made strides integrating evolutionary biology into our understanding of government. A similar process has occurred in anthropology. These disciplines are uniquely positioned to begin the delicate task of bringing the biological conception of interests, both ultimate and proximate, into political theory, ethics, and policy studies.

A tentative acknowledgement of genetic interests has been advanced in political science. J. Beckstrom[34] recommends that the property of individuals who die without leaving a will should be distributed to their closest genetic kin, since in the absence of other information, this is the best predictor of a person's wishes. But, overall, the social sciences have been so far removed from biology that most of political and sociological theory has not even come to terms with familial genetic interests, a generation after Hamilton's mathematical kin selection theory swept the field of theoretical biology. It is time for us to do a little catching up. In addition to analysis, evolutionary approaches to politics have a vital role in ethics and policy formulation by helping individuals identify their vital interests—both ultimate and proximate—and devising fair strategies for protecting those interests. Evolutionists are uniquely equipped to identify genetic interests and formulate principles of biological fairness because only neo-Darwinian theory identifies the ultimate causes of social behaviour.

Incorporating genetic interests into social theory will be a large undertaking, which this essay can initiate but hardly complete. One central question, which I try to answer in Chapter 7, is how various political systems affect familial and ethnic inclusive fitness. The answer will require a systematic analysis that begins with hunter-gatherer societies, passes through tribal, big-man and chieftain systems, thence kingdoms, city states, agrarian empires, ending with an analysis of modern forms, including democracy, fascism, communism, constitutional monarchy, republics, commercial states, and theocracies. It was relatively easy to discover that familial genetic interests are most unequal in despotic societies, since these facilitate polygyny,[35] and that relatively egalitarian societies facilitate monogamy, which can also be socially imposed.[36] The more difficult question is, which political system best defends genetic interests contained by ethnic groups? Some indirect guidance will come from the analysis of patriotic altruism from the perspective of kin selection.[37] A more direct approach will be to analyse state formation, as undertaken by Masters,[38] though he did not consider ethnic genetic interests. Masters argues that the original state emerged from a process of negotiation between rulers and ruled in such a way that the latter did not sacrifice their inclusive fitness. In this view, the lower ranks in the proto state not only benefited from the increases in absolute fitness due to the state's territorial expansion, but retained *relative* fitness with the higher ranks. Whatever the plausibility of such accounts,[39] they show that biopolitics has concerned itself with familial genetic interests in sophisticated ways. Ethnic genetic interests await treatment.[40]

To this end, in the following chapter I describe the distribution of individuals' genetic interests in family, ethny and species, review and formulate strategies used or usable to defend them, and assess the ethical status of those strategies.

*Notes*

1 Darwin's theory has been augmented to arrive at the modern synthesis, so called because it incorporates knowledge of genetic transmission of characteristics, the study of which was initiated by the Austrian monk Gregor Mendel in 1866. The theory of natural selection was actually co-authored with Russell Wallace, the American naturalist, and presented in 1858 to the Royal Society, but has been identified with Darwin, partly because he had originally formulated it 20 years earlier in notebooks, but also for pragmatic reasons of Darwin's greater fame, powerful connections, painstaking experimental investigation of the theory, and incomparable scientific output.
2 Hamilton (1964); E. O. Wilson (1975); Dawkins (1976; 1982).
3 Dawkins (1976, p. 21).
4 Maslow (1954).
5 Alexander (1987, p. 81).
6 Williams (1997, pp. 41, 45).
7 Alexander (1985/1995, p. 182).
8 E. O. Wilson (1975, p. 3).
9 Alexander (1985/1995, p. 183).
10 Dawkins (1978, p. 69).
11 *Ibid.*, p. 61.
12 The notion of genetic interests is rarely made as emphatically as by Alexander (1985/1995, pp. 182–3). References to the genetic interests of individual organisms are

scattered throughout the literature, for example Godfray (1995, p. 133), states that an individual has a 'genetic interest in its siblings', and G. C. Williams (1997, p. 49) observes that a honeybee serves its 'genetic interests' by promoting its mother's reproduction.

13 G. C. Williams (1997, p. 42).
14 *Ibid.*, p. 140.
15 Eibl-Eibesfeldt (1994).
16 Eibl-Eibesfeldt (1998).
17 Hutchinson and Smith (1996, p. 6).
18 Trivers (1971); Corning (1983).
19 Wiessner (2002b).
20 Wiessner (1998/2002).
21 MacDonald (1991).
22 Degler (1991).
23 Eibl-Eibesfeldt (1982, p. 14); Alexander (1985/1995, p. 182).
24 For example, B. F. Skinner (1948; 1971).
25 B. F. Skinner (1984).
26 Degler (1991); Shipman (1994).
27 Shipman (1994).
28 Gould (1981/1996); for a critique of Gould's book see Rushton (1997).
29 Sociobiology Study Group of Science for the People (1976/1978); see Wilson's defence (1978b).
30 Rushton (1994, p. 1).
31 Alcock (2001).
32 E.g. Chagnon (1980); Daly et al. (1997); van den Berghe (1981).
33 Arnhart (1995); Caton (1988); Flohr and Tönnesmann (1983); G. Johnson (1986; 1987); Masters (1989; 1998/2002; in press); Sanderson (2001; in press); Somit (1976); G. Schubert and Masters (1991); J. Schubert (1983); and see the journal *Politics and the Life Sciences*, published since 1982.
34 Beckstrom (1993).
35 Betzig (1986).
36 Alexander (1979); MacDonald (1995).
37 E.g. Johnson (1986); Salter (2002); Shaw and Wong (1989).
38 Masters (1989, pp. 189—90; 1998/2002).
39 Salter (1995, p. 21).
40 The omission of ethnic genetic interests in biosocial science is remarkable, because inclusive fitness theory drives much research into social behaviour of humans as well as of non-human animal species. Why then not study human politics as it affects reproduction? The notion that an ethny (or tribe) is an aggregate genetic interest for its members is at most implicit, even in studies of group strategies that posit self-sacrificial altruism (e.g. MacDonald 1994). Similarly, texts on human ecology can demonstrate familiarity with modern evolutionary theory yet fail to raise questions about genetic fitness in discussing stratification and the state. In their broad ranging undergraduate text, Richerson et al. (1996, p. 353) assert that inequality is a moral dilemma of complex societies, but do not discuss the inclusive fitness consequences of stratification or state formation, whether relative or absolute, individual or tribal.

# 2. Concentrations of Kinship: Family, Ethny, Humanity

*Summary*
Genetic interests are in principle quantifiable, since they occur in individuals and groups with measurable concentrations. These concentrations decline in concentric circles of relatedness from self (and identical twins), to family, clan, ethnic group, geographic race, whole species, and even other species. Genetic interests in kin are well studied, but ethnic genetic interest (ethnic kinship) is poorly studied and as a result not previously quantified. Hamilton's theory of inclusive fitness, elaborated by Harpending to accommodate data from population genetics, allows the estimation of ethnic kinship. Kinship with other members of one's ethnic group (co-ethnics) is zero if that group is considered in isolation, without genetic competitors. But in fact the world population consists of many ethnies, some of which are more closely related than others. Some ethnic groups are so distantly related that randomly-chosen co-ethnics are closely related by comparison. Emerging evidence from population genetics indicates that in most situations individuals have a larger genetic stake in their ethnic groups than in their families.

Where do our genetic interests lie? The conceptual answer is provided by inclusive fitness theory as developed by Hamilton in 1964. The theory has at its heart the idea that within a species an individual's reproductive success—his or her fitness—depends partly on the reproductive success of other individuals who share some of the individual's distinctive genes. The theory has been applied mainly to kin groups, families in particular, and is often referred to as 'kin selection', though this is actually a special case of inclusive fitness theory. Kin selection was the first conceptual breakthrough in solving the problem of altruism, first recognized by Darwin. Darwin could not explain how members of neuter casts, as found in social wasps, bees, and ants, could have evolved. After all, they did not have offspring, instead devoting themselves to helping the queen reproduce. The answer found by Hamilton is that these sterile individuals do reproduce, but via the queen, since the queen shares a large fraction of their genes.

Hamilton's answer provided a new way of looking at parental reproduction. Parenthood, Hamilton realized, is just one way for an individual to get its genes into the next generation. Because parenthood is the most common way organisms reproduce, previous naturalists including Darwin had been led to believe that the individual is the unit of selection. However, as noted in the first chapter, individuals never in fact reproduce except in asexual species. We have never cloned ourselves, ultimately because it isn't healthy for a lineage to do so; cloning eliminates the variation needed to continue the perpetual evolutionary arms race against pathogens and, more generally, against changing environmental conditions. No variation means no adaptation which means, sooner or later, death.

An individual's genetic interest in a particular group is the number of copies of the individual's distinctive genes carried within the group by reproducing individuals. An individual's genetic interest in a group has two dimensions: (1) the average per person concentration of the individual's genes existing in the group; and (2) group size.

It is uncontroversial that individuals' distinctive genes are concentrated within their families. The proportion of shared genes, denoted by $r$ (for relatedness) in Hamilton's original formulation, declines by 50 percent for each generational step. An individual shares half his or her genes with each offspring, a quarter with each grandchild, an eighth with first cousins, and so on. The formula for calculating $r$ between two individuals is simply half raised to the power of the number of generational steps $n$ separating them. The generational steps must go via a common ancestor or ancestors. Consider the case of first cousins set out in Figure 2.1.

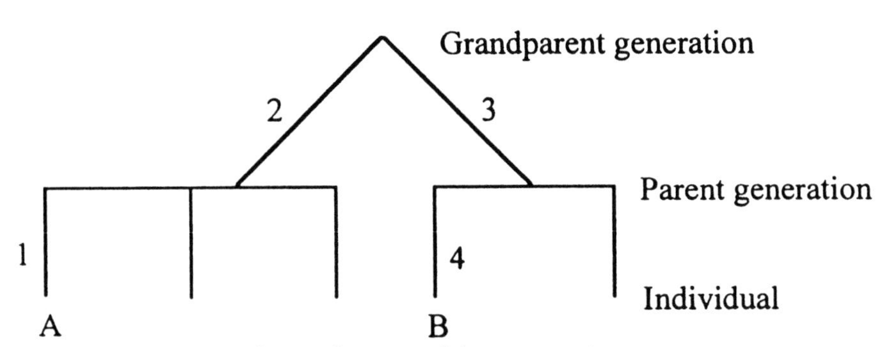


*Figure 2.1. Family tree showing lineage path between cousins.*

In Figure 2.1, consider the relatedness $r$ of cousins A and B. The method for determining $n$ is to count the generational steps to the common ancestor, going up to the grandparent generation then back down to the other individual. There are four steps separating A and B, so $r$ is one half raised to the fourth power, which is one sixteenth, or 0.0625. Since first cousins share two grandparents, this result needs to be doubled to yield the relatedness of one eighth, or 0.125.

The aggregate size of familial genetic interests varies with size of family. In a society with average family size of three children the nuclear family will contain 1.5 copies of each parent's genes (3 x 0.5). Nieces and nephews add another 1.5 genetic copies (6 x 0.25) and first cousins add another 1.5 copies (12 x 0.125). Together these three sets of relatives add 4.5 copies of each parent's genes. The familial inclusive fitness is less in typical western societies whose average family size is below the replacement number of 2.1 children per woman. Assuming 2 children per family, then children (2 x 0.5), nieces and nephews (2 x 0.25), and cousins (4 x 0.125) add to 2 copies. In a rapidly growing population with average family size of 4, the same calculation yields 6.5 copies. Beyond first cousins family ties in Western societies are weak. Extended families in industrial societies are not highly interdependent and generally do not act as corporate groups as is typical in hunter-gatherer bands and tribal societies. The high mobility and individualism of modern Western societies has weakened the clan as a distinct structural element. For this reason I count relatives beyond first cousins as part of the ethny when calculating the distribution of genetic interests.

An individual's familial genetic stake is readily calculated. But individuals' genetic interests in their ethnies have not been calculated by demographers or population geneticists, the disciplines most likely to possess the needed data. With rare exceptions, scholars in neither discipline have been interested in issues of altruism. The lack of interest shows in neglect of the genetical theory of altruism formulated by W. D. Hamilton, R. Dawkins, R. Alexander, E. O. Wilson and others. My search for relevant theory turned up little useful in theories within demography or population genetics, and led instead to ethology, anthropology, and psychology. Finally, I found the crucial quantitative theory able to link genetic assay data and inclusive fitness in the work of Hamilton himself and Henry Harpending, an anthropological geneticist who applies Hamilton's theory.

### *Ethnic genetic interests: qualitative theory*

Qualitative reasoning about ethnic kinship goes back at least to Darwin's 1871 book *The Descent of Man and Selection in Relation to Sex* in which he advanced a group selectionist theory of human social evolution. Darwin was trying to explain the evolution of morality, a theme taken up in the same group-selectionist mode by Sir Arthur Keith[1] after the Second World War. Both scientists were limited in their theorizing by incomplete knowledge of the genetic basis of heredity. Recent theories of group selection, though consistent with the existence of ethnic genetic interest, do not quantify it or give it much attention.[2] Instead these theories focus on the evolution of behaviours and other traits.

A generation later anthropologist Pierre van den Berghe more directly implied the existence of ethnic genetic interests in his analysis of family systems. '[R]eproduction is almost always a bad *economic* bargain for parents. Consequently, there must be more in it for them, and that "more" is, of course, fitness.'[3] In his classic 1981 sociobiological analysis of ethnicity, *The Ethnic Phenomenon*, van den Berghe argued that ethnic groups constitute extended families, both subjectively in the perception of their members and objectively in the genetic commonality produced by common descent. Since in this view ethnies are large families, van den Berghe's theory implies that they represent a store of their members' distinctive genes.

Similarly, the ethologist Eibl-Eibesfeldt's[4] theory of the evolutionary function of ethnic solidarity entails a reproductive interest residing in ethnies. According to this theory, genetic group selection (in the sense of extended kin selection) was made possible by tight solidarity found in hunter-gatherer bands and tribes. Through indoctrination and the discipline imposed by mutual monitoring, bonds in these groupings can become so strong that they motivate self-sacrificial behaviour in the context of intergroup conflict, and facilitate territorial conquest and differential population growth. Eibl-Eibesfeldt[5] observes how tribes adopt cultural group markers, including language and dialect that heighten in-group identification and solidarity. He notes that linguistic and genetic group boundaries roughly coincide, as shown by Cavalli-Sforza,[6] and concludes that ethnies are in part the result of group competition in which genetic interests were served by national solidarity. Whether or not group selection has been a factor, in Eibl-Eibesfeldt's theory members of proto-ethnic group were closely related and thus shared a 'common genetic interest'.[7] He notes that this appears also to be true of some modern nations, citing van den Berghe's 1981 analysis.[8] Eibl-Eibesfeldt was thus among the first to refer to ethnic genetic interests.

Another recent precursor of the concept of ethnic genetic interests comes from the psychologist J. Philippe Rushton, who formulated genetic similarity theory (GST).[9] He applied kin selection theory to ethnies, implying a genetic interest shared by co-ethnics. Rushton also discussed scenarios in which group solidarity could increase the fitness of the genes shared by members of an ethny. In particular he suggested that if competition between ethnies resulted in replacement, this would constitute a form of group selection (in the sense of extended kin selection), since ethnies share more genes within than between groups.

GST is mainly directed to explaining assortative choice of mates and friends. Rushton noted the large data set showing that people choose mates and friends who are similar in ethnicity and other variables, and argued that this increases fitness. GST has been criticized,[10] and the relevance of this debate to ethnic genetic interests is discussed in Chapter 8 (pp. 257–280).

In a 1989 paper Rushton argued that genetic similarity can render altruism adaptive between nonkin, by boosting inclusive fitness of their shared genes. In their commentary on Rushton's paper, evolutionary psychologists John Tooby and Leda Cosmides rejected the possibility that similarity detection could facilitate kin selection because, they argued, in Hamilton's theory the responsible genes must not only be similar, but identical by recent common descent. Rushton's theory is proofed against this criticism in two ways. First, the difference between GST and kin selection theory disappears if genetic similarity is interpreted to mean genes shared from a common *distant* ancestor, which is in fact the cause of intra-ethnic genetic kinship. GST is then better conceptualized as extended kin selection theory. The second line of defence concerns the distinction between similarity and genealogy. Pedigree data are rarely adequate to allow us to estimate the proportion of genes shared between two distantly related individuals. Hamilton's original 1964 theory defined relatedness by pedigree. By the early 1970s he had revised his theory.[11] The revision was more general and based on genetic similarity, operationalized in terms of genetic variation. Geneticist Alan Grafen concludes that 'Hamilton's rule is true no matter how genetic similarity arises . . . '.[12] In this revised theory kin selection becomes a special case of inclusive fitness, and the latter can be measured from gene assay data independent of knowledge about genealogy. After reviewing Hamilton's revisions, Pepper states that, '[u]nder certain simplifying assumptions the two definitions coincide, but when they do not it is the modern statistical definition rather than the original genealogical version that makes inclusive fitness theory work . . .'.[13] As set out in the next section, this gene-based definition means that kinship coefficients can be directly measured from gene frequencies, thus circumventing the impractical pedigree method.

By the 1990s population genetic theorists had adopted the term 'genetic similarity' and developed methods for measuring it.[14] Rushton's view that ethnicity predicts genetic similarity turns out to be conceptually valid and terminologically up to date. While constituting an important contribution to the evolutionary analysis of ethnicity, Rushton's analysis falls short of capturing the idea that an ethny is a store of genetic interests for each of its members. But this step is completely consistent with Rushton's analysis, as it is with Eibl-Eibesfeldt's and van den Berghe's. Indeed, taken together with Hamilton's revised theory, this body of work implies the existence of ethnic genetic interests. The remaining problem is to quantify that interest.

*Ethnic genetic interests: quantitative theory*

A starting point for thinking quantitatively about the genetic interest contained within ethnies is a 1975 paper by Hamilton.[15] The paper is a theoretical treatment of relatedness within and between semi-isolated 'towns', which can be taken as corresponding to primordial tribes. The model presented by Hamilton indicates that the relatedness between members of an isolated tribe will rise to some maximum due to the growing web of kinship within it. A steady trickle of migration between tribes does not very much affect the ceiling of within-tribe relatedness, which can rise as high as 0.5, or that between siblings in a panmictic population. This high figure and the model that generates it are not to be found in population genetic texts.[16]

Hamilton's model is startling, for it raises the possibility that ethnic genetic interests exist on such a massive scale that they dwarf the largest familial inclusive fitness. If this is true, then self-sacrificial altruism on behalf of the ethny might indeed be adaptive.[17]

Hamilton's analysis contradicts the view that in the Pleistocene inter-tribal conflicts were conducted by groups with very low within-tribe relatedness.[18] In his model, a random individual has a large genetic interest vested in the local population.[19] As noted earlier, Hamilton's revision of his 1964 theory dispensed with genealogical kinship and redefined relatedness as a statistical measure of genetic similarity. In the 1975 paper Hamilton summarized the change thus:

> Because of the way it was first explained [by Hamilton], the approach using inclusive fitness has often been identified with 'kin selection' and presented strictly as an alternative to 'group selection' as a way of establishing altruistic social behaviour by natural selection. But . . . kinship should be considered just one way of getting positive regression of genotype in the recipient, and that it is this positive regression that is vitally necessary for altruism. Thus the inclusive fitness concept is more general than 'kin selection'.[20]

Hamilton's revised theory is of the first importance in quantifying ethnic genetic interests. Genealogical data are unavailable for the many millennia of prehistory during which populations split one from the other and, due to selection, mutation, and genetic drift, acquired their genetic differences. For an ethny to constitute a large family, distant kin must carry genetic interests for each other. Hamilton addressed this issue when he pointed out the pervasive web of relatedness that can result in high levels of interrelatedness. '[C]onnections which the remote townsman does not so easily know of make up in multiplicity what they lack in close degree'.[21]

Dawkins has also emphasized that an individual's distinctive genes are carried in distant kin,[22] an idea he has expressed in characteristically gripping imagery:

> Individuals do not, in an all or none sense, either qualify or fail to qualify as kin. They have, quantitatively, a greater or less chance of containing a particular gene. . . . [T]he post Hamilton 'individual' . . . is an animal plus 1/2 of each of its children plus 1/2 of each sibling plus 1/4 of each niece and grandchild plus 1/8 of each first cousin plus 1/32 of each second cousin . . . Far from being a tidy, discrete group, it is more like a sort of genetical octopus, a probabilistic amoeboid whose pseudopodia ramify and dissolve away into the common gene pool.[23]

Dawkins's organism metaphor helps convey the diffuse, distributed nature of genetic interests, but also fits the idea of a 'superorganism'. It has been suggested by some theorists that a bee swarm or a religion constitutes an individual organism.[24] Hamilton disagreed with this conceptualisation, preferring to take the perspective of the genetically-interested individual, the position I take in this volume.[25] Genetic interest residing in a population is a collective good that belongs, as it were, to its individual members. Any reference to an ethnic group's genetic interest in this text is shorthand for this meaning.

An alternative to Dawkins's metaphor is a landscape of kinship. The highest mountain peak represents ego's relatedness to himself of 100 percent (identical twins are twin peaks). One cannot trace a smooth slope from this peak to the common gene pool. Instead it recedes in a series of plateaux. From the peak there is an abrupt fall of 50 percent relatedness to the base camp of the nuclear family, perched on a narrow ledge, though broader that the lonely summit. The next fall is only half as steep and comes to rest at a still considerable elevation, for a time, on a wider plateau shared by grandparents, grandchildren, nieces and nephews. Successive slopes are ever gentler but the sequence of plateaux never disappears altogether; modern conditions can produce surprise cliffs. Slight ethnic gradients mark the boundaries of ancient tribes; genetic ridge lines are still palpable after many centuries of intermarriage. More significantly for the present argument, these lower slopes lead to very wide plateaux. The ancient tribes that once lived at low population densities have grown in size by several orders of magnitude, swapping genes with neighbouring groups but at a rate inhibited by geographic and cultural barriers. Such gradients mark ethno-linguistic boundaries even within Europe despite millennia of migrations and invasions (Map 2.1). Although these inclines can be abrupt, they are hardly something to trip over unless approached *en masse*. But together they form a grand staircase branching and winding its way between continents. When science or transcontinental migration allows us to look down the stairwell, we behold a dizzying abyss of exponen-

*Map 2.1. Thirty-three steep genetic gradients in Europe based on an assay of 60 neutral alleles (Barbujani and Sokal 1990). Many more gradients between smaller adjacent populations were found, but were not as steep. Most (31) gradients correspond to linguistic boundaries, themselves often corresponding to geographical barriers.*

tially receding kinship. In the relativistic universe of ethnicity, all our many landings represent, to each of us, the highest flight of kinship. Yet from whichever angle they are viewed, kindred ethnies remain clustered only one or two steps removed. This brings us to the evidence from population genetics.

To summarize this section, Hamilton's revised theory of inclusive fitness theory applies not only to relatives, but to any individuals sharing distinctive genes, even if their relatedness is remote and cannot be documented. This revised theory frees the analyst to measure genetic interests directly from genetic assay data.

### *Population genetic theory and ethnic kinship*

Two difficulties attend translating genetic assay data into measures of ethnic genetic interests. First, the theory from population genetics that connects Hamiltonian theory to genetic assay data does not deploy Hamilton's central concept of relatedness. Instead it refers to kinship. Secondly, genetic assays do not typically measure kinship, but genetic variance or genetic distance. I shall deal with these in turn.

Population geneticists usually refer to kinship rather than to relatedness, since the latter is not as well defined mathematically (see Appendix 1). The coefficient of kinship $f$ resembles Hamilton's coefficient of relatedness $r$ except that in most cases $2f = r$. The kinship coefficient is the probability that a gene found in one individual's genome at a particular locus is identical to one found in another individual at the same locus. When these individuals are randomly sampled from different groups, the kinship coefficient is a measure of the kinship between the groups. This new definition allows pairs of individuals to have negative kinship, meaning that they share fewer genes than is typical for the population, as well as positive kinship, when they share more genes than is typical.

The coefficient of kinship was used by population geneticists until the 1980s, and is of special utility in genetic epidemiology.[26] However, worldwide assays of kinship across large numbers of genes have not been conducted. Data are limited to assays of one or a few genes, and typically measure within-population kinship, or kinship within a regional cluster of populations. What is needed for present purposes is an assay of kinship between different populations, including those that are geographically close and distant. Unfortunately, no world assay of genetic kinship at the population level exists.

A global genetic assay performed in the 1980s by Cavalli-Sforza, Paolo Menozzi, and Alberto Piazza measured not kinship but variance or 'genetic distance', expressed as the coefficient $F_{ST}$,[27] Fortunately, Harpending[28] shows that kinship can be expressed in terms of variance:

$$f_o = F_{ST} + (1 - F_{ST})[-1/(2N - 1)] \quad \ldots 2.1$$

where $f_O$ is the local kinship coefficient, $F_{ST}$ the variance of the metapopulation, and N the overall population. When N is large, as it usually is with modern ethnies, a good approximation for the above equation becomes, simply:

$$f_o = F_{ST} \quad \ldots 2.2$$

This means that genetic distance, $F_{ST}$, is both a measure of genetic variance between two populations and a measure of kinship within each of them.[29] Hamilton came to the same conclusion in a 1971 paper.[30] Harpending explains the implication of this equivalence thus: 'This will mean that helping behavior within the subdivision [e.g. an ethny] will be selected against locally, because kinship is negative locally, but it may be positively selected within the species because kinship between donor and recipient is positive with reference to the global base population.'[31]

This is a finding of considerable importance, because it allows the estimation of average kinship coefficients between human populations based on $F_{ST}$ measures, as will be done in the next chapter. In the remainder of this section I discuss the meaning of Harpending's simple equation.

To correctly interpret this equation it is necessary to understand the relativistic nature of kinship, as explained with regard to ethnicity by Hamilton in the quote below. The kinship of two randomly chosen individuals in a population is zero. In the same context, two siblings have a kinship of 0.25 (equivalent to Hamilton's relatedness *r* of 0.5; see Appendix 1). The relativity of kinship can be illustrated with a version of J. B. S. Haldane's famous hypothetical example of altruism between kin. Haldane argued that it is adaptive to give one's life to save two drowning siblings or eight drowning cousins.[32] What he omitted to make explicit was a background assumption—that these rescues occur in the context of a population of zero relatedness between random pairs. Cousins have a relatedness of one eighth not in an absolute sense, but *in comparison to* this zero relatedness. Hamilton defined the kinship of two random members of a population as zero, but his own 1975 model indicated that when the variance of the meta-population is taken into account it is possible for high relatedness of random pairs within a subdivision (local population). (Recall that Hamilton's model allowed for intratribal relatedness to rise as high as 0.5 [equal to kinship of 0.25], which is the level of siblings.) Hamilton expressed this point in an earlier paper, pre-empting much of Harpending's analysis:

> If there is free mixing within subdivisions [e.g. an ethny] an encounter concerns a randomly selected pair from the subdivision. The correlation of gametes from such a pair is zero with respect to their subdivision. Thus an altruistic trait expressed in random encounters is certainly counterselected within the subdivision. The correlation of gametes with respect to the population is $F_{ST}$, which is always greater than zero, depending on the degree to which the gene frequencies of the isolates have differentiated. Thus if there is a gain to inclusive fitness on the basis of [a coefficient expressed in terms of $F_{ST}$] the genes for the trait are positively selected in the population as a whole.[33]

Recently D. S. Wilson[34] has made a similar point about the relative nature of fitness. He connects groups and the evolution of altruism in a similar manner to Hamiltonian inclusive fitness theory.[35] He argues that the absence of genetic competition between groups increases the payoff from competition within groups, increasing the maladaptiveness of altruism between random group members. The main difference is that D. S. Wilson does not emphasize the genetic gradients at group boundaries, instead arguing that any trait can delineate a group, including culturally caused traits. This is a general approach that captures all possible types of group selection, but the most likely and hence common type of group selection has probably operated between extended kin groups or, more accurately, between groups separated by a significant genetic distance. We know that bands and populations speaking the same dialect were such groups, and that these bounded the species' primordial social organization.

What if the world consisted of cousins? The kinship between random pairs would be zero, so that it would not be adaptive for them to show altruism towards each other. In this hypothetical case, there are no competing individuals or groups against which cousins have an elevated level of kinship. Adaptive altruism would then be limited to the nuclear family, where relatedness is four times higher than between cousins (eight times higher for identical twins).

An ethny is analogous to a population of cousins. If the world consisted only of that ethnic group (or cousins), the relatedness of random pairs would be effectively zero and therefore the ethny as a whole would hold zero genetic interests for its members. Then adaptive altruism would only be possible between relatives of closer degree than cousins. But in fact the world consists of a great many ethnies. Taking the whole world population together, the kinship of random pairs is zero. The question is, in this situation what is the kinship of random pairs chosen from an ethny? Is it realistic to speak of 'ethnic cousins'?

As we shall see in the next chapter, the data from population genetics indicates that aggregate ethnic kinship is at least several times larger than that contained by a nuclear family, and between populations drawn from different geographic races can be many orders of magnitude larger. Before considering some of these quantities in the case of contemporary ethnies in the next chapter, I want to discuss some objections to the reality of ethnic genetic interests.

*Objections to the existence of ethnic genetic interests*

The 'genetic unity of mankind' has often been cited as a reason to reject the biological importance of race and ethnicity (see also Chapter 4, section e). It is often implied that humans are too closely related for ethnic or racial nepotism to be adaptive. After all, some scholars argue, according to the 'out-of-Africa' or 'Eve' theory of human origins, all existing populations derived from a single population of anatomically modern humans that emerged perhaps 200,000 years ago in East Africa. According to this theory, now the majority view among evolutionists, colonists from Africa completely replaced all existing hominids in the rest of the world, beginning about 100,000 years ago (Map 2.2). Supporting data come mainly from genetic analysis based on assumptions about mutation rates.[36] A related argument is that since leaving Africa, populations have mixed so much that none is a pure lineage; the species forms one large population. For example, Cavalli-Sforza argues for the likely admixture of African and Asian genes about 30,000 years ago in the ancestral European gene pool, and uses this to criticize de Gobineau's belief in European racial purity.[37] The argument successfully dispels the mystical notion of a pure racial essence. Clearly purity is a relative concept better conceptualised as degree of homogeneity or heterogeneity.[38] The question raised by Cavalli-Sforza that is relevant here is: Could two populations have distinct genetic interests if they once partially interbred?

Accepting the Eve theory of the recent origin of races, and accepting subsequent partial mixture, presents no difficulty for the concept of genetic ethnic interests, since we know from direct measurement that significant genetic variation exists between populations. Neither should we expect any difficulty theoretically. The small groups that left Africa to colonize the world were almost certainly bands comprised of closely related individuals, and thus already genetically differentiated to some degree from the remainder of the species. According to theory expounded by Cavalli-Sforza himself, genetic drift within these small bands would have rapidly increased the genetic distance between and the kinship within them.[39] That is without factoring in selective forces of changed environment. Consistent with this time scale, E. O. Wilson has argued that microevolution of human populations can occur within 1000 years, major adaptations taking perhaps 2000 years, and speciation 40,000 years.[40] Lumsden and Wilson's 'thousand year rule' formalized this position.[41] Anthropologists have tested culture-led models of evolution in a small-scale society and found evidence of mul-

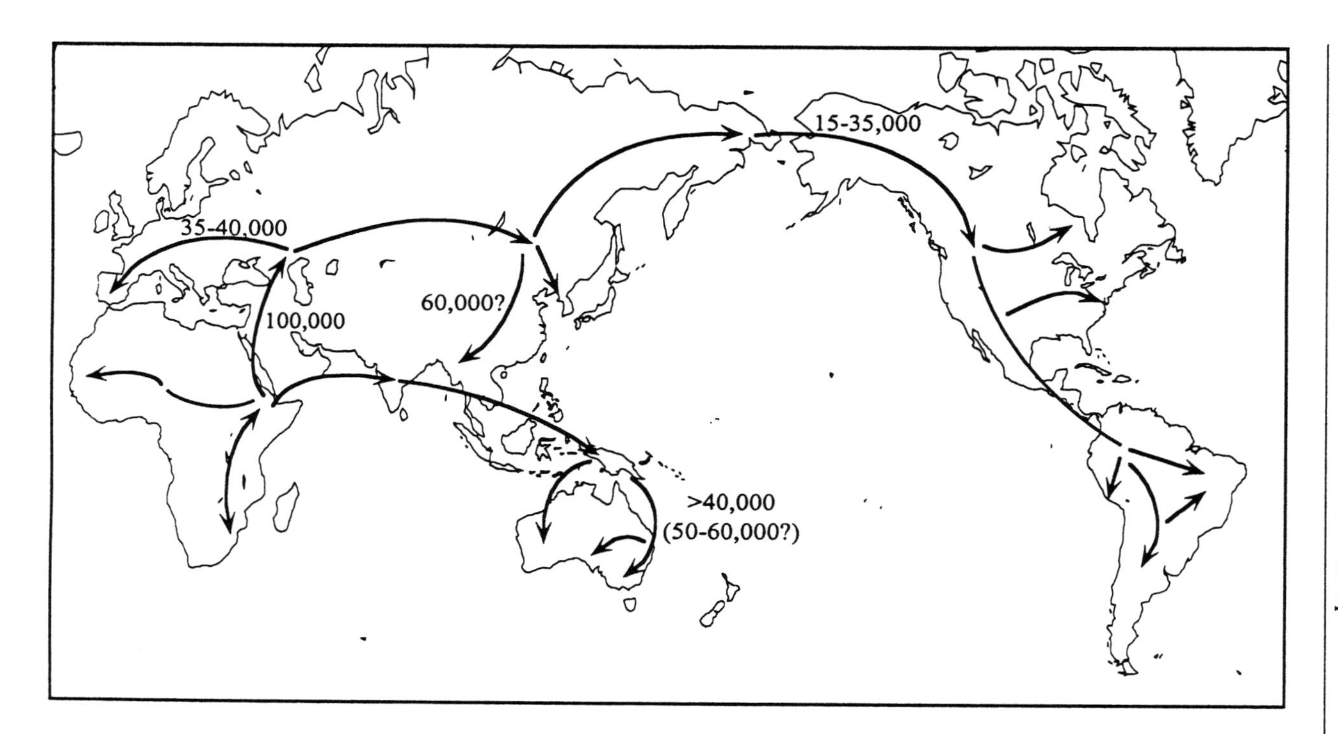


*Map 2.2. The 'out-of-Africa' theory of human origins. Anatomically modern humans originated in East Africa about 200,000 years ago, and began migrating out of the continent about 100,000 years ago (based on Cavalli-Sforza and Cavalli-Sforza [1995, p. 122]).*

tiple group replacements over periods of as little as a few centuries.[42] As discussed earlier, Hamilton argued mathematically that over many generations tribal kinship can reach high levels despite a constant trickle of interbreeding with other tribes.[43] It follows that a 30,000 year separation, whatever the initial admixture, is plenty of time for significant genetic distances between populations to evolve. Some difference might be selected by cultural divergence between adjacent groups within a period of a few centuries, assuming restricted gene flow.

Note that the Eve theory offers the weakest evolutionary foundation for assuming the importance of ethnic kinship. The competing theory of human origins, the multiregional theory,[44] holds that modern humans are the result of ancient hominid populations evolving in parallel. In this theory, the many populations around the world retained membership in the one species through gene flow caused by migration and interbreeding. There almost certainly were some cases of replacement as one group displaced another, but the multiregionalists point to evidence of racial continuity going back much further than 100,000 years. A recent test of the replacement theory found that archaic human skulls resemble modern skulls from the same region to an extent that appeared to be incompatible with the theory of complete replacement.[45] If the multiregional theory is correct, poorly calibrated measures of genetic distance between living human populations might be masking the existence of ancient lineages. It is possible that gene frequencies alone do not tell us everything about group differences in epigenesis, the process of interaction between genes and environment that guides development to produce anatomy and behaviour.[46] If so, the ethnic genetic interests of geographically distant populations—i.e. races—could be much greater than indicated by gene frequencies alone. However, majority opinion among specialists favours the 'out of Africa' theory. Whichever theory is correct, improvements in genetic sequencing techniques and in our understanding of epigenesis will provide more accurate estimates of the genetic interests bound up in ethnies.

A more serious objection to the existence of ethnic genetic interests is Cavalli-Sforza's finding of cross-cutting gene frequencies. For example, he points out a broad cline (slope) of genes running from the Near East to Northwestern Europe, and another running from Southwestern to Northeastern Europe. The implication is that individuals do not in general have more of their genes in the local population than in distant ones. Yet Cavalli-Sforza's own life's work contradicts any such implication because he finds net genetic distances between ethnies. As Cavalli-Sforza himself points out, in calculating the genetic distance between populations it is not sufficient to consider the frequencies of one or a few genes, but 'essential to average the distance between two populations over many genes if one wants reproducible conclusions'.[47] Applying Harpending's analysis to the assay of 120 selectively neutral alleles conducted by Cavalli-Sforza and colleagues[48] indicates that even the various European populations, much closer genetically than continentally separated races, have a genetic dimension, and that the lineage tree connecting those populations is what a reasonably well informed historian or demographer or tourist would expect.

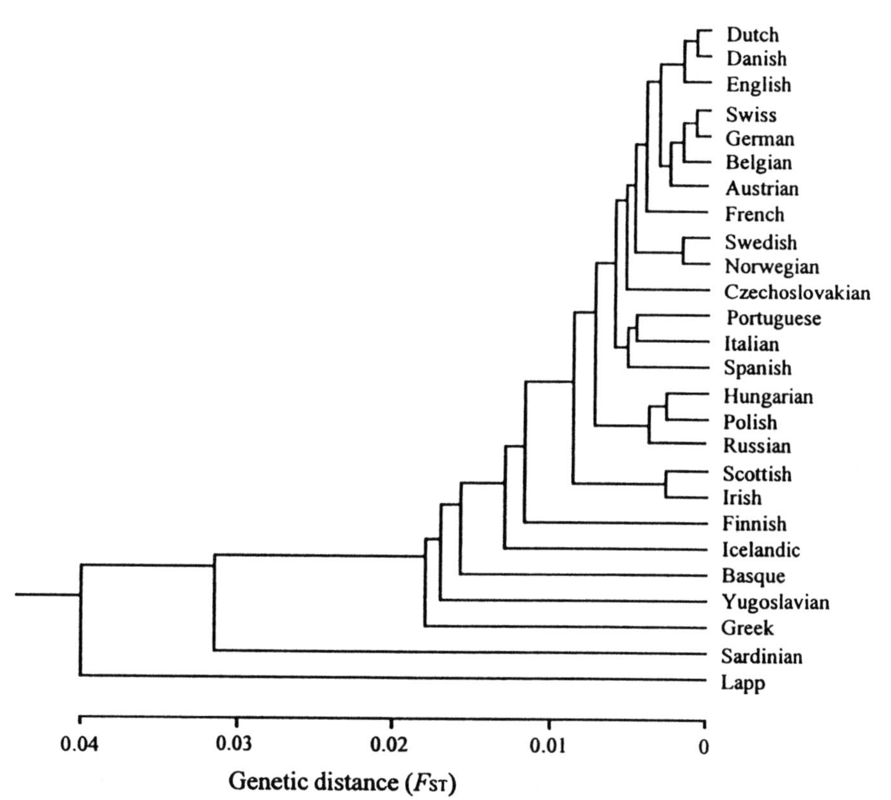


*Figure 2.2 Genetic tree of 26 European populations based on $F_{ST}$ measures estimated from an assay of 88 genes (from Cavalli-Sforza et al. 1994, p. 268).*

Consider for example the genetic tree of 26 European populations presented by Cavalli-Sforza et al.[49] It shows the Swiss and Germans as belonging in the same cluster, with Swedes and Norwegians in another. The groupings are not always intuitive. For example, Portuguese and Italians are closer together than either is to Spanish, though the differences are slight and all three speak romance languages. The French, though speaking a romance language, are closer to the Germanic cluster than they are to the Italian-Portuguese-Spanish cluster. The latter exception to the broad correlation between linguistic and genetic similarity is explained by Cavalli-Sforza et al. as follows.[50] The French language originated from the Roman occupation of Gaul, populated by at least 74 mostly Celtic tribes with some German admixture. However, Celtic peoples were themselves the result of multiple ethnic admixtures within central and northern Europe, and the whole region is relatively genetically homogeneous. Hence French genes are, on average, more similar in frequencies to the genes of central and northern Europe than they are to frequencies found in Italy or the Iberian peninsular, where both Celtic and German settlement came later and in smaller numbers. Another exception to the language-gene correlation is Hungary, whose genes are closer to those of Poles than the latter are to Russians', even though Poland and Russia both speak Slavic languages and Hungarians speak a Uralic language closest to Finnish, Lapp, and Estonian. Cavalli-Sforza et al. explain that Hungarian was imposed by the conquering Magyars who originated in the Urals. But the Magyar genetic contribution was not enough to alter the population's predominant genetic similarity to its neighbours.[51] Despite linguistic boundaries being weakly correlated with genetic boundaries, European ethnies nevertheless have a fuzzy genetic distinctiveness, contrary to Cavalli-Sforza's summation.

Another reason to doubt the view implied by Cavalli-Sforza—that populations are so mixed that ethnic kinship is effectively zero—is offered by British geneticist Bryan Sykes. Based on comparison of mitochondrial DNA, a genetic sequence inherited from the maternal line, Sykes disconfirmed Cavalli-Sforza's theory that the European gene pool was replaced by Neolithic farmers who entered Europe about 8,000 years ago. The 'demic-diffusion' theory is the reason Cavalli-Sforza emphasizes the cline running from South-East to North-West Europe. According to his theory, the discovery of farming in the Near East led to a demographic explosion and an advancing wave of migrating farmers who arrived in Europe in such large numbers that they swamped the indigenous hunter-gathers who had settled there tens of thousands of years earlier.[52]

If Cavalli-Sforza's theory were correct then Europeans could not trace much of their ancestry back to the Palaeolithic and to the Cro-Magnons who replaced the Neanderthals. The theory is subversive of belief in European distinctness, since it amounts to the view that Europeans are relatively recent settlers and largely of Near Eastern origin. The slight but noticeable regional differences in colouration, physiognomy, and physique would be surface characteristics produced by only a few thousand years' evolution acting on a Near Eastern population, instead of the result of tens of millennia of environmental and social selection operating on indigenous peoples. However, in 1996 Sykes and colleagues advanced an analysis of mitochondrial DNA indicating that only 15–20 percent of the European gene pool is derived from the Near East.[53] Moreover, regional clusters of mitochondrial strains pointed to six founding populations—at least on the maternal line—dating from between 45,000 and 15,000 years ago to the original Cro-Magnon settlement of various different regions of Europe (a seventh founding population was the early farmers who entered Europe in the Neolithic). For example, the mitochondrial analysis of Basque genes indicated that this population was not a rare remnant of the original European hunter-gatherer population, but a representative European population.[54] As Sykes expressed the implications of his finding for European identity, 'These were not the faint whispers of a defeated and sidelined people but a resonant and loud declaration from our hunter-gatherer ancestors: "We are still here.".'[55]

Sykes's mitochondrial analysis of European origins was confirmed by an assay of Y-chromosomes of 1007 European males conducted by an international team that included Cavalli-Sforza.[56] This chromosome, which determines male sex, is passed down the paternal line. This study confirmed the 20 percent admixture of Near Eastern genes in European populations and found evidence of ten founding tribes in the Palaeolithic. Thus clusters of Europeans, while showing some relatively recent admixture from outside Europe, are a largely autochthonous people who can trace their origins back for several tens of thousands of years to a small number of founding tribes. Cavalli-Sforza's broad cross-cutting clines do exist, but have not in the main reduced regional distinctiveness. Ethnic genetic interests exist within a region even as criss-crossed with migration routes as Europe.

Cavalli-Sforza emphasizes the fuzziness in the genetic map of Europe. It is likely that some group differences are so small that only a marginal genetic interest exists in the home group. The ethny has little or no genetic interest for those individuals whose characteristic genes occur at low frequencies in the population and who have few blood relatives within it. While this should not be ignored, neither should the fact that ethnic genetic interests are usually larger than those found within families. For the majority the interest is real *vis-à-vis* most other groups, so long as the group is a descent group, which most ethnies are.[57] As is clear from Cavalli-Sforza et al.'s world-wide data reproduced in Table 3.1, genetic differences can be relatively sharp when two groups are brought together due to transcontinental migration. In this case the hundreds of intervening clines concertina to form a deep fault line. The underlying genetic distance between such populations is often marked by pronounced physical differences of colour, stature and physiognomy, as well as many other differences, including proneness to different diseases and responses to medication.[58] Such racial differences, especially when occurring together in sets, are reliable indicators of large genetic distance.

Improvements in genetic sequencing methods are allowing ever more accurate surveys of the genetic plateaux of human relatedness, measured in gene frequencies and summarized as genetic distances. As techniques and data become more fine-grained, more people should be able to identify their ethnic genetic interests. If present scientific trends continue, they will discover that their ethnies are super families containing large deposits of their distinctive genes. In Chapters 6 and 7 I shall argue that ethnic kinship can be safely mined using the right social technologies. Like the fossil fuels that still power our industry, ethnic kinship was laid down over the ages. Its volatility risks explosion and depletion. It recovers slowly from admixture. Yet it resembles a renewable resource when properly conserved. Ethnic kinship is the ultimate form of social capital, an adaptive basis for altruism and thus trust and cooperation.

To conclude, the theory surveyed in this chapter indicates that concentrations of genetic interests are highest in families, declining to the ethny, thence to geographic race, and declining further in concentric circles of relatedness to the whole species. In the absence of threats to the species as a whole, individuals have no genetic interest in it. However, because of their size, ethnies carry large genetic stakes for their members. Just how large ethnic genetic interests can be is calculated in the next chapter, where I express these interests in units of kinship.

Notes

1 Keith (1947/1968).
2 Boyd and Richerson (1985); Sober and Wilson (1998).
3 van den Berghe (1979, p. 179 fn).
4 Eibl-Eibesfeldt (1982).
5 Eibl-Eibesfeldt (1998/2002, pp. 26–41).
6 Cavalli-Sforza (1991).
7 Eibl-Eibesfeldt (1982, p. 14).
8 Eibl-Eibesfeldt (1989, p. 100).
9 Rushton et al. (1984); Russell et al. (1985).
10 E.g. Grafen (1990, pp. 50–51).
11 Hamilton (1970; 1972); for a discussion see Pepper (2000, pp. 355–6).
12 Grafen (1990, p. 46).
13 Pepper (2000, p. 356).
14 E.g. Grafen (1990).
15 Hamilton (1975, especially pp. 141–50). This paper was called 'reductionist, racist, and ridiculous' by the anthropologist S. L. Washburn in 1976 (quoted by Hamilton 1996, p. 317). The offending passage (pp. 149–50) speculates that barbarian inva-

sions introduce genes coding for altruism into old civilizations, revitalizing them after the passage of a few centuries of assimilation. Hamilton, like Fisher (1930/1999, Chapter 12), thought that civilization was dysgenic, though for different traits. (Fisher was concerned mainly with the decline of traits for economic competence. However, like Hamilton he argued that heroism and clan bonding were found mainly in primitive tribes—heroism often to an excessive degree—and were selected out once such a tribe became civilized [1930/1999, pp. 245–50]). The idea of a decline of gene-based altruism within civilizations is a biological version of the Islamic historian Ibn-Khaldun's (1332-1406) famous thesis set out in *The Mugaddima* (Hill 1993). To the end of his life Hamilton defended his speculations as plausible and worth making. Regarding the paper as a whole, 'I have hardly changed my opinions and certainly haven't with regard to what Washburn indicated as the most offending passage' (Hamilton 1996, p. 317). But he did allow that he should have included the word 'cultural' at one point, to reinforce his already expressed view that there are cultural as well as genetic factors at work in forging altruism. 'However, I have never seen any evidence that a genetic interpretation or a genetic component is out of court and the idea in general continues to be justified to my mind' (ibid.). Hamilton did not think the idea insulted or threatened any race.

16 It is remarkable that leading population geneticists such as L. Cavalli-Sforza have not discovered or used Hamilton's work. Hamilton (1975) and Cavalli-Sforza and Bodmer (1971/1999) appear to have conducted their investigations in ignorance of one another. This is not surprising given that Hamilton's theory took until the mid 1970s to acquire fame among evolutionary biologists (Segerstråle 2000, p. 54). The paper's reputation was enhanced by E. O. Wilson's exposition of the theory in his 1975 text *Sociobiology* and by Dawkins's popularization in *The Selfish Gene* (1976). However, the question remains as to why mainstream population genetics continued to ignore such an important theory. Bodmer and Cavalli-Sforza (1976, p. 554) briefly refer to kin selection theory without offering references, while Cavalli-Sforza et al. (1994) cite none of the sociobiological theorists. Sociobiological research is not cited in Cavalli-Sforza's books (1991; 1995; 2000). Even one of his books that dealt with cultural transmission between kin failed to incorporate inclusive fitness theory (Cavalli-Sforza and Feldman 1981). Part of the explanation is that Cavalli-Sforza has been mainly interested in explaining the evolution of physical traits and culture rather than social behaviour. However, he has repeatedly commented on issues of nationalism, racism, and migration, issues intimately bound up with altruism and kinship. Yet based on Cavalli-Sforza's work alone one would never guess that parents had a genetic interest in their children. By 1975 Hamilton had redefined inclusive fitness theory such that it could estimate inclusive fitness from gene assay data, surely a major step for population geneticists seeking to understand social evolution. The puzzle of Hamilton's absence from Cavalli-Sforza's work is deepened by the fact that both admired the evolutionary theory of R. A. Fisher (e.g. Cavalli-Sforza 2000, p. 22) and had adopted his gene-centred theory. Until Hamilton modified his theory, both adopted essentially the same definition of the coefficient of kinship. Hamilton can be forgiven for failing to reference population genetics literature published years after his pathbreaking paper appeared, but the reverse is perplexing. A scientist with the scholarly sweep of Cavalli-Sforza must have been aware of Hamilton's theoretical work.

17 Another remarkable result of Hamilton's model is that, again contrary to intuition, the ceiling for intra-tribal kinship is not sensitive to either the size of tribes or the pro-

portional rate of migration. What counts is the absolute number of migrants per unit time (1975, pp. 142–3). Low rates of inter-tribal migration (exogamy) would have allowed the retention of high intra-group relatedness, and thus maintained the tribe as a large store of genetic interests.

18 E.g. Tooby and Cosmides (1989, p. 544).

19 It should be noted that Tooby and Cosmides acknowledge the existence of such a tribal genetic interest by arguing that 'an individual's decisions on coalition formation, . . . when summed over the members of the two groups, would often have aggregated into substantial inclusive fitness effects' (1989, p. 544). This misses the evolutionary origin of rising group kinship over hundreds of generations, but is nevertheless consistent with that part of Rushton's (1989b) argument dealing with ethnic group competition. At the proximate level there is some evidence of the importance of genealogical knowledge in primordial group processes. For example, in contemporary primitive societies kinship strongly affects group fissioning, alliance formation, and extended exchange networks (e.g. Chagnon 1980; Koch 1974; Wiessner 2002b).

20 Hamilton (1975, pp. 140–41; [p. 337 in the 1975/1996 reprint]). See also Michod and Hamilton (1980/2001, pp. 108–9), where they state that the 'coefficient of relatedness' and 'regression coefficient of genotype of recipient on genotype of altruist' are the same except when the interactants are inbred to different degrees. Michod and Hamilton (1980/2001) reviewed alternate definitions of relatedness formulated to that date and concluded that they were all equivalent.

21 Hamilton (1975, p. 142).

22 Dawkins (1979, p. 187).

23 Dawkins (1978, p. 67).

24 D. S. Wilson and Sober (1989); D. S. Wilson (2002).

25 Hamilton (1975/1996, p. 360).

26 E.g. Morton (1982); Gudmundsson et al. (2000).

27 Cavalli-Sforza et al. (1994, p. 29).

28 Harpending (1979, p. 624).

29 Scattered statements by Cavalli-Sforza and colleagues, when taken together, indicate that $F_{ST}$, the measure of genetic distance, is also a measure of the kinship coefficient within subdivisions, but I could not find an exposition of the relationship in their work. Cavalli-Sforza and Bodmer (1971/1999, p. 399 ; and see p. 451) state that '[t]he mean kinship coefficient $f$ can be equated to Wahlund's variance'. Cavalli-Sforza et al. (1994, p. 29) denote Wahlund's variance by '$F_{ST}$', the measure of genetic distance used in their world assay. See also Wright (1951).

30 Hamilton (1971, p. 89).

31 Harpending (1979, p. 624); and see Appendix 1 by Harpending.

32 Haldane's (1955).

33 Hamilton (1971, p. 89).

34 D. S. Wilson (2002, pp. 37–9).

35 Hamilton (1975); Harpending (1979).

36 Hedges et al. (1991); A. C. Wilson and Cann (1992).

37 Cavalli-Sforza (2000, pp. 75–6).

38 A genetically-informed de Gobineau might have limited himself to claiming that Europe is the most homogeneous of continents, as do Cavalli-Sforza et al. (1994, p. 122). One might interpret medical geneticists' search for homogeneous populations as

a search for racial purity (Gulcher et al. 2000). Populations that have remained endogamous for many generations are prized by researchers because they facilitate identification of genes responsible for various illnesses. Geographically isolated populations in Finland and Iceland as well as religiously isolated populations such as Jews are examples. These populations are relatively homogeneous, but none is ethnically pure in any absolute sense (e.g. Helgason, A., *et al.* 2000; Thomas et al. 2002).

39 Genetic drift causes kinship between random group members to increase each generation at a rate of $1/2N$, where $N$ is the breeding population (Cavalli-Sforza and Bodmer 1971/1999, p. 707). Assume that ancestral human groups were as large as contemporary tribes and hunter-gatherer dialect groups, about 500, each with 300 reproducing individuals, so that $N = 300$. Then each generation within-population kinship increased by about 0.0017. Assume a generation time of 25 years. Then every thousand years kinship increased by about 0.067, which is slightly greater than the kinship of first cousins. At that rate, in as little as 3,750 years the genetic distance between separated populations would be so great that random group members would have a kinship of 0.25, or that between parent and child.

40 E. O. Wilson (1971, pp. 204–8; 1975, p. 569).

41 Lumsden and Wilson (1981).

42 E.g. Soltis et al. (1995).

43 Hamilton (1975).

44 Thorne and Wolpoff (1992); Wolpoff and Caspari (1997).

45 Wolpoff et al. (2001).

46 Lumsden and Wilson (1981).

47 Cavalli-Sforza et al. (2000, p. 22).

48 Cavalli-Sforza et al. (1994, p. 73).

49 *Ibid.*, p. 268—Figure 2.4.

50 *Ibid.*, pp. 280–85.

51 *Ibid.*, p. 263.

52 'If the population of Europe is largely composed of farmers who gradually immigrated from the Near East, the genes of the original Near Easterners were probably diluted out progressively with local genes as the farmers advanced westwards. However, the density of hunter-gatherers was probably small and the dilution would thus be relatively modest' (Cavalli-Sforza and Bodmer 1976, p. 548, quoted by Sykes 2001, pp. 151–2). This quote was followed by a qualification, to the effect that the degree of demic diffusion of Neolithic Europe by Near Eastern genes was 'not yet settled' and required further research.

53 Richards et al. (1996); Sykes (2001).

54 Sykes (2001, p. 143).

55 *Ibid.*, p. 150.

56 Semino et al. (2000).

57 van den Berghe (1981).

58 Satel (2002).

# 3. Immigration and the Human Dimension of Ethnic Genetic Interests

*Summary*
In the global village all humans become each other's genetic competitor, regardless of intent. Ethnic kinship (ethnic genetic interest) only exists in relation to other ethnies as genetic competitors. Ethnic genetic interest is lost when all or part of an ethny is replaced by another ethny. Immigration effectively replaces part of the native population when it does not permanently raise the territory's carrying capacity by an amount sufficient to allow for the immigrants. Hamilton's theory allows us to express ethnic genetic interest as an equivalent number of children. Asymmetric immigration between closely related ethnies replaces native children in the receiving population by a small amount, while immigration from genetically distant populations has large effects.

In the previous chapter I presented evidence indicating that ethnic genetic interests are usually of significant proportions and often large. 'Ethnic genetic interest' is a cold analytic concept. What does it mean in terms of human values? Familial genetic interest is readily grasped to be a vital interest, The death of a child or other family member has a great emotional impact. Rises and falls in family genetic interests are powerfully registered in the joy of a new birth and the grief of a loved one's passing. But ethnies are composed of anonymous multitudes. The ebb and flow of an ethny do not impress themselves on members as reliably as do fluctuations in family size. A family member who dies in his sleep is mourned, but the decline of an ethny due to low birthrates and peaceful immigration from other ethnies can escape notice, even though the damage is much greater. One might arouse a sense of interest by applying a tribal metaphor, for example by referring to 'the nation' or 'the people', but this large unit is unsuitable for measuring our genetic stake in relatively small numbers of fellow ethnics.

One way to put a human face on ethnic genetic interests would be to count it in equivalent number of family members. In the previous chapter I reported the insight of the great geneticist J. B. S. Haldane, that it is adaptive to give one's life for two siblings or eight cousins, but not for fewer. We are now in a position to apply Haldane's human metric to ethnic kinship, since ethnies are, genetically, extended families. For how many drowning co-ethnics is it adaptive to risk one's life? The answer is at hand, using the theory reported in the previous chapter. Answering Haldane's question is the goal of the present chapter.

Before getting to numbers, it is necessary to identify the modern events affecting ethnic genetic interest. This is ultimately a matter of population size, which can be directly reduced through warfare, genocide, and the loss of limiting resources such as territory. The fact that a 30 percent loss of population is a 30 percent loss of ethnic genetic interest is obvious. But competition can have powerful effects without any behaviour that is aggressive in the usual sense of the word. The prime examples in the contemporary world are peaceful migration between states and high rates of reproduction of one ethnic group within a multiethnic state. Group competition can involve peaceful as well as violent means. Examples of peaceful means are 'competitive breeding'[1] and discrimination, for example in economic affairs. The tribes within which humans have spent so much of their history were territorially based and policed their borders. The same is true of modern states. The special quality of a defended territory is that it insulates the population from the vicissitudes of demographic disturbances in the metapopulation, namely the connected phenomena of uneven population growth and migration. When an ethny controls the borders of a territory that is large enough to support the population, loss of numbers relative to other ethnies is not necessarily fatal, i.e. it need not lead to replacement. Territory adequately defended guarantees continuity and the chance to ride out a temporary downturn in numbers.

Thus a fundamental, though by no means unique, threat to ethnic genetic interests is loss of the ethnic monopoly of a territory. The great advantage of this monopoly is that it facilitates continuity of the tribe's gene pool by boosting the group's ability to defend against mass immigration, whether violent or peaceful. Defence of a territory is a basic ethnic group strategy, a cooperative group effort among members of the same ethny to defend themselves from or compete with members of other ethnies. The strategy is as adaptive today as in primordial environments. A decimated, defeated, or impoverished population has a chance to recover if it retains control of its territory. But a large-scale influx of genetically distant immigrants has the potential permanently to reduce the genetic interest of the original population.

Mass migration between diverse populations combined with the existence of public goods in wealthy societies such as low cost medical support and other forms of welfare have produced effective ethnic competition within many Western states. For example the founding European-derived ethnies of the United States are set to become a minority in that country by the middle of the 21st century, falling to 40 percent by 2100.[2] A similar though slower pattern of replacement is occurring in some other Western societies, due to the combination of below-replacement birth rates and immigration from genetically distant populations. The typical pattern is for the native born population to be falling in numbers due to an excess of deaths over births, but the overall population to be relatively stable due to immigration. This is the case in the following countries (with excess of deaths over births in brackets): Italy (11 percent), Sweden (7 percent), Germany (14 percent), and Austria (0.6 percent).[3] In Britain, government forecasts indicate that immigration alone will account for half the growth of the population in the first two decades of the twenty-first century. Moreover, non-white immigrants usually have higher fertility than the white British population. If the non-white minority continues to grow at the high end of its 2–4 percent annual growth rate reported since 1950, whites will be in the minority in the British Isles by 2100. On current trends, whites will be in the minority in London by 2010.[4] The Dutch port city of Rotterdam has a population of 540,000 people, 45 percent not of Dutch descent.[5] If present trends continue the Hague will be minority Dutch by 2020 and Israelis of Jewish descent will become a minority in Israel by about 2030 due to higher Arab fertility and immigration in the family reunion category.[6] Russia's Far East, explored and settled in the seventeenth century by a Slavic population, is being swamped by Chinese migrating north. Unless the trend is reversed, ethnic Russians are set to lose about 40 percent of their territory or, if they retain it, to be significantly replaced in Russia as a whole.[7] These challenges are real enough for majorities but minorities have usually fared worse as diaspora peoples have discovered through the trials of centuries. Not to control a territory creates risks of repeated group subjugation, displacement, and marginalization. For all of past human experience and still today, control of a territory is a precious resource for maintaining ethnic genetic interests in the long run.

### *Territory and population carrying capacity*

It might seem that [mates and land] would not repay the expected cost of [warfare], but it has to be remembered that to raise mean fitness in a group either new territory or outside mates have to be obtained . . . (W. D. Hamilton).[8]

The continuing relevance of ethnic genetic interests in an age of mass immigration is indicated by ecologist Garrett Hardin's analysis of global and national population carrying capacity.[9] Hardin begins by noting that in the modern world most habitable spaces have been colonized. Moreover, the earth's surface has a limited carrying capacity, as do its parts. This is the maximum population beyond which some value, such as freedom from hunger or overcrowding, is lost.

The most basic carrying capacity is the number beyond which population growth is self correcting, because any further growth is cancelled out by die-offs. Technological advance can increase carrying capacity, but only in finite steps.[10]

E. O. Wilson uses a different analysis to make essentially the same point: the earth is full and its present population is probably unsustainable.[11] His formulation is based on the 'ecological footprint', which is 'the average amount of productive land and shallow sea appropriated by each person in bits and pieces from around the world for food, water, housing, energy, transportation, commerce, and waste absorption'. The developing world has a per capita ecological footprint of about one hectare, while that for the United States is 9.6 hectares. If every human being were to consume at the average level of the United States with existing technology, four more planet earths would be needed. Populations and levels of consumption appear set to continue growing for the time being, but only at the cost of lost biodiversity and a collapse in the earth's capacity to renew ecosystems. Population must level off at some point, whether through design or accident. Either way, average family size must fall to that of zero population growth, about 2.1 children.

Hardin points out that capping global population growth requires that every state limit its numbers to the carrying capacity of its ecological resources.[12] Each society's ecological footprint must not, in the long run, exceed that of the territory it controls. However, a society practising such self discipline is vulnerable to immigration, because any net intake will reduce the relative size of the native population. Hardin does not discuss ethnic genetic interests, but does remark that in this situation the cost of immigration would fall 'most heavily on potential parents, some of who[m] would have to postpone or forego having their (next) child because of the influx of immigrants'. To imagine such an extreme future world one need only contemplate the present situation in China, where the government administers a strict policy limiting families to one child. Note, however. that China's overpopulation is due to past organic population growth, not immigration. Hardin thus only considers individual fitness, though this is an important advance on most analysts. As we saw in the previous chapter, the damage done to the native borns' ethnic genetic interests can be much larger than the effect on their individual fitnesses.

Immigrants must affect a country's capacity to hold the native population. If the immigrants contribute to the economy in ways that the native population cannot, the carrying capacity is raised. If they are a drain on resources or average productivity, they lower that capacity by taking the place of potential native born. In the present example, let us assume that immigrants have equal capacities to the native born, and let us consider immigrants in lots of 10,000. Such a number of immigrants will lower the effective carrying capacity of a country by

10,000, more or less; more if the immigrants have a higher birth rate than the native population; less if their birthrate is lower. To simplify further, assume that birthrates are equal, in which case the loss of effective carrying capacity is 10,000. If the immigrants and native born have the same ethnicity the native population loses no ethnic genetic interests. Kin are being replaced by kin of similar degree. But if the immigrants are from different ethnies, especially genetically distant ones, there will be a loss of genetic interests for each member of the native population. How large is that loss measured in units of kinship?

*Ethnic genetic interest expressed as numbers of kin*

At first sight the data on inter-ethnic kinship might appear rather counterintuitive. Consider the data from Cavalli-Sforza et al. (see Table 3.1).[13] Assuming that these figures accurately reveal overall patterns of ethnic kinship, they indicate that intra-ethnic kinship coefficients range from 0.0021 (random English in relation to Danes) to 0.43 (random Australian Aborigines in relation to Mbuti, a Pygmy population in Central Africa). Yet one would expect that kinships between autochthonous populations within the one region to be closer than those between continentally-separated populations. This seeming paradox is resolved by the realization that there is an inverse relationship between kinship within and between populations. This rule can be illustrated using Table 3.1. Consider the two examples just mentioned. Recall from the discussion in the previous section that if the world population were wholly English then the kinship between random pairs would be zero. But if the world consisted of the English and Danes, then two random Englishmen would have a slightly positive kinship of 0.0021 (kinship coefficients are multiplied by 10,000 in Table 3.1). This is slightly closer than the kinship of eight linear generations separation, or a descendant to his or her great great great great great great grandparent. In other words, because English and Danes are very close genetically, there is only marginal kinship between randomly chosen English (or Danes). A random Englishman would not lose many copies of his distinctive genes if another random Englishman were replaced by a Dane.

Since the kinships within and between populations stand in an inverse relation, an Australian Aborigine loses many more genes if a random fellow ethnic is replaced by an African, such as a Mbuti. This is because Australian Aborigines and the Mbuti Pygmies are very distant kin according to Cavalli-Sforza's data. In a world consisting of Aborigines and Mbuti, two random Aborigines (or Mbuti) are very *close* kin, almost as closely related as identical twins, with a kinship of 0.43 (identical twins have a kinship of 0.5). Such a replacement reduces the Abo-

| | BAN | EAF | WAF | SAN | MBU | IND | IRA | NEA | JPN | KOR | MNK | THA | MNG | MAL | FIL | NTU | SCH | BAS | DAN | ENG | GRK | ITA | CAM | ESK | PLY | AUS |
|---|---|---|---|---|---|---|---|---|---|---|---|---|---|---|---|---|---|---|---|---|---|---|---|---|---|---|
| Bantu | 0 | | | | | | | | | | | | | | | | | | | | | | | | | |
| E. African | 658 | 0 | | | | | | | | | | | | | | | | | | | | | | | | |
| W. African | 188 | 697 | 0 | | | | | | | | | | | | | | | | | | | | | | | |
| San | 94 | 776 | 885 | 0 | | | | | | | | | | | | | | | | | | | | | | |
| Mbuti | 714 | 1232 | 801 | 1495 | 0 | | | | | | | | | | | | | | | | | | | | | |
| Indian | 2202 | 1078 | 1748 | 1246 | 2663 | 0 | | | | | | | | | | | | | | | | | | | | |
| Iranian | 2241 | 1060 | 1796 | 1267 | 2588 | 154 | 0 | | | | | | | | | | | | | | | | | | | |
| Near Eastern | 1779 | 709 | 1454 | 880 | 2138 | 229 | 158 | 0 | | | | | | | | | | | | | | | | | | |
| Japanese | 2361 | 1345 | 2252 | 1905 | 3089 | 718 | 1059 | 1056 | 0 | | | | | | | | | | | | | | | | | |
| Korean | 2668 | 1475 | 1807 | 1950 | 2996 | 681 | 905 | 933 | 137 | 0 | | | | | | | | | | | | | | | | |
| Mon Khmer | 2446 | 1538 | 1951 | 1977 | 2766 | 866 | 1282 | 987 | 961 | 946 | 0 | | | | | | | | | | | | | | | |
| Thai | 3364 | 1602 | 2480 | 2064 | 3872 | 852 | 1155 | 1023 | 743 | 814 | 99 | 0 | | | | | | | | | | | | | | |
| Mongol-Tungus | 2882 | 1423 | 1733 | 1398 | 2568 | 509 | 681 | 827 | 218 | 170 | 1093 | 957 | 0 | | | | | | | | | | | | | |
| Malaysian | 1658 | 1216 | 1365 | 1434 | 1743 | 1130 | 1489 | 1173 | 1175 | 1001 | 264 | 455 | 1251 | 0 | | | | | | | | | | | | |
| Filipino | 2913 | 1770 | 2299 | 1922 | 3776 | 872 | 908 | 909 | 1020 | 1218 | 552 | 625 | 737 | 485 | 0 | | | | | | | | | | | |
| N. Turkic | 2486 | 1386 | 2163 | 1448 | 2989 | 638 | 821 | 710 | 627 | 732 | 1259 | 1225 | 728 | 1189 | 1044 | 0 | | | | | | | | | | |
| S. Chinese | 2963 | 1664 | 1958 | 2231 | 3384 | 847 | 1092 | 983 | 541 | 498 | 254 | 105 | 705 | 635 | 315 | 1109 | 0 | | | | | | | | | |
| Basque | 1474 | 922 | 1299 | 1307 | 1965 | 418 | 285 | 246 | 1481 | 1063 | 1831 | 1726 | 1049 | 1784 | 1634 | 903 | 1675 | 0 | | | | | | | | |
| Danish | 1708 | 909 | 1458 | 1025 | 1462 | 293 | 179 | 238 | 1176 | 947 | 1463 | 1390 | 680 | 1628 | 1279 | 820 | 1306 | 184 | 0 | | | | | | | |
| English | 2288 | 1163 | 1487 | 1197 | 2373 | 280 | 197 | 236 | 1244 | 982 | 1100 | 1143 | 896 | 1275 | 1117 | 866 | 1152 | 119 | 21 | 0 | | | | | | |
| Greek | 1479 | 892 | 1356 | 1068 | 1735 | 272 | 70 | 129 | 1175 | 904 | 1482 | 1355 | 735 | 1482 | 1109 | 794 | 1095 | 231 | 191 | 204 | 0 | | | | | |
| Italian | 2292 | 1234 | 1794 | 1181 | 2931 | 261 | 133 | 208 | 1145 | 936 | 1446 | 1382 | 905 | 1599 | 1136 | 949 | 1236 | 141 | 72 | 51 | 77 | 0 | | | | |
| C. Amerind | 2237 | 1475 | 2293 | 2143 | 3499 | 1089 | 1199 | 1037 | 658 | 790 | 1522 | 1323 | 970 | 1731 | 1527 | 859 | 1192 | 1539 | 1266 | 1246 | 1271 | 1198 | 0 | | | |
| Eskimo | 3251 | 2116 | 2693 | 2217 | 3329 | 940 | 1234 | 1225 | 791 | 843 | 1595 | 1417 | 545 | 1617 | 1597 | 796 | 1304 | 1637 | 1180 | 1185 | 1254 | 1135 | 903 | 0 | | |
| Polynesian | 2649 | 1414 | 1992 | 1940 | 3136 | 927 | 1142 | 869 | 823 | 890 | 860 | 589 | 969 | 849 | 650 | 1147 | 508 | 1406 | 1210 | 991 | 1096 | 1215 | 1312 | 1415 | 0 | |
| Australian | 3272 | 2131 | 2694 | 2705 | 4287 | 1176 | 1546 | 1408 | 321 | 850 | 1699 | 1314 | 781 | 1665 | 1300 | 1580 | 1081 | 1949 | 1400 | 1534 | 1498 | 1413 | 1360 | 1230 | 1145 | 0 |
| | BAN | EAF | WAF | SAN | MBU | IND | IRA | NEA | JPN | KOR | MNK | THA | MNG | MAL | FIL | NTU | SCH | BAS | DAN | ENG | GRK | ITA | CAM | ESK | PLY | AUS |

*Table 3.1. $F_{ST}$ distances x 10,000 between selected populations around the world (sampled from Cavalli-Sforza et al. 1994, p. 75; standard errors are omitted).*

rigine's inclusive fitness almost as much as eliminating an identical twin or two children.

Taken together with the limits imposed by carrying capacity, these kinship coefficients mean that, other factors being equal, asymmetric immigration is more harmful to the receiving population's genetic interests the more genetically distant the immigrants. The $F_{ST}$ distances shown in Table 3.1 allow us to estimate the loss in genetic interest caused by cross-migration of 10,000 individuals between 26 native populations. Appreciation of the genetic interests involved is aided by converting this loss to child-equivalents. Losses of genetic interest will not be counted in units of random fellow ethnics, but in the larger unit of offspring. As will be discussed in greater detail in Chapter 5 (p. 117–133), in Hamilton's town model the kinship of random co-ethnics could rise as high as that between parent and child in outbred populations.[14] Hamilton noted that when this happened, actual parent-child kinship would rise significantly higher, though he did not specify how high. Harpending (see Appendix 1) offers a formula for that higher figure, based on the $F_{ST}$ between populations.

$$f = 0.25 + 3F_{ST}/4 \qquad \ldots 3.1$$

Let us apply this formula in considering the impact on the genetic interests of a random Englishman of 10,000 ethnic Danes replacing 10,000 ethnic English (or vice versa). To simplify, let us assume that this is a neat replacement, so that over succeeding generations all the immigrants survive to reproduce.[15] We also assume that the Englishman loses no genealogical kin in the process. Replacement involves two effects, the removal of 10,000 Englishmen and the introduction of 10,000 Danes who, according to the population genetics formula we are using, have negative kinship to the English population. Recall that the English-Danish $F_{ST}$ is 0.0021. Then removal of 10,000 Englishmen in this case reduces the genetic interest of our random Englishman by 10,000 x 0.0021 = 21 kinship units. The replacing Danes bring a negative kinship of the same magnitude. Subtracting the latter from the former gives a loss to the Englishman's genetic interests of 42 kinship units. But what do those units mean in human terms? We can express those units as a number of children by dividing 42 by the parent-child kinship of the English. The latter is

$$f = 0.25 + (0.0021 \times 3/4)$$
$$= 0.2516$$

The number of children lost due to the immigration of 10,000 Danes is therefore $42/0.2516 \approx 167$. This is a large family indeed. Due to the loss of fellow ethnics alone the Englishman's genetic interests would be lowered by the equivalent of 167 children (or siblings). Repeating the scenario with Bantu immigrants, the loss to English genetic interests of replacement of 10,000 English is 10,854 children (or siblings).[16] Bantu suffer the same loss from 10,000 English immigrants to a Bantu territory.

This last figure is puzzlingly high. How can the loss due to replacement exceed the number replaced? Actually it does not, because we are counting genes, not individuals. The result only looks strange when gene counts are converted into child-equivalents, though the conversion is valid. Random members of an ethnic group are concentrated stores of each other's distinctive genes, just as children and cousins are concentrated stores. Some ethnies are so different genetically that they amount to negative stores of those distinctive genes. So migration has a double impact on fitness, first by reducing the potential ceiling of the native population, and secondly by permanently replacing those lost individuals' familiar genes with exotic varieties. Referring to carrying capacity also helps understand the dramatic loss of fitness caused by asymmetrical interracial immigration. In a large nation the loss of 10,000 fellow ethnics, say due to a natural catastrophe or war, could be made up in a generation, as the population rose to the country's carrying capacity. Filling their places with immigrants reduces that loss by an amount equal to the immigrants' kinship with the natives. When Danes replace English, the loss is almost completely reduced because the two ethnies are closely related. But when Bantu replace English (or vice versa) the loss is exacerbated, because these populations have large negative kinship with one another.

The loss is not diminished by somehow being spread across the entire ethny. For the native it is a collective loss in the same way that collective goods are shared without being diminished. It applies to every randomly-chosen member of the native ethny, wherever he or she may live. Children *per se* are not lost. We are assuming that there will be the same number of children in the society. Neither is it a symbolic loss for people everywhere, like some humanitarian disaster. After all, the immigrants produce replacement children. For them the process of ethnic replacement increases fitness. The loss is limited to the native ethny in a very personal way. For a native woman it is equivalent to the loss of *her* children and grandchildren, for a native man it is equivalent to the loss of *his* children and grandchildren, though on a much larger scale. The large scale of these ethnic genetic interests means that the loss is only slightly mitigated if these individuals' own children are not replaced.

It becomes clear from the above analysis that ethnic genetic interests are usually very large compared to familial genetic interests. The mathematics on which these estimates rely would need to be in gross error for this not to be true. Indeed, for inter-racial immigration the losses would still be large if ethnic kinship were one hundredth the values estimated above. Neither would inaccuracies in other assumptions of the analysis alter the result by much. For example, consider the mitigating effect of immigrants increasing the receiving country's long-term carrying capacity. I have assumed nil economic benefit (and cost), but the genetic loss will still be large even if the economic benefit mitigates half or even 90 percent of the loss. The result is also robust for all except radical modifications of the assumption of a neat replacement of natives by immigrants, since partial replacement still causes large losses of ethnic kin. Finally, two-way migration between ethnies must be almost perfectly symmetrical for costs and benefits to cancel.

Ethnies are indeed super families as van den Berghe argued. Although being more dilute stores of genetic interest than families, ethnies can number in the millions and so are often orders of magnitude more precious. If immigrants replaced one quarter of the English nation of approximately 50 million people, the remainder would suffer a very large loss even if their own relatives were not affected. If 12.5 million Danes and similar peoples moved to England, the genetic loss to the remaining English would be the equivalent of 209,000 children. The corresponding loss due to the same number of immigrants from India would be 2.6 million children, and due to Bantus over 13 million children. All these losses apply in the reverse direction, if there was mass English immigration to Northern Europe, India, or Bantu Africa.

Large ethnic genetic interests make public charity and self-sacrificial heroism directed towards one's ethnic group potentially adaptive. As we see from the above estimates, ethnic altruism is most adaptive when it aids fellow ethnics in the face of competition from genetically distant ethnies, such as those belonging to different geographic races. Figure 3.1 shows the relative genetic distances of the major races, based on the world-wide assay by Cavalli-Sforza et al.[17] The genetic distance measures between these races are shown in Table 3.2. Subjects were individuals who could trace their ancestry from indigenous populations, the autochthonous peoples that inhabited a region before the great migrations of the modern era began to mix geographical races from about 1600.

Like other racial differences, the English-Bantu genetic distance is large because the populations have been separated for many thousands of years. The genetic distance between English and Bantu is so great that, on the face of it, competition between them would make within-group altruism among random English (or among random Bantu) almost as adaptive as parent-child altruism, if the altruism were in the service of that competition. Thus it would appear to be more adaptive for an Englishman to risk life or property resisting the immigration of two Bantu immigrants to England than his taking the same risk to rescue one of

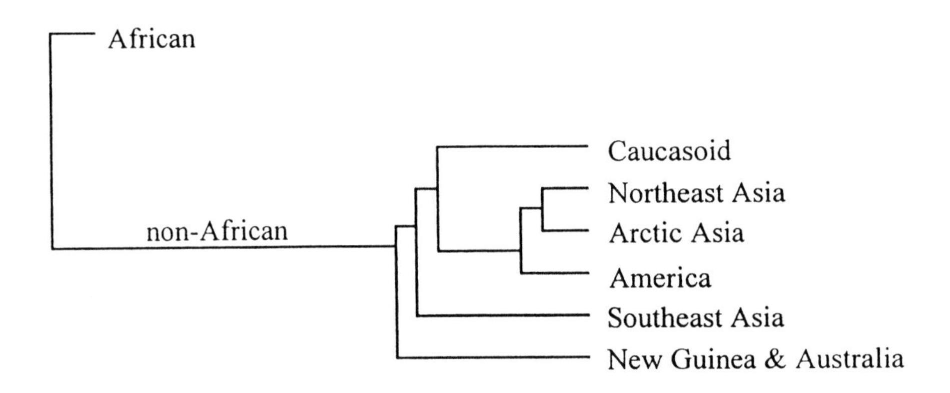


*Figure 3.1. Genetic distances between seven geographical races. Clustering is apparent (drawn from Cavalli-Sforza et al. 1994, p. 79).*

| | AFR | NEC | EUC | NEA | ANE | AME | SEA | PAI |
|---|---|---|---|---|---|---|---|---|
| Africans | 0.0 | | | | | | | |
| Non-European Caucasoids | 1340 | 0.0 | | | | | | |
| European Caucasoids | 1656 | 155 | 0.0 | | | | | |
| Northeast Asians | 1979 | 640 | 938 | 0.0 | | | | |
| Arctic Northeast Asians | 2009 | 708 | 747 | 460 | 0.0 | | | |
| Amerindians | 2261 | 956 | 1038 | 747 | 577 | 0.0 | | |
| Southeast Asians | 2206 | 940 | 1240 | 631 | 1039 | 1342 | 0.0 | |
| Pacific Islanders | 2505 | 954 | 1345 | 724 | 1181 | 1741 | 437 | 0.0 |
| New Guineans and Australians | 2472 | 1179 | 1346 | 734 | 1013 | 1458 | 1238 | 809 |
| | AFR | NEC | EUC | NEA | ANE | AME | SEA | PAI |

*Table 3.2. Racial kinship coefficients of nine geographical races x 10,000. Racial kinship between random individuals varies according to competition (e.g. potential immigration) from different races. Random co-ethnics have zero kinship when the ethnic group is considered in isolation. (From $F_{ST}$ distances provided by Cavalli-Sforza et al. 1994, p. 80; rounded to nearest integers; standard errors omitted).*

his own children from drowning, unless the immigrants were bringing qualities of such economic value that they would permanently raise the Island's carrying capacity. The same applies in the reverse direction; two Englishmen migrating to Bantu Africa constitute a greater loss of long-term genetic interest than does a random Bantu losing a child. But the genetic distance between English and

Danes is so small that in the equivalent situation it would be adaptive to take only a small risk in resisting small numbers of immigrants. Despite the potentially large payoff, intra-ethnic altruism can be maladaptive when it benefits the reproduction of free riders within the group. This problem and institutional counter-strategies are discussed in Chapter 6.

Cross racial immigration is most damaging to ethnic genetic interests. This becomes apparent if one considers the number of immigrants needed to reduce the ethnic genetic interest of a random native by the equivalent of one child (assuming as before that immigrants are economically equivalent to natives and that there is a finite carrying capacity for each country). Table 3.3 shows replacement numbers for immigration between races distributed around the world. Table 3.4 shows the numbers for immigration within Europe, a generally homogeneous region. The latter are usually about an order of magnitude greater than the former, meaning that immigration within a race is usually less harmful to ethnic genetic interests than immigration between races. There are exceptions. According to these data, immigration of non-European Caucasoids to Europe (and the reverse) is less harmful than Lapps or Sardinians immigrating to England (or the reverse) (see Table 3.4). Note, however, that this exception occurs within regional subsets of the broad Caucasoid racial group. The overall pattern is clear. Within local regions of Europe the genetic impact of immigration declines up to another order of magnitude. Immigration between ethnies of the same race can still be maladaptive for the receiving population, but the threshold is typically 10 to 100 times that of inter-racial immigration.

| Immigrants/host | AFR | NEC | EUC | NEA | ANE | AME | SEA | PAI |
|---|---|---|---|---|---|---|---|---|
| Africans | ∞ | | | | | | | |
| Non-European Caucasoids | 1.3 | ∞ | | | | | | |
| European Caucasoids | 1.1 | 8.5 | ∞ | | | | | |
| Northeast Asians | 1.0 | 2.3 | 1.7 | ∞ | | | | |
| Arctic Northeast Asians | 1.0 | 2.1 | 2.0 | 3.1 | ∞ | | | |
| Amerindians | 0.9 | 1.7 | 1.6 | 2.0 | 2.5 | ∞ | | |
| Southeast Asians | 0.9 | 1.7 | 1.4 | 2.4 | 1.6 | 1.3 | ∞ | |
| Pacific Islanders | 0.9 | 1.7 | 1.3 | 2.1 | 1.4 | 1.1 | 3.2 | ∞ |
| New Guineans and Australians | 0.9 | 1.4 | 1.3 | 2.1 | 1.6 | 1.2 | 1.4 | 1.9 |
| | AFR | NEC | EUC | NEA | ANE | AME | SEA | PAI |

*Table 3.3. Number of immigrants between nine geographical races needed to reduce the ethnic genetic interest of a random native by the equivalent of one child.*

| | BAS | LAP | SAR | AUT | CZE | FRE | GER | POL | RUS | SWI | BEL | DAN | DUT | ENG | ECE | IRI | NOR | SCO | SWE | GRK | ITA | POR | SPA | YUG | FIN |
|---|---|---|---|---|---|---|---|---|---|---|---|---|---|---|---|---|---|---|---|---|---|---|---|---|---|
| Basque | ∞ | | | | | | | | | | | | | | | | | | | | | | | | |
| Lapp | 2.4 | ∞ | | | | | | | | | | | | | | | | | | | | | | | |
| Sardinian | 5.2 | 2.2 | ∞ | | | | | | | | | | | | | | | | | | | | | | |
| Austrian | 6.8 | 4.4 | 4.6 | ∞ | | | | | | | | | | | | | | | | | | | | | |
| Czech | 8.2 | 3.0 | 4.2 | 35.1 | ∞ | | | | | | | | | | | | | | | | | | | | |
| French | 13.8 | 3.9 | 4.8 | 33.3 | 17.7 | ∞ | | | | | | | | | | | | | | | | | | | |
| German | 7.8 | 4.4 | 4.2 | 66.2 | 24.4 | 46.7 | ∞ | | | | | | | | | | | | | | | | | | |
| Polish | 8.9 | 3.5 | 4.8 | 17.7 | 19.9 | 19.3 | 27.0 | ∞ | | | | | | | | | | | | | | | | | |
| Russian | 9.3 | 4.2 | 5.1 | 19.9 | 17.0 | 21.6 | 21.2 | 42.0 | ∞ | | | | | | | | | | | | | | | | |
| Swiss | 8.0 | 3.7 | 3.9 | 104 | 40.7 | 54.7 | 125 | 21.2 | 16.4 | ∞ | | | | | | | | | | | | | | | |
| Belgian | 12.1 | 4.1 | 5.3 | 78.5 | 29.4 | 39.4 | 83.7 | 31.6 | 24.9 | 89.7 | ∞ | | | | | | | | | | | | | | |
| Danish | 7.2 | 4.1 | 4.0 | 46.7 | 23.5 | 29.4 | 78.5 | 18.5 | 16.0 | 66.2 | 59.9 | ∞ | | | | | | | | | | | | | |
| Dutch | 11.0 | 4.0 | 4.4 | 33.3 | 19.3 | 39.4 | 78.5 | 23.5 | 22.3 | 78.5 | 104 | 139 | ∞ | | | | | | | | | | | | |
| English | 10.9 | 3.5 | 4.1 | 23.1 | 21.2 | 52.5 | 57.2 | 18.2 | 16.2 | 45.0 | 83.7 | 59.9 | 73.9 | ∞ | | | | | | | | | | | |
| Icelandic | 6.0 | 2.9 | 3.5 | 8.5 | 7.6 | 8.9 | 12.2 | 9.1 | 7.8 | 11.2 | 16.4 | 14.6 | 12.8 | 16.8 | ∞ | | | | | | | | | | |
| Irish | 9.0 | 2.6 | 3.6 | 11.2 | 11.1 | 13.8 | 15.3 | 8.7 | 8.2 | 14.9 | 17.0 | 18.8 | 16.8 | 42.0 | 13.0 | ∞ | | | | | | | | | |
| Norwegian | 6.8 | 4.3 | 3.3 | 20.9 | 16.8 | 22.7 | 59.9 | 21.9 | 14.3 | 38.3 | 52.5 | 66.2 | 59.9 | 50.4 | 17.3 | 16.2 | ∞ | | | | | | | | |
| Scottish | 8.9 | 3.2 | 3.9 | 17.3 | 12.4 | 20.5 | 24.0 | 10.7 | 10.1 | 21.6 | 21.6 | 31.6 | 26.4 | 46.7 | 11.6 | 43.5 | 21.9 | ∞ | | | | | | | |
| Swedish | 7.8 | 4.1 | 3.7 | 16.0 | 14.3 | 16.4 | 32.4 | 15.6 | 11.7 | 23.1 | 37.1 | 35.1 | 30.9 | 34.2 | 12.2 | 13.7 | 69.8 | 17.3 | ∞ | | | | | | |
| Greek | 5.8 | 4.4 | 7.0 | 14.9 | 10.3 | 9.9 | 9.1 | 7.4 | 8.1 | 8.8 | 12.5 | 6.9 | 6.7 | 6.5 | 4.7 | 4.7 | 5.7 | 5.3 | 5.8 | ∞ | | | | | |
| Italian | 9.2 | 4.1 | 6.0 | 29.4 | 16.6 | 37.1 | 33.3 | 19.9 | 17.0 | 28.8 | 42.0 | 17.7 | 19.9 | 24.9 | 9.1 | 9.8 | 14.6 | 11.5 | 13.5 | 16.6 | ∞ | | | | |
| Portuguese | 9.0 | 4.2 | 4.1 | 26.4 | 27.5 | 26.4 | 24.9 | 19.6 | 13.1 | 24.0 | 40.7 | 16.6 | 21.2 | 8.9 | 8.8 | 11.2 | 17.5 | 13.3 | 16.4 | 12.5 | 28.8 | ∞ | | | |
| Spanish | 12.4 | 3.1 | 4.6 | 18.5 | 19.6 | 32.4 | 18.5 | 11.1 | 10.6 | 29.4 | 30.1 | 16.0 | 16.8 | 27.0 | 8.0 | 11.4 | 13.3 | 12.9 | 13.0 | 8.1 | 20.9 | 26.4 | ∞ | | |
| Yugoslavian | 7.5 | 2.6 | 4.6 | 11.7 | 12.8 | 10.5 | 11.0 | 9.5 | 7.7 | 10.8 | 25.4 | 8.3 | 9.6 | 8.2 | 4.3 | 5.0 | 7.6 | 5.4 | 6.2 | 6.2 | 10.9 | 9.4 | 7.6 | ∞ | |
| Finnish | 5.7 | 6.3 | 4.1 | 16.6 | 7.5 | 12.1 | 16.6 | 9.4 | 8.5 | 11.5 | 20.2 | 13.4 | 10.5 | 11.2 | 8.3 | 6.0 | 13.7 | 7.9 | 15.6 | 8.7 | 13.7 | 10.9 | 8.2 | 5.4 | ∞ |
| Hungarian | 8.5 | 4.1 | 4.9 | 31.6 | 18.5 | 18.2 | 27.5 | 50.4 | 42.0 | 22.3 | 24.4 | 16.4 | 18.0 | 18.2 | 7.6 | 8.6 | 16.6 | 10.5 | 13.0 | 14.6 | 20.9 | 20.2 | 11.0 | 9.6 | 11.2 |
| | BAS | LAP | SAR | AUT | CZE | FRE | GER | POL | RUS | SWI | BEL | DAN | DUT | ENG | ECE | IRI | NOR | SCO | SWE | GRK | ITA | POR | SPA | YUG | FIN |

*Table 3.4. Number of immigrants between 26 European ethnies needed to reduce the ethnic genetic interests of a random native by the equivalent of one child. (Based on $F_{ST}$ genetic distances provided by Cavalli-Sforza et al., 1994, p. 270).*

There is a striking contrast between the replacement effects shown in Table 3.3 and 3.4. It requires only 1.1 African immigrants to depress the European genetic interest by the equivalent of one child (or vice versa). But it takes 59.9 Danish immigrants to have the same effect on an English population, or 27 Polish immigrants on Germans, or 42 English immigrants on Irish (and all vice versa). The infinity symbol '∞' is meant to show that a large number of immigrants from the same ethny can immigrate without reducing ethnic genetic interests. In fact too large a rise in population numbers from any source will exhaust resources and endanger long-term interests.

The replacement effects shown in Tables 3.3 and 3.4 put in a harsh light the widespread assumption that ethnic competition is no longer adaptive.[18] Take the United States as an example, a country frequently held up as the unambiguous economic beneficiary of growing ethnic diversity. The United States is also the most powerful state in the world, with the largest economy and most effective armed forces. Yet like undeveloped peripheral economies that bore the brunt of colonial expansion from 1500, it is failing to defend the genetic interests of the majority of citizens.

Earlier in the chapter I mentioned the rapid decline in white American's relative fitness. That trend warrants closer inspection and consideration of its impact on other ethnies. Since the immigration reform bill of 1965, Americans of European descent have fallen rapidly in relative numbers. In 1960, the white population was 88.6 percent.[19] By the 2000 census the non-Hispanic white population was down to 69.6 percent of the population.[20] While the white population kept growing, the higher birth rate of minorities and the large immigration influx of almost one million legal immigrants per year plus many illegals, caused its proportion of the population to decline by 21 percent within two generations. The US Census Bureau projection is that by 2050 the white proportion of the population will be 52.8 percent.[21] The United States' founding population is heading towards minority status by 2060.

By the same token, the immigrants are benefiting their own genetic interests. Their home countries typically accept no immigrants and have much higher fertility rates and populations than do Western societies. So the process is a boon for the genetic interests of immigrant ethnies.

Many post-1965 immigrants with professional qualifications have arguably boosted the economy above the load they exert on it (but perhaps not the country's long-term carrying capacity). However, most have been a net burden both on the public purse and the jobs of native Americans.[22] Also the great majority of immigrants are genetically distant from European and African Americans, coming mainly from Mexico and East Asia. Cavalli-Sforza et al. estimate the genetic distance between Europeans and American Indian as 0.1038, which has been re-

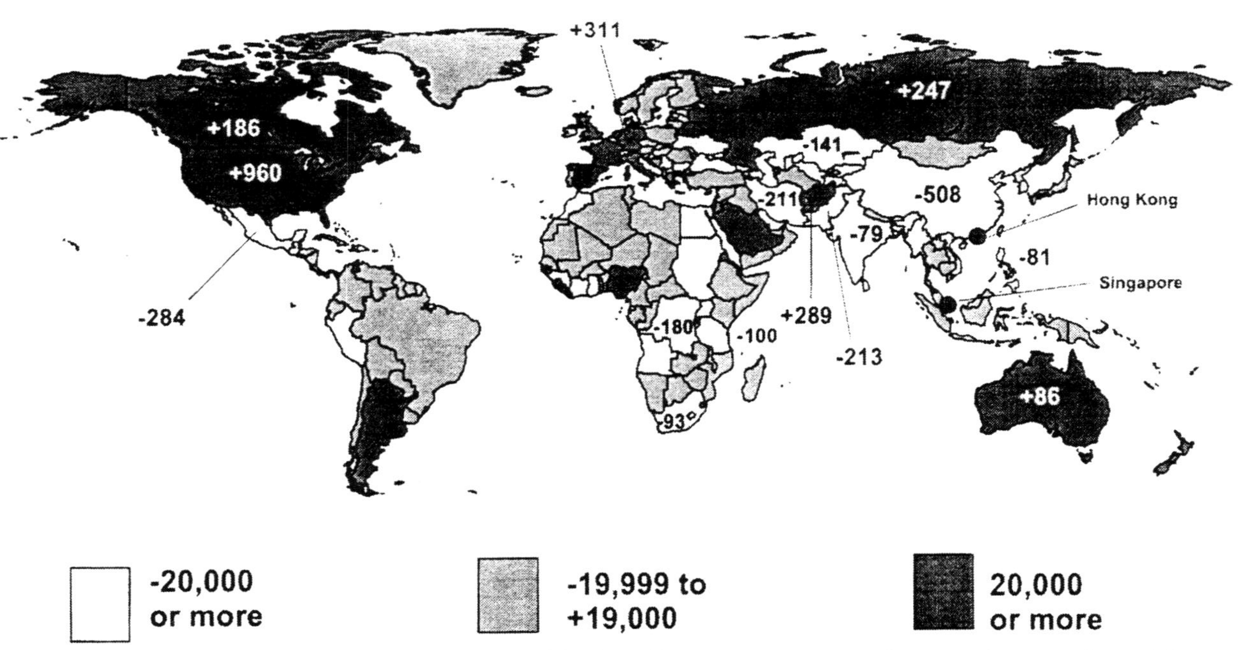


*Map 3.1. Global net annual migration by country, 1996–2001. Most western societies show significant intakes, while non-western societies are usually neutral or show outflow (based on U.S. Bureau of Census estimates and projections). Numbers indicate average annual net migration in thousands for selected countries. Many of the smaller countries are classified as having little or no net migration, although reliable data are not available. Printed with permission of Rodger Doyle.*

duced somewhat due to European admixture, while the distance between Europeans and Northeast Asians is 0.0938 (see Table 3.2 above; all figures in the Tables are multiplied by 10,000).[23] This means that Mexican and Northeast Asian immigration has effectively replaced a great many native European- and African-American genes as a proportion of the country's long-term carrying capacity. Prospectively, Americans of European and African descent have lost, and are losing, the equivalent of millions of children due to post-1965 immigration.[24] The same is happening to many other societies. Map 3.1 shows asymmetries of international immigration in the second half of the 1990s.

Despite the above demographic trends, it is sometimes maintained that ethnic competition no longer pays. If one takes ethnically-restrictive immigration policy to be a form of inter-group competition, then Americans of European and African descent would have benefited by conserving their genetic interests had they engaged in competition of this form, for example by maintaining the quota system enacted by Congress in 1924. This system was aimed at maintaining ethnic proportions within the country that existed in the late nineteenth century. Conversely, any efforts by minorities to overturn that quota system and widen the 1965 open door constituted an act of competition with the ethnic majority that has harmed majority genetic interests more than any terrorist attack or any war.

As noted earlier, economic effects can indirectly benefit the native population by increasing their country's carrying capacity, thus helping all resident groups increase their absolute numbers. But this argument can be taken to absurd lengths. For example, if the members of an immigrant ethny are generally more productive than the natives, complete replacement would result in the carrying capacity being greatly increased, but this would still mean extinction for the natives. Alternatively, native numbers might be increased in *absolute* terms by immigrant contributions to the economy. But if this results in them falling in *relative* numbers to minority status, they are likely to lose control of state policy and be unable to prevent their further marginalization. Clearly there will be an optimum level of immigration rather than a maximum, defined by countervailing costs and benefits. When genetic interests are taken into account, immigration of genetically distant groups will be most adaptive when kept to the minimum necessary to fill vital economic needs, rather than pushed to a maximum in pursuit of short-term economic gain.

Let me illustrate the previous point with an historical example. Imagine that in 1600 American Indians had been offered an informed choice between two futures in the year 2000. One future is the present United States with a level of economic development unattainable without the efforts of millions of settlers and immigrants from Europe. The other is one or more Indian civilizations in possession of the present area of the United States but with economies less developed than at present. Which would they have chosen if they had valued their ethnic genetic interests? A temporary delay, even one of decades or centuries, in acquiring some skill or institution would seem a weak excuse for sacrificing the future of all succeeding generations. In the modern era of automation the rush to import manual labour is much more short sighted. Even if the economy and with it the population failed to grow at maximum speed, this would be a small price to pay for retaining a continent for one's ethnic kin.

In this chapter I have tried to express ethnic genetic interests in terms of equivalent numbers of close kin. When the abstract formulas and data of population genetics are expressed in these human terms, it becomes easier to understand that people have important genetic stakes in their ethnies as well as in their families. Some will respond, 'Noone cares about genetic interests!'. In the next chapter I consider this and other objections to the concept of genetic interests.

*Notes*

1 Parsons (1998).
2 *Statistical Abstract of the United States: 2001*, Table 15. This is probably an overestimate of the white proportion, because it includes Hispanics of partly Amerindian descent who classify themselves as of wholly European descent. For 2100 projections, see US Census Bureau (2000, internet data, Table NP-T4-H, middle series projection). The total US population is projected to be 571 million in 2100, with the 'white' population at 404 million and the Hispanic population at 190 million. The resulting projection of 40 percent non-Hispanic white is probably an underestimate because the Hispanic population is partly of European ancestry.
3 See CIA World Factbook, January 2002, http://www.cia.gov/cia/publications/factbook/.
4 Browne (2000).
5 *International Herald Tribune* 23–24 March 2002, p. 3.
6 Beaumont (2002).
7 In 2002 only one third of Russia's frontier-crossing posts had immigration checkpoints (*Moscow News* 33, 28 Aug. –3 Sept. 2002, p. 4).
8 W. D. Hamilton (1975, p. 148).
9 Hardin (1968; 1974/2001; 1993).
10 The concept of carrying capacity has been criticized because of the difficulty, and perhaps impossibility, of precisely quantifying that capacity for any particular territory. The variables that determine carrying capacity are subject to human choice, for example lifestyle as constituted by use of technology and types of foods preferred (Cohen 1995; Daily and Ehrlich 1992). That does not imply the absence of limits to growth (Clarke 2002), only that the analysis of those inevitable limits is difficult and that, where error-prone, should err on the side of conservation; it is safer to underestimate than overestimate carrying capacity.
11 E. O. Wilson (2002, p. 72).

12 Hardin (1974).
13 Cavalli-Sforza et al. (1994, p. 75).
14 Hamilton (1975).
15 In fact, if die off due to exceeding the carrying capacity were randomly distributed, the immigrants or their descendants would bear their share of the general population decline. In the case of 10,000 immigrants to a population of 50 million, that share would be 2 individuals (10,000 - 10,000 x 50,000,000/50,010,000). Complete replacement becomes a poorer approximation when the number of immigrants is a large fraction of the native population.
16 English parent-child kinship in the context of a population consisting of English and Bantu is f = 0.25 + (0.2288 x 3/4) = 0.4216. 10,000 English removed = 10,000 x 0.2288 = 2288 aggregate kinship lost. 10,000 Bantu entering = 10,000 x – 0.2288 aggregate kinship lost. Total loss = 4576 aggregate kinship. Equivalent number of English children = 4576/0.4216 ≈ 10,854 children lost.
17 Cavalli-Sforza et al. (1994).
18 E.g. Rubin (2000).
19 *Statistical Abstract of the United States: 2001*, Table 10; Hispanics were not yet counted separately.
20 US Census Bureau website, May 2002, Table PCT002. The white population is somewhat larger than stated, because a minority of Hispanics are of predominantly European descent.
21 *Statistical Abstract of the United States: 2001*, Table 15.
22 Borjas (1999); Smith and Edmonston (1997).
23 Cavalli-Sforza et al. (1994, p. 80).
24 Assuming that European admixture has reduced the European-Mexican genetic distance by 50 percent to 0.052. Nevertheless 2.8 Mexican immigrants replace a European-American child, and 1.7 Northeast Asian immigrants do the same. Even fewer will replace an African American child—only 0.9 Mexican and 1.0 Northeast Asian immigrants (based on genetic distances between unmixed populations).

# 4. Who Cares? . . . and other Objections to the Concept of Genetic Interests

*Summary*
(4a) Objection from lack of human motivation: Who cares?
(4b) Objection from the teleological nature of genetic interests.
(4c) Objection from levels of analysis: Do only genes have genetic interests?
(4d) Objection from competition between genes in the same genome.
(4e) Objection from the genetic unity of mankind.
(4f) Objection from random genetic drift.
(4g) Objection from non-genetic replicators.
(4h) Objection from inequality in ethnic genetic interests.

In this chapter I try to anticipate objections to the notions that genetic fitness is an interest and that it is the only ultimate one. Some of these objections are plausible, at least initially, while others can be readily dispensed with. The idea that cultural group strategies (section 4g) are ultimate interests warrants serious attention, though I am critical of the idea. I very much doubt that this chapter will resolve all objections in the minds of all readers, but my aim is the more modest one of simply identifying some major lines of criticism and defence and thus facilitating future debates about genetic interests.

The six criticisms I discuss in this chapter are: (a) Objection from lack of human motivation: Who cares?; (b) Objection from the teleological nature of genetic interests; (c) Objection from levels of analysis: Do only genes have genetic interests?; (d) Objection from competition between genes in the same genome; (e) Objection from the genetic unity of mankind; (f) Objection from random genetic drift; (g) Objection from non-genetic replicators; and (h) Objection from inequality in ethnic genetic interests.

### *(4a) Objection from lack of human motivation: Who cares?*

Perhaps genes are not interests, if interests are defined as conscious wants. The *Oxford English Dictionary* offers one of many definition of interests as a kind of feeling: '7.a. The feeling of one who is concerned or has a personal concern in any thing; hence, the state of feeling proper to such a relation'. Life, liberty and happiness qualify as interests by this definition because all are valued in mind and action. Motivation-based interests are to be expected from the perspective of neo-Darwinian evolutionary theory because motivations are usually a good guide to adaptive living. Life, liberty and property are individual phenotypic interests as well as being valued ends in themselves. What could be more evolutionary than to be motivated to protect our own lives and those of kin? Talk of genetic interests is nothing more than metaphysics, the objector might continue. The concept has no implications for the real world. People look after their genetic interests, or not, without being explicit about it. People have been living, breeding, and dying since the species began. Populations have waxed and waned, mixed and moved, and we're still here. What counts is the happiness and rights of human beings, not the destiny of tiny unfeeling fragments of DNA. Compared to feelings the gene is a highly abstract conception of little interest, since it involves mere sequences of nucleic acids with no human (phenotypic) qualities. It does not by itself elicit any emotion, except sometimes curiosity. How can this be construed, it might be asked, as an interest in the usual sense of the word? Even the creator of inclusive fitness theory found it difficult to identify with the genetic level of analysis: '[O]ne thing has not changed—this is my dislike for the idea that my own behaviour or behaviour of my friends illustrates my own theory of sociality or any other. I like always to imagine that I and we are above all that, subject to far more mysterious laws. . . . [Yet my theory] has turned out very successful.'[1] Alexander's remarks on the possibility for deliberate pursuit of genetic interests can be interpreted as calling into doubt the meaningfulness of genetic interest, even though he sees this as the ultimate interest:

> We can easily assume that the capacity for culture has allowed (as an incidental effect) various degrees of uncoupling of human behavior from patterns that would maximize genetic reproduction. In modern urban society, such uncoupling is rampant. It is my strong impression . . . that the use of self-reflection to contemplate the *raison d'être* of genes is not an evolved function of these genes but an incidental effect of their action which owes its existence to the rise of technology. . . . I suspect that we would have to undergo considerable genetic evolution before reflection on our genetic history would be likely to cause us to maximize our reproduction.[2]

On the one hand Alexander recognizes the existence of genetic interests, and notes that conscious efforts to advance them would be adaptive. He is not claiming that, to be an interest, it must be apprehended as such and actively striven for. Neither must some proximate interests. The insane and retarded still have an interest in maintaining themselves in cleanliness and dignity, even if they do not recognize the difference. The lazy and modest still have an interest in some of the things money can buy. But Alexander maintains that humans will not take genetic interests seriously until they evolve a special mental module like the appetite system or the emotions. In the parlance of evolutionary psychology, he thinks that without domain-specific mental capacities, maintenance of genetic interests must be left to other instincts such as parent-child bonding and sexuality, which are not general strategies able to take into account changing conditions. If he is right, if humans are not evolved consciously to pursue genetic interests even after reflecting on their genetic history, then the concept of genetic interest might be hollow. Perhaps if this interest cannot motivate protective action it must remain a descriptive idea unless and until humanity evolves to the extent that people can get excited about it.

Surely Alexander is mistaken. In our modern world many interests are not intrinsically motivating, only being valued when we understand their significance. Would keys to a castle be more than a curio to hunter-gatherers unaware of the wealth and prestige they can unlock? It is in our material interest to hold keys and analogous devices (degrees, credit cards, pass words, bank account numbers) despite their abstractness. Recognizing something as an interest requires background knowledge, sometimes quite sophisticated, of the contexts in which it becomes valuable. Such recognition depends on general intelligence, what evolutionary psychologists call domain-general capacities. General intelligence is more ponderous than the domain-specific modules that process information subconsciously with such alacrity. But our civilizations are largely built on the scientific and technical advances accumulated from conscious pondering.

It might be countered that objects and codes are not interests in themselves. They only attain value because they allow access to things we all intuitively value, that we have feelings about, such as status and resources. In this account keys are not intrinsic interests. It is objects, states of being and other individuals that we consider valuable—that are intrinsic interests. Nothing is an interest that does not unlock such valuables. This is a plausible view, but hardly a criticism of the notion of genetic interests. Genes produce myriad effects in the real world, including health and kinship, that are intrinsically valuable. Thus genes have always been valuable, even before they and their actions were discovered. Recall from Chapter 1 that Alexander maintains that genetic interests exist even if not consciously grasped.[3]

On the whole, serving genetic interests upholds human proximate interests. Many of the values we hold most dear are preserved down the generations because individuals strive to preserve their genetic interests, even when those interests are vaguely apprehended or not apprehended at all. The maternal affection shown by mothers in all mammalian species directly serves their genetic fitness by promoting the continuation of their genes. Conversely, if maternalism, with its painstaking devotion, sacrifices and risk, did not serve genetic fitness, the genes coding for that costly set of behaviours would ultimately be selected out of the gene pool. The same applies on the mass level, in relations between ethnies. An obvious example is attachment to ethnic identity, which indirectly motivates defence of genetic continuity. Another is the self-sacrificial patriotism of warriors in societies primitive and modern. This is not to suggest that pursuing proximate interests is pleasant to all concerned. The aggressive content of human nature has been adaptive for negotiating dominance rank and acquiring and defending resources. The fact remains that proximate interests such as status and wealth are paths to genetic fitness. A less obvious case is the compassion and mutual aid shown by welfare societies. The welfare state is without exception the product of political developments in relatively ethnically homogeneous societies.[4] When rising ethnic diversity leads to ethnic stratification, as found in the United States, Canada, and Britain, welfare tends to decline.[5] Wealthier taxpayers who disproportionately foot the welfare bill, are more willing to subsidize the needy from their own ethnies than those from others. Welfare and compassion are positively valued, as are privation and exclusion negatively valued. These values are affected on a mass scale by ethnic nepotism, and that phenomenon serves the genetic interests of the discriminating taxpayers. To object that altruism should be indiscriminate, equally helping the needy of all groups, is not to deny that nepotism is better than no altruism at all. Once again, serving genetic interests upholds human values.

The point should be emphasized that genes *only* become interests when part of the reproductive chain of life; when they contribute to the creation of humans and influence their development; or when such function is in prospect. If it were possible to manufacture billions of copies of one's genome in the form of powdered protein, and disperse them in the world or in outer space, that would hardly be in one's genetic interests. But it does serve genetic interests to have part of one's genome help form a new human. Critics of the idea of genetic interests must somehow get around the core biological idea that it is adaptive to reproduce.

The objective sense of an interest is repeatedly implied as a usage of the word. The *Oxford Dictionary* again:

> Interest . . . I. 1. The relation of being objectively concerned in something, by having a right or title to, a claim upon, or a share in. 2. a. The relation of being concerned or affected in respect of advantage or detriment; esp. an advantageous relation of this kind. . . . b. That which is to or for the advantage of any one; good, benefit, profit, advantage. . . . c. *in the interest (interests) of*: on the side of what is advantageous or beneficial to. . . . 8. The fact or quality of mattering or being of importance (as belonging to things); concernment, importance.

What if conscious appreciation is made a criterion for something to constitute an interest? We can conceptualise copies of our genes in phenotypic terms: as shared blood, as family, and as familiar appearance and behaviour. Valuing such characteristics is to value the genes that contribute to them, assuming that the perceiver is equipped with sufficient general intelligence to understand the causal link and that the culture in which he or she lives carries information about that link.

Human intelligence need not be supplied with scientific knowledge about genes to spur us to fitness-enhancing action. We know that metaphors such as shared blood can motivate adaptive behaviour because this is how people have often conceived of their families and tribes from time immemorial.[6] Hamilton pointed out that the blood metaphor is an effective guide to relatedness, often to distant branches of pedigrees.[7] In practice phenotypic analogies need not have any truth content. Such options include the mystical, such as ethnic relatedness constituting a god's seed, or some special quality such as moral superiority. Any phenotypic analogy is functional if it motivates behaviour protective of genetic interests, and it is in the nature of analogies deployed by general intelligence that they can be hung on many hooks, including abstract concepts such as DNA.

Genetic interest could motivate as a token of success. It is conceivable that individuals aware of life's evolutionary dimension can treat genetic fitness as a safety indicator. The assumption would be that if they or their groups are not sustaining their genetic line, for example by monopolizing a territory, something is wrong and should be put right. Similarly, pride could motivate concern for genetic interests, in which case adaptiveness would be associated with winning and maladaptedness with losing, not an unreasonable interpretation. Not many people like to think of themselves as losers. Winning and losing are matters of status, and keeping up with the Joneses absorbs large resources of time and money. The result is not always aesthetically pleasing, and the rat race repels as well as attracts, but there is movement, and arguably the injection of values promoting genetic fitness would tend to stabilize genetic interests.

A critic might concede that familial genetic interests are closely bound up with human values, but maintain that ethnic genetic interests are not. Ethnic kin, he might contend, are strangers, and humans evolved to value familiar others. In response to this criticism I refer again to the idea that ethnic genetic interest can be used as a token of success. It is a powerful token indeed, one that predicts success in securing a range of values. Group genetic interests track cultural values. Defence of an ethny simultaneously preserves its genes and culture, including religion, cosmology, political culture, and any other acquired traits passed on through socialization within the family and ethnic institutions. This is generally true, although ethnies can acquire elements of each other's culture, such as language, religion and technology. Also, since adaptive ethnic altruism is homologous with extended kin selection (see p. 40 above), one is also defending all group qualities with high heritability, such as cognitive profile, personality, and physiognomy. Those differences are moderate within geographical races, but can be large between races.[8] Moreover, there is growing evidence that differences in traits such as IQ have large effects on standard of living, both between individuals and ethnies within the one society[9] and between societies.[10] So even if our critic rejects the value of ethnic kinship in its own right, he might embrace it if he cares about preserving for future generations the proximate values linked to it, such as culture and wealth; but probably not, because that would require concern for the fate of future generations of his ethny as distinct from the species in general.

An effective counter to the view that humans cannot be motivated by genetic interests, even indirectly, is that they are and always have been. The cooperative defensiveness shown by band and tribal peoples is bound to have boosted inclusive fitness, because it is universal and ancient, thus likely to have been an evolutionarily stable strategy. Other forms of group spirit, including patriotism and nationalism and religious solidarity, have been powerful motivators of group continuity. Even in present day Western societies where ethnic sentiment is often considered passé by the ruling elites and where whole populations are being displaced by mass immigration, indirect concern over genetic interests lives on in one place or another. A common underpinning of group defensiveness is identity. Many people feel a strong affinity for their ethnic identities, and many more are prone to do so. Social identity processes tend to prioritize ethnicity, a marker for genetic interests. Voices raised against assimilation, replacement migration, and under-replacement birth rates might be marginalized,[11] but they tend to disprove the proposition that humans cannot be motivated by genetic interests. When those voices talk about ethnic and racial descent, they are a direct refutation of that proposition.[12]

The 'who cares?' objection can be delivered from a more pragmatic direction. As already argued, profound social change, driven largely by technological developments, can undermine the adaptiveness of social motivations. For illustrative purposes I have chosen two cases, one real, the other imaginary. It is a fact that across all developed economies the so-called demographic transition led to an inverse relationship between socio-economic class and fertility.[13] This trend began in Europe in the nineteenth century, when middle class families began to limit family size. The trend became accentuated in the 1960s when the contraceptive pill became available to women. The wealthiest classes, those most able to afford children, have the fewest, while the poorest members of society have the most. There are several likely causes, one promising candidate being the greater cost of rearing children in high status families, due to the growing demand for education and training needed to equip children to compete in modern labour markets.[14] Another possible cause is class differences in conscientiousness. The ready availability of efficient contraception makes conceiving a matter of intention or accident. If poorer individuals make more accidents, including failing to properly self-administer contraceptives, then they will tend to conceive more children. This could not be a stronger renunciation of genetic interests, since those who can *don't* and those who don't intend *do*. If genetic interests are of overriding importance, perhaps we should all strive to be poor?

My reply begins by agreeing that below-replacement fertility runs counter to genetic interests, and by accepting the reality of the inverse correlation between socio-economic status and individual fertility in developed economies. But neither trend is a problem for the *theory* of genetic interests. The biosocial literature carries persuasive accounts of these phenomena, to the effect that in past evolutionary environments the genome did not need to program an autonomous imperative to maximize fertility. Instead it coded for a set of independent adaptations such as sexual and nurturant behaviour which together reliably caused individuals to reproduce, in the context of environmental conditions including a lack of efficient contraceptive technology. The availability of such technology allows evolved motivations to produce suboptimal reproductive behaviour.

There is more to the problem than contraception, namely the motivation to use it to prevent conception. This is due to a number of factors, including economic competitiveness and status criteria for which children can be a handicap. From the biological perspective these factors are maladaptive when they result in suboptimal family size. Correcting them would require alterations to economic criteria and to status criteria, such that both placed greater value on children. Taking that step requires the assertion of reproductive values, a step that cannot be logically or empirically (theoretically) impelled. To assert otherwise would be to commit the naturalistic fallacy, the attempt to deduce values from facts. Conveniently, the initial criticism reduces to the assertion that genetic interests will fail to become overriding imperatives while people continue to avoid bearing children, which hardly affects my basic claim that reproduction is the ultimate interest. I continue to avoid the naturalistic fallacy in Chapter 9 where I discuss some ethical issues of pursuing genetic interests. I conclude that it is certainly not a theoretical truth that one ought to defend one's genetic interests. However it is immoral to prevent those who do value their reproductive interest from nurturing it. When such individuals belong to our families and ethnies, as they often do, our reproductive behaviour affects their perceived interests. Harming ourselves also harms our relatives. Duty to one's family is thus a reason not to behave maladaptively.

The second illustration of the way that changing environments, such as new technology, can undermine the adaptiveness of social motivation is imaginary, based on the allusion to robotic eagles in the opening paragraph of the book. This might seem a fanciful example, but bear with me. An example that radically separates phenotypes and genes is helpful because it shows how important an explicit comprehension of genetic interests might be.

Recall the criticism of the genetic interest concept that began this section: genes have no emotional significance while phenotypes do, so why not care exclusively about peoples' behaviour and appearance and let the genes take care of themselves? The toy industry cares exclusively about the outward behaviour and appearance of its products. The equivalent of robotic bald eagles are set to take the consumer market by storm, providing children with toys, everyone with poop-free pets, and someday households with 24-hour-a-day non-unionized maids. Will humans ever care for their robots the way they care for biotic life? Perhaps—if robots can be built that mimic the releasing stimuli to which our species is evolved to respond. These stimuli, though innate, are not confined to humans. We find young birds and mammals cute and feel protective towards them.

Robots are already being designed with the crude ability to show human-like facial expressions, such as the 'face robot' from the Hara-Kobayashi Laboratory in Tokyo.[15] In late 2000 an American toy manufacturer began marketing a robot baby doll designed by iRobot corporation in conjunction with Rodney Brooks, a scientist at the Artificial Laboratory at the Massachussetts Institute of Technology. The doll asks to be fed, babbles, smiles, coos, cries, and plays games, partly in response to how it is being handled.[16] Again, will humans ever care for their robots the way they care for biotic life? The answer will become a definite yes if human-like interpersonal signals sent by robots overwhelm our abstract knowledge that robots share none of our genes, that they do not belong somewhere on our extended family tree. My guess is that for a great many individuals abstract knowledge of relatedness will matter less than the immediacy of interpersonal signals, and that man-machine emotional bonding will result.

Now we get to the point about maladaptive motivations. Brooks believes that should robots be constructed with humanlike intelligence and consciousness it will be unethical to treat them as slaves. 'You get into the moral question—would it be okay to breed a race of subhumans? Essentially, enslavers thought they were dealing with subhumans. If that's not okay, will it be okay to deliberately build subhuman machines? And certainly we feel now it's okay. We don't feel any empathy for the machines but that may be a consideration ultimately. . . .'.[17] This position combines vivid psychological insight with poor biology. Brooks thinks it would be wrong to have any entity be our slave that could elicit our empathy, arguing from the *lack* of empathy slaveholders once felt for their human slaves. If the slaveholders were wrong in casting their slaves as subhuman, he implies, then robot owners would similarly be wrong to cast their robots as subhuman. The syllogism makes sense only if divorced from the most basic understanding of biology, and from a concept of genetic interests, implicit or explicit. Human slaves of any race were as human as their masters. It was *false belief* that designated them subhuman, but a similar belief about robots would not be false. Introducing an inkling of biological interests, of a sense of loyalty to humanity as a whole, helps set priorities. In that light the issue is not whether a slave is subhuman, an inferior type of humanity, but whether it is human at all. From the perspective of genetic interests, we owe more empathy to our fellow humans, even to fellow mammals, than to any robot, no matter how well the robot amuses us or endears itself to us by imitating selected human characteristics. Humans have more genetic interest in their own species' survival than in the survival of artificial entities that replicate elements of the human phenotype, however perfectly. All this is probably an excellent case for banning the construction of human-like robots, since bonding with objects that should be serving our needs is bound to become an obstacle to their instrumental treatment.

But this discussion is drifting from the point. If we care more about phenotypes than genotypes, then 'who cares?!' will often be an effective repost to any evangelising call to preserve genetic interests. One either feels protectively about genetic interests or not. But individuals who do manage to feel protective about their genetic interests, in however an indirect or counterintuitive manner, stand a better chance of behaving adaptively in the face of rapid demographic and technological change. Perhaps a principled concern with adaptiveness—motivated by affiliation for family and community—can link the intellectual realization of genetic interests to the will to act.

### *(4b) Objection from the teleological nature of genetic interests*

I have encountered criticisms of the idea of genetic interests based on rejection of teleological explanation.

a. *Objection*: Human behaviour is often directed towards goals, such as acquiring food or mates, but it is fallacious to portray humans as deliberately striving to maximize their reproductive fitness. Fitness might or might not be an outcome of our behaviour, but with rare exceptions it is not a conscious goal. *Reply*: The present essay is not primarily a theory of human behaviour, but of interests. Rather than being a work of explanation, this is mainly an exercise in political theory dealing with what people are *able* to do if they want to behave adaptively. It also contains some ethical ideas about how to make one's adaptive strategies universalisable, in the Kantian sense (Chapter 7) and compatible with general welfare (Chapter 9). As to whether humans consciously strive to maximize reproduction, that is an empirical question. Clearly many individuals want children and care for them, and many feel defensive about their tribe or ethny. It is doubtful, however, whether this constitutes deliberate strategizing to maximize reproduction. That is not a premiss of the present analysis.

b. *Objection*: Reference to 'ultimate causes' and 'ultimate interests' is teleological in the same way that Aristotle proposed 'final causes' to explain why something exists or behaves in a particular way. Most egregiously teleological is the notion that all life has evolved towards preserving genetic continuity, it might be alleged. This leads to the view that love and hate exist because they function to preserve genes. *Reply*: The present book does indeed commit teleological reasoning, but so does modern biology in its quest to identify ultimate as well as proximate causes. It is quite appropriate to ask and answer the question 'Why are people?' from such a perspective, as does Dawkins.[18] Natural selection 'has made us almost all that we are'.[19]

c. *Objection*: Discussing genetic interest implies that evolution can become teleological in the sense of an outcome of conscious human agency. Does this not conflict with the mainstream mechanistic tradition in evolutionary thought? *Reply*: This is the view taken by Dawkins, who claims that 'evolution is blind to the future', so blind in fact that people can do nothing to save their societies from extinction should evolutionary mechanisms pull them in that direction.[20] This has certainly been true for much of evolutionary history, but does not account for the agency of human intelligence. I would go further, and suggest that it is absurd in the case of contemporary humans because we live in an era of widespread knowledge of evolutionary mechanisms and advancing technologies for manipulating genetic and social processes. Animal and plant breeders have exercised human agency on non-human evolution for millennia, and do so at an accelerating pace. In the human domain, something as everyday as immigration policy affects a population's gene frequencies. The regulation of immigration is an obvious way in which human agency can influence human evolution.

### *(4c) Objection from levels of analysis: Do only genes have genetic interests?*

Assuming as valid the notion of objective interests, independent of motivations or even awareness, it could be argued that neo-Darwinian theory emphasizes the genes' phenotypic interests, not phenotypes' genetic interests. From the replicator's vantage point phenotypes exist for the convenience of genes. This line of thinking might conclude that if phenotypes have any interests they must bear on their own phenotypic needs. A rough guide to these needs is striving behaviour but includes the objective need of the organism to survive and flourish. Put differently, phenotypes might have only proximate interests, not ultimate ones. The latter type of interests might only adhere to replicators, not vehicles.

This argument fails to account for what Alexander calls 'the direction of striving of the phenotype', quoted earlier. Predictably from the evolutionary perspective, phenotypic needs and motivations usually point to the reproductive interests of their genes. Phenotypes are, after all, genes' survival vehicles, to use Dawkins's term.[21] Genes are our ultimate interests because they are the basic units of selection, partially defined by Dawkins as 'active replicators', those that positively influence their probability of being copied.[22] All functional genes are active replicators because they help shape the organism, thus affecting the organism's reproductive chances, and thus the genes' likelihood of being copied from one generation to the next. The other criterion making genes units of selection, already mentioned, is that they are germ-line replicators. Active germ-line replicators, such as functional genes, are units of selection and hence ultimate interests. The general mutuality between genetic and phenotypic 'striving' in the Environment of Evolutionary Adaptedness indicates that even if we count only phenotypic needs and motives as interests, these are strongly identified in that environment with genetic interests *as the genes' interests.*[23] Hamilton nicely expresses this confluence of phenotypic and genetic interests:

> [T]he idea of the inclusive fitness of an individual is . . . a useful one. Just as in the sense of classical selection we may consider whether a given character expressed in an individual is adaptive in the sense of being in the interests of his personal fitness or not, so in the present sense of selection we may consider whether the character or trait of behaviour is or is not adaptive in the sense of being in the interests of his inclusive fitness.[24]

So while Hamilton thought it analytically useful to consider traits as an interest of inclusive fitness, he also indicated that the individual organism also has inclusive fitness ('his inclusive fitness') and thought there was a confluence of individuals' characteristics and inclusive fitness, in the sense that the former was evolved to advance the latter. But it cannot be concluded that we are servants to our genes. Genes code for phenotypic traits or the capacity for them that we value as human beings both in ourselves and others: capacity for love; trust; beauty; sexual attractiveness; curiosity; playfulness; intelligence. These traits, plus some like aggression and hatred about which we are ambivalent, are simultaneously the rent we pay for genetic fitness and a labour of necessity by our genes on behalf of their survival machines.

Genetic and phenotypic interests are not easily distinguished in naturally selected species. But that distinction is clearly made by identifying behaviours that sacrifice the individual in the interests of other individuals; hence the evolutionist's interest in altruism. Surely the primacy of phenotypic (or vehicular) interests cannot be maintained when so many phenotypes in so many species give highest priority to their genetic interests; when selfishness and altruism are shown convincingly to be strategies for ensuring genetic continuity.

### *(4d) Objection from competition between genes in the same genome*

How can an individual have genetic interests if there is competition between genes within the genome? Can non-functional DNA sequences be relegated to non-interest status, as I have assumed? Negative answers to these questions would subvert my earlier contention that all functional genes contribute to overall genetic interests. I shall attempt an answer by starting with another subversive question: If all functional genes add to our genetic interests, does this mean that genetic adaptation is impossible, since this entails substitution of at least one allele by another? If so, my definition of genetic interests is incompatible with evolution, while purporting to be inspired by it. The way out of this apparent paradox is the escape clause contained in the words 'functional genes'. The genome is not unitary but a set of mostly cooperating elements. This is the case for both additive effects and synergistic ones. Additiveness was assumed by Charles Darwin and in the 1930s by R. A. Fisher who visualized the units of heredity constituting a parliament of genes in which all members acted to produce a fitter phenotype. Synergistic interaction of genes was first postulated by Sewall Wright, also in the 1930s, and is known by the technical name of epistasis.[25] All complex adaptations rely on synergistic interplays of genes, where the effect of an allele at one locus relies on the action of alleles at other loci. As a rule genes interact to produce complex adaptations such as limbs and brains, and interact in such a way that the overall effect is greater than the sum of individual gene effects. That is synergistic, not additive. The genome is seen not so much as a parliament of genes but an internal combustion engine, with complex interdependencies of the parts producing an overall effect qualitatively different from the effect of a single gene. There are both additive and epistatic genetic effects, and in both conditions natural selection is possible via substitution of alleles. It follows that overall genetic interest can be preserved by sacrificing one or a few parliamentary members or engine parts.

One way to look at this is by taking the gene's-eye view, by considering the genes' interests and how these correspond to the organism's genetic interests.

Phenotypes can lose inclusive fitness through individual alleles in their genomes becoming maladaptive, due to mutation or to changes in the environment. Those dysfunctional alleles no longer serve the interests of the majority of the genes comprising the genome and thus the individual's genetic interests would be preserved or increased by substitution of maladaptive alleles.

To complete this section I want to distinguish genetic interests from genes for altruism. It might be inferred from sociobiological theory that our ultimate interests consist only of genes coding for altruism (hereafter 'altruistic genes'), since according to that theory, altruism is a strategy for propagating only altruistic genes. This notion is vulnerable to the following *reductio*. If 'altruistic genes' were the only ultimate interest, then individuals and whole species that lack altruism would have no interest in survival or propagation. Yet there are many species including many in the reptile class that live and reproduce without helping conspecifics or even nurturing their young.[26] Is it plausible that a nonaltruistic organism has no interest in perpetuating its line? Inclusive fitness theory was invented by W. D. Hamilton, a man fascinated by altruism from his student days.[27] The theory is meant to explain only helping behaviour that carries a net cost for the helper; it is not a theory of the function of all genes. There are strategies apart from altruism by which genes promote their own continuation in the next generation. All entail coding for adaptive phenotypic traits, and include building organs, regulating the functioning of other genes, organizing behaviour patterns for finding food and mates, even manipulating the behaviour of other species.[28] Building and maintaining an organism can hardly be considered less important than giving it some particular behavioural properties. It follows that all mutually functional genes, all genes that contribute to adaptive phenotypic traits, are part of the individual's genetic interests. In Chapter 5 (pp. 120–123) I discuss the related distinction between genetic interests and inclusive fitness, noting that the former is the *frequency* of an individual's distinctive genes while the latter is the *effect* the individual has on the reproduction of those genes.

### *(4e) Objection from the genetic unity of mankind*

Another challenge raises what might be called the 'universalist paradox', an early criticism of kin selection theory, which goes as follows. It is well known that humans share over 98 percent of their genes with chimpanzees. Commonality within species is much higher, so that even the most genetically distant humans share almost all their genes. One estimate is that human genetic variation occurs against a backdrop of 99.9 percent genetic similarity. That is, only one in every thousand human genes differs between individuals. There may be as few as 3 million base pairs differing between randomly chosen individuals, from a total of 3 billion in the genome. Put differently, if there are about 30,000 genes in the genome (each averaging 100,000 base pairs), the maximum difference between individuals amounts to 30 complete genes. Based on this apparently low figure it is sometimes argued that group differences must be vanishingly small. The point is often combined with the assertion, rejected in Chapter 2 (pp. 48–50), that not enough time has elapsed since humans emerged in East Africa 200,000 years ago for significant genetic differences to have evolved.

The argument is mainly applied to functional characteristics, such as intelligence. It is asserted that since genetic differences are trivial, any phenotypic differences between populations must be caused by nongenetic factors, such as climate, diet and culture. The issue is raised to address what is thought to be the major issue of race relations, invidious comparisons between groups. However, genetic interests would exist unscathed if humans everywhere were identical in personality, intelligence, athletic prowess, and disease resistance. Our children and ethnies are precious by virtue of kinship, not because they can outperform other children and ethnies. The argument concerning group differences in functional characteristics is thus of secondary relevance to the present thesis and I shall treat it cursorily. It is also predicated on a false premiss and reverses the scientific order of reasoning.

It is a false premiss that genetic differences are trivial. On what grounds can we assume that 30 genes are unimportant? Geneticists believe that just one regulatory gene, the testis determining factor on the Y chromosome, is responsible for all sex differences. The Human Genome Project's discovery of fewer genes than anticipated was interpreted by the experts in the field as meaning that genes must have more functions than previously believed. Genes can perform multiple functions by producing different kinds of proteins that circulate around the body cueing development, behaviours, and other processes. It follows that a small genetic change can have large effects, a proposition that seems reasonable when one considers the markedly different structure and behaviour of humans and chimpanzees despite sharing over 98 percent of their genes. Not only can individual genes produce a variety of proteins, but the typical gene comes in many versions, increasing the effective genetic variation. There could be up to half a million gene variants making up the functional repertoire in the global population. Also, human diversity consists of more than 30 genes. The 3 million base pairs are spread over a great many genes. As dramatically demonstrated by genetic diseases such as phenylketonuria, myotonic dystrophy, and Huntington's chorea, a mutation of just one base pair in a gene can radically alter its function.

Racial differences between humans are also striking, not only in appearance but in functional characteristics such as athletic performance and temperament, and some of these differences have innate characteristics, for example being universal, insensitive to environmental change, and appearing early in development. Rushton offers a compendium of evidence on genetic group differences.[29] Let me discuss one concrete example to convey the flavour of that evidence. Note that this evidence has been available for several decades. Freedman and Freedman found that the well-known Chinese characteristic of calmness (compared to Caucasians and Africans) is apparent soon after birth, even for babies born to parents of Chinese ancestry raised in the United States.[30] This is good evidence that this trait has a genetic component, since so many environmental factors were controlled (mothers and babies had the same maternity hospital and same postnatal treatment).

The gene-centred argument against the importance of innate differences reverses scientific reasoning. The function of theory is to explain and predict facts, not obscure them. Since it is known beyond reasonable doubt that there are significant genetically-caused physical and behavioural differences between individuals and races (though with much clarification still needed), the finding that a relatively small number of genes distinguish individuals and groups implies that those few genes pack a large punch, like the sex-determining gene on the Y chromosome. It is simply bad reasoning to assume that small genetic differences cannot, in principle, code for large phenotypic effects.

*Universalist paradox*

The more interesting challenge posed by the new genetic findings strikes at the heart of sociobiological theory. If all humans share 99.9 percent of their genes, is it not adaptive to be altruistic towards everyone? The theoretical flaw in this reasoning was first perceived by Hamilton.[31] Hamilton's mathematical reasoning is clarified in verbal form by Dawkins[32] using Maynard Smith's concept of evolutionary stable strategies, in the following way. A universal altruist, one who distributed resources randomly, would be outbred by a kin altruist,[33] one who restricted generosity to kin. So within a few generations the gene that caused universal altruism would have fallen in frequency in the population and be slipping towards complete replacement by genes that directed altruism towards relatives.

Four years before his death, Hamilton repeated the universalist challenge and his refutation with great clarity. '[S]urely it is open to humans and other intelligent animals to realize that, if the genes are so alike, it really only makes sense to treat relatives and non-relatives as alike to the self to the degree of the measured similarity—in other words, to be extremely generous and loving towards everyone.'[34] This is the view implied by Cavalli-Sforza and other geneticists who deemphasize or ignore neo-Darwinian theory. But Hamilton points out that this is false reasoning, because

> [e]ven when it is near to complete monomorphism the evolved nepotistic programme of behaviour is still at work in maintaining its base. Simply by being there it is beating back all mutant 'cheat' genes that fail to perform altruism but still reap its fruits from the acts of others. . . . Since there is hardly an easier mutation to imagine than one that causes a constructive behaviour like altruism *not* to be performed, so creating a passive and successful receiver for the altruism of others, the point is serious.[35]

In Hamilton's formulation kinship has been an effective criterion for directing altruism towards individuals carrying one's distinctive genes, even when genetic differences are vanishingly small. The same is true of ethnic genetic interests. The historical record and extended family histories are the approximate equivalents of genealogies. The more reliable an ethnic genealogy, the more it is adaptive to favour one's ethny over other populations even in the absence of genetic assay data. Even if gene surveys could find no genetic differences, it would be more adaptive to direct one's public altruism to one's ethny than to the world at large, as a means of protecting against free riding mutants.

Hamilton's analysis immediately falsifies the widely-circulated argument by geneticist Richard Lewontin that the race concept should be abandoned as of no scientific value since 'only' 10–15 percent of genetic diversity exists between populations while 85–90 percent exists within populations.[36] However, as we saw in Chapter 3, a 12.5 percent genetic variance between two populations implies within-population kinship equivalent to that found between grandparent and grandchild or between aunt and nephew. Lewontin's genetic estimate is not only compatible with the existence of high ethnic kinship, it is a rough measure of it. However the debate might be resolved concerning race differences in functional traits, Lewontin's argument fails to show that altruism directed at members of one's ethny is necessarily maladaptive. If it were, then nepotism shown by grandparents and aunts would be malaptive due to the genetic variation found within all families. Lewontin's own data indicate that ethnic nepotism can have selective effects, and is therefore of scientific as well as practical importance.

Genealogy is not an end in itself. The sequencing of genomes allows us, in principle, to bypass kinship and tribal instincts. We can discover intellectually who is and who is not genetically related in ways more reliable than feelings of attachment, on the one side, and strangeness on the other. The absence of objective genetic knowledge until recently has forced reliance on genealogical information, which has been effective enough during most of human experience but has recently broken down due to the rapid change of our man-made environment.

Genetic science will be valuable as an adaptive tool for determining the average kinship of whole populations.

*Animal liberation*

Let us return to the original universalist paradox, or rather a particular version of it which might be called the 'animal liberation' version. The title comes from P. Singer's 1981 book of the same name. The argument goes thus. Since we share over 98 percent of our genes with chimpanzees, is not a random chimp essentially as precious as a human? Should we not recognize large genetic interests in much of the animal kingdom? It is currently estimated that mice share more than 90 percent of human genes, dogs and pigs 86 percent, rattlesnakes 79 percent, the fruit fly 50 percent, and yeast 36 percent.[37] The genetic isolation that defines species does not prevent great genetic similarity. As between human populations, genetic variation within some species can be greater than variation between them.[38] Since all species represent variations on an original life form that evolved billions of years ago (Figure 4.1), do not these represent human genetic interests in various concentrations? And since the genetic distances between individual humans are very small compared to those between humans and chimpanzees, let alone between humans and yeast, should we not feel protective about our genetic interests residing in all fellow human beings without regard for ethnicity, race or indeed kinship? Note that this challenge does not doubt the concept of genetic interests but it does doubt that defending these interests entails any discrimination between humans or much between humans and other primate species. It follows that discrimination in its many forms does not and cannot serve genetic interests, especially between humans, thus rendering the latter concept moot in the domains of politics and social interactions in general.

My response is partly to accept the implication, based on gene frequencies, that humans have an interest in their genes in whatever species they happen to be propagated. I might never develop an emotional bond to yeast beyond its role in producing beer, but genetic interests do, in principle, extend in concentric circles to an interest in all biotic life, in proportion to genetic proximity as well as utility. Dawkins makes a similar point, but argues that despite a shared genetic heritage, genetic similarity has not influenced natural selection processes, except between close kin. This is presented as a critique of group selection. 'Lions and antelopes are both members of the class Mammalia, as are we. Should we then not expect lions to refrain from killing antelopes, "for the good of the mammals"?'[39] This *reductio ad absurdum* is effective against the straw-man theory of group selection implied by Dawkins, but it is poor prognosis in the case of humans, since he does not account for the potential agency of high intelligence

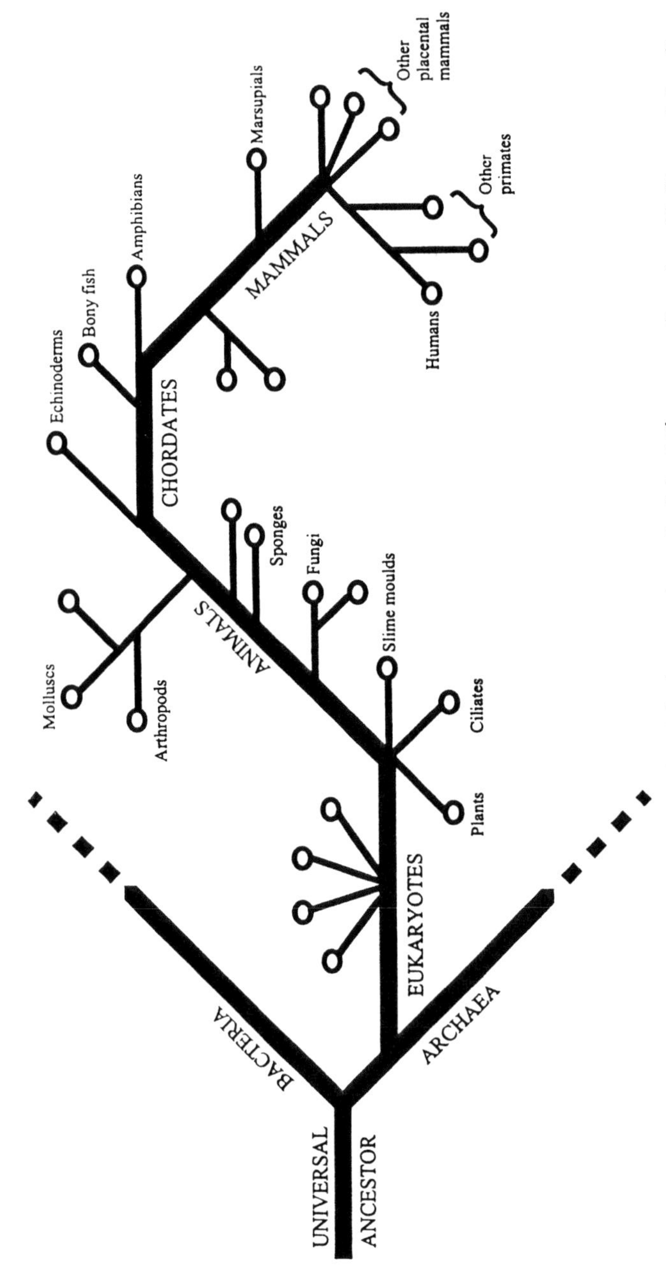


*Figure 4.1. The tree of life according to accumulated findings at the end of the 20th century (based on Collins and Jegalian 1999, p. 54). Humans share most of their genes with other primates.*

applied to finding adaptive solutions to rare contingencies. Faced with some types of threats, an adaptively-minded intelligent lion might spare antelopes, even if this meant its own demise. A mind experiment helps make this point in the case of humans.

Imagine that a cataclysm destroys all human life except for some remote tribe, say a group of Aborigines living in Central Australia. That tribe's survival and eventual repopulation of the globe would represent the restoration of most of the genes of every human who had been individually destroyed. Knowing this outcome ahead of time might be a source of comfort for doomed individuals who took genetic interests seriously. The Aboriginal survivors would have greatest reason to rejoice, apart from their avoiding an unpleasant death, because their individual survival would translate into the survival of their distinctive genes. African Pygmies' (Mbuti) joy might be slightly muted if they were aware of the great kinship distance between them and Aborigines. Let us extend the hypothetical devastation. If life as a whole were threatened by an asteroid impact, the survival of a plant species or even better a species of animal, would rescue a large fraction of our genetic interests. In the case of Aborigines we can emotionally appreciate some of the import of our survival through them. We respond to the imagined scenes with thanks that people like ourselves survived and prospered. With a little intellectual effort it might be possible to extend this affection for mankind to a broader affection for all biotic life, as postulated by E. O. Wilson,[40] albeit less enthusiastic than for our fellow humans. If so, this would indicate that we can, at least in principle, appreciate genetic kinship wherever it resides. In the unlikely event that the continuation of the species or the genera or of life of any kind was absolutely conditional on individual self sacrifice, that would be adaptive.

In the absence of universal threats of the kind postulated above, distinctive genes define genetic interests. A gene that is so widely distributed that it is present in predator and prey and in competing conspecifics, has no interest in influencing its phenotypes to cease predating or competing, because it survives whatever the outcome. Such genes become part of the background environment, their status as interests (between competing organisms) cancelled by ubiquitousness. Only distinctive genes have an interest in one phenotype prevailing over another. Genes coding for altruism sometimes express that interest by influencing their phenotype to be less competitive towards other phenotypes bearing copies of themselves. But many more distinctive genes do not code for altruism. They are still 'selfish' genes in Dawkins's famous phrase; still acting to perpetuate themselves, though not by coding for altruism; and they still represent genetic interests.

Relatedness is critical because it greatly improves the chance of sharing distinctive genes. Biophilia does not logically prevent me from feeling much greater affection for species and individuals that are more closely related, even if the gene frequency differences involved are quantitatively small. The intellectual justification for this prejudice is Hamilton's free rider argument quoted above. Another relevant argument comes from Dawkins.[41] Relatedness alone does not make selective altruism adaptive, but *contrasts* in relatedness do. This is the point of the 'nation of cousins' analogy in Chapter 2. As pointed out by Trivers, it is adaptive for mothers preferentially to care for their own offspring because of shared genes, but even mothers and children can have conflicts of genetic interest, for example over the timing of weaning and subsequent births.[42] If competition continues within the family—in which occurs the closest genetic relationships—how much more can we expect competition where relatedness is much lower? Genetic interests are thus served by cooperation and competition, sometimes within the same relationship, depending on circumstance.

### *(4f) Objection from random genetic drift*

It might be argued that Darwinism is an incomplete guide to genetic interests in light of the theory of random genetic drift advanced by Sewall Wright.[43] It is now generally accepted that random sampling of parental genes produces genetic change within small isolated populations. The process occurs in small populations of band and tribal size because in large populations most alleles are represented in so many individuals that it is most improbable that a random sample will not retain them, and in the same proportion as in the parental generation. Random genetic drift was not part of Darwin's theory, which emphasized Malthusian shortages leading to a perpetual competition for existence. In fact much of the genetic distinctness of populations is not due to adaptation, but to random drift. How meaningful, then, is talk of ethnic genetic interests? Why should we value something that is the result of chance?

The short answer to this objection is that random processes are fundamental to evolution and are even built into the way we produce our children. Sexual species have as part of their basic design the random sampling of parental genes during meiosis, to pass onto the next generation. If the random element in ethnogenesis is a reason to doubt the genetic value of ethnic kin, it must also be a reason to devalue one's offspring as a store of genetic interests. Conversely, if our children are a genetic interest despite being random samples of our genome, then our ethnic families are also a genetic interest despite being aggregate random samples of parental generations.

Random drift has contributed greatly to the genetic differentiation of separated populations, and hence to the evolution of ethnies as super families. Drift is therefore not a problem for the concept of ethnic genetic interest, but a partial explanation and predictor of it. It is perhaps ironic that the same process splits populations, producing genetically distant descendant populations that can come into competition.

### *(4g) Objection from non-genetic replicators*

Perhaps replicators other than genes are ultimate interests. Are not characteristics such as behaviours and somatic traits also vital interests? And are not genes merely the code by which such traits are replicated? Certainly characteristics are reproduced in sexual reproduction, albeit imperfectly and often blended with other characteristics. And natural selection operates directly on traits, not genes. Finally, characteristics such as child-like features and the appearance of close kin can directly elicit protective motivation; we treat them as if they were interests. A problem with this view is that traits are not replicators while genes are. Traits recur in lineages not because they replicate themselves but because the genes coding for them do. As for the eliciting of protective motivation, the neo-Darwinian emphasis on genes does not usually conflict with a focus on the characteristics for which they code. As argued in section 4a above, our ultimate interests are the transgenerational causes of valued traits, rather than the traits themselves.

This leads to a further query. If genes are ultimate interests because they code for adaptive characteristics across generations, then perhaps other long-term influences on phenotypes are ultimate interests as well. Environment, including culture, influences phenotypes. Many of our genes have been naturally selected by local conditions, including light intensity, temperature, predators, and climate-sensitive diseases such as malaria. Others have been selected by cultural environment, for example the lactose tolerance found in populations with a history of dairy farming. Thus populations owe their gene frequencies partly to environmental conditions. Does this make the environment an ultimate interest? There is clearly a reproductive interest in maintaining healthy environments, but these cannot be ultimate interests unless they are replicators. This is not true of climate. No matter how long-lasting an environment, it is not a replicator. However, non-human species are, and we depend on them for sustenance. Love of living things, E. O. Wilson's biophilia again, might reflect an ultimate interest lying beyond human genes. But apart from making life possible, climatic and biotic environments do not provide the detailed information needed to sculpt human nature. Whatever caused us to inherit our functional genes, they are our replicators and continue as such in different environments, unless and until selected out of the gene pool. In principle humans could thrive on new worlds surrounded by extraterrestrial life forms that performed the same life-supporting functions as life on earth. Unlike human genes, the biosphere is in principle if not in practice a replaceable interest. Our only permanent genetic interest in other species is the genes we share with them. The rest is utility.

Memes are a more serious challenge to the genes' claim to a monopoly as bearers of irreplaceable ultimate interests. Memes are units of cultural replication postulated by Dawkins that might qualify as ultimate interests.[44] Like genes, memes are conceptualized by Dawkins to be replicating entities that use phenotypes as disposable survival vehicles, just as genes do. For example, he sees religion as a memetic mind virus, some varieties of which have been contagious.[45] One property of religion memes is that belief, for example in gods or hell-fire, is self-perpetuating, without serving any adaptive function, because our brains are adapted in such a way that they are vulnerable to infectious invasion by such memes. These memes serve only their own interests. 'God exists, if only in the form of a meme with high survival value, or infective power, in the environment provided by human culture.'[46]

Any ultimate interest must be replicated down the generations and must code for human phenotypes with sufficient fidelity to retain their properties for many generations. If memes meet this qualification then it follows that preserving culture might be as much an ultimate interest as preserving genes. Certainly cultural information shapes the human phenotype in many important ways, including cultural identities and languages. Are memes an ultimate interest, and if so, do they compete with genetic interests?

Memes do qualify as replicators in one important sense of possessing a degree of reproductive fidelity, which Dawkins emphasized as central to the concept of replicator.[47] He was defending gene-centric neo-Darwinian theory against Bateson's attempted reductio of the idea that individuals are the genes' way of making copies of themselves.[48] Bateson noted that succeeding generations of birds build nests and the nests protect each generation of hatchlings. It follows, he argued, that birds are a nest's way of making new nests. But this wrongly defines nests as replicators, Dawkins replied. 'A nest is not a true replicator because a "mutation" which occurs in the construction of a nest, for example the accidental incorporation of a pine needle instead of the usual grass, is not perpetuated in future "generations of nests".' Since changes to memes are replicated in future minds that they colonize, they act as germ-line replicators.

Do memes qualify as active replicators? The analogy of memes favoured by Dawkins is that of a virus, a mind infection that passes from person to person.[49] Dawkins summarizes the qualities of a good replicator: 'longevity, fecundity, fidelity.'[50] Yet as Alexander points out, memes do not replicate but are selected and replicated by phenotypes in a manner not analogous with genetic reproduction. 'In whatever sense, and to whatever extent, learning is not a "blank slate" phenomenon in usual human environments—and surely it is not—then culture is a vehicle of the genes . . . .'[51] Genes but not memes are active replicators. Alexander goes on to argue thus:

> Cultural novelties do not replicate themselves or spread themselves, even indirectly. They are replicated as a consequence of the behavior of the vehicles of gene replication. Only if the decisions or tendencies of such vehicles of gene replication (individuals) to use or not to use a cultural novelty are independent of the interests of the genetic replicators can it be said that cultural change is independent of the differential reproduction of genes.[52]

When a minority of many alternate memes is selected to be learnt and passed on to other individuals, and the selection is influenced by genetic predispositions, these memes cannot be said to be pure germ line replicators but expressions of the 'extended phenotype', to use Dawkins's phrase.[53] Genes code for phenotypic predispositions to acquire certain types of memes. But memes cannot affect genes except by affecting mate choice and other aspects of reproduction, i.e. via the ultimate process of selection, in competition with other environmental and selective influences. Long-lived memes, those that might qualify as germ-line replicators, increase their reproductive fitness by increasing individuals' genetic fitness. In this sense too genes are prior to memes. Most memes are better conceptualized as tools serving the organism, as argued by psychologist Del Thiessen.[54] Religion is a good example. Religious ideas are sometimes contagious but as D. S. Wilson has recently argued, the spread and persistence of religions over many centuries is more plausibly due to their functional role of enhancing practitioners' genetic interests by, for example, integrating communities, resolving conflicts, and serving sectional interests.[55] Historian William H. McNeill comes to the same conclusion from an ecological perspective: '[I]t is arguable that for most of the people most of the time, the moral injunctions and the hope for a better future that the teachings of the higher religions inculcated conduced to survival. Had this not been the case, the new religions surely would not have spread and survived as they did.'[56] McNeill counts 'great world religions' on strictly statistical grounds, as being Christianity, Buddhism, Islam and Hinduism, but also mentions 'less numerically successful rivals—Judaism, Manichaeism, and Zoroastrianism'.[57] He identifies aristocracies and bureaucracies as the main 'macroparasites' that have exploited whole populations. Evolutionary psychologist Kevin MacDonald's analysis of the Medieval Church complements this view by showing how Catholicism acted as the vanguard of popular resistance against the aristocracy by enforcing monogamy on elite males.[58] Allowing for exceptions, religious culture and human genes are probably coevolved, with the former selected to benefit the latter. One recent hypothesis formulated along these lines is that religion is built upon biologically-evolved signs of commitment that are difficult to fake. Unconditional commitment helps build the trust needed to produce collective goods, in turn of great adaptive value to group members. According to anthropologist William Irons, natural selection has favoured the psychological predisposition to form and communicate religious commitments, including the religious kind.[59] Thus, in the main, the analogy of the organism as a vehicle used for the convenience of memes stands reality on its head. In Thiessen's view the challenge for the memeticist is to find at least one example of memes as architects of the organism. Then and only then should we take seriously the meme-gene analogy.

One approach to meme theory that fails Thiessen's test is that of Susan Blackmore in her book, *The Meme Machine*.[60] Blackmore contends that memetic fitness usually augments genetic fitness. But at one point she suggests that memetic evolution can take over at the expense of the genes. This becomes possible, Blackmore thinks, once genetic evolution produces a species adept at imitation. She thinks that human imitative behaviour has put our memes beyond genetic selection: '[O]nce the genes have given us imitation they cannot take it back.' The resulting extreme behaviour, such as potlatch in which individuals compete in giving away their most precious possessions, can be 'like a parasite that may, or may not, kill its host, while most of our altruistic behaviour is symbiotic or even beneficial'. Putting aside the possibility that potlatch is a form of status competition with real payoffs in status and mates, Blackmore's vision of runaway control by memes does not appear to be evolutionarily stable. In a society where ruinous giving was the norm, genes for constraining imitation in ways that helped individuals retain resources would sooner or later become common. If there is some gene-based variation in imitative behaviour, as there is with most behaviour patterns studied by behavioural geneticists,[61] then individuals who refrain from giving away all their possessions will have more offspring and spread their less compulsive ways. Genetic fitness would constrain memetic fitness.

Most memes fail to qualify as ultimate interests. However Robert Boyd and Peter Richerson go some way towards meeting Thiessen's challenge by developing and testing a theory according to which groups can adopt meme packages (beliefs; rituals; modes of organization; modes of economic production) that establish new conditions that then benefit particular phenotypes and result, over generations, in changed gene frequencies.[62] Memes can indeed be architects of the organism as agents of selection. Boyd and Richerson call the meme packages 'cultural group strategies'. If modern humans have been genetically shaped by these strategies then the meme complex comprising the strategy comes prior to the genes selected by the complex, including some genes coding for certain somatic and behaviour traits. Does it follow that these memes constitute an ultimate interest of at least equal rank to genes? There are adherents to the idea that cultural survival is more important than genetic survival. Here is Australia's senior demographer, who recognizes the genetic dimension of ethnicity but subordinates that genetic interest to cultural values:

> Some people think that a steady replacement of Anglo-Celts by other ethnic groups is highly desirable. . . . Personally, [replacement of Anglo-Celts] does not worry me so long as "Australian values" remain: free speech; freedom of religious worship; equality of the sexes; reasonable equality between social classes (i.e. no aristocracy); and so on.[63]

Any priority of memes over genes should become apparent by considering conflicts between memetic and genetic interests. Richerson and Boyd describe several examples of the selective effects of cultural group strategies.[64] These are lactase tolerance, group ideologies, dialect evolution, and diffusion of innovation. Of these, only some group ideologies do not clearly serve genetic interests. I have in mind belief systems that induce celibacy or early death in war or transfer of resources to a chief or other genetic competitor. Could ideology be an example of competition between memetic and genetic fitness? And if so does it mean that memes are ultimate interests? There is an apparent conflict because an ideology that directs loyalty to the group can be induced in the laboratory among unrelated subjects.

Social psychological experiments show that humans have a marked tendency to identify with groups of all kinds, and to develop ethnocentric attitudes concerning these groups. Tajfel reports a study of group identification in which negative stereotyping of outgroups developed spontaneously, in the absence of intergroup competition.[65] Tajfel randomly assigned subjects to groups that had no conflict of interest and no history of inter-group hostility. Even when subjects were informed of the random composition of the groups, they still attempted to maximize group differences, apparently in an attempt to outcompete other groups. At least in Western cultures, group competition is thus easily triggered. Laboratory ethnocentrism is intensified when there is conflict of interest between groups. The seminal research in this area was conducted by Sherif, who randomly assigned boys to groups and then set these groups against one another in a series of competitions.[66] Group membership became a salient aspect of the boys' personal identities. Furthermore, the boys negatively stereotyped members of other groups and acted on these evaluations with aggressive conduct.[67] Emotional attachment to ingroup and rejection of outgroup members is implicated in these processes. Ethnic-like factors produce an exaggerated group identification, but are not prerequisites. This is consistent with an evolutionary history in which memetic interests have won out over genetic interests, as hypothesized by Boyd and Richerson.[68]

If ethnocentrism can attach itself to random groups, why have people throughout history shown greater tribal allegiance than allegiance to vocational categories such as hunters or gatherers or farmers? Why have ethnic loyalties usually taken priority over class loyalty? Recent insights provided by psychologist Lawrence A. Hirschfeld provide a persuasive solution to this puzzle.[69] Hirschfeld conducted a series of psychological experiments with young children, finding that knowledge of race is not derived from observations of physical difference. Nor does it develop in the same way as knowledge of other social categories. Three-year-old children distinguish inherited characteristics from uninherited ones. Hirschfeld concludes that ethnic thinking is the product of a special-purpose cognitive ability—a 'domain-specific competence'—evolved for understanding and representing heritable traits. One might call this competence an innate descent-group module, a concept seemingly made to order to help explain the reoccurrence of ethnic nepotism. A common set of abstract principles has underlain all systems of ethnic thinking throughout history. Hirschfeld's findings indicate that descent groups, both families and ethnies, have been so important to fitness in human evolution that the species is hardwired with special mental equipment for identifying, categorizing, and learning about them. Neuroscientists are homing in on the brain regions responsible for processing facial characteristics, including racial differences.[70] Once again, memes are selected by genes. Cultural group strategies cannot be simply designated in any clear cut way to be prior or external to genetic influences.

Social identity processes combined with our innate descent-group mental module prime us to sort our social environments according to kinship and ethnicity and to identify with the descent groups to which we belong. However, identification with non-kin groups is easily induced, as demonstrated by social psychology experiments and everyday experience. Does this mean that in identifying with groups humans sometimes give priority to memetic interests over genetic interests? This would follow if such group identification led to altruism in the sense of unreciprocated giving. If so this would be a misfiring of a behavioural tendency, perhaps because of an artificial environment that confuses mechanisms evolved to serve genetic continuity. MacDonald notes the compatibility of social identity theory with Boyd and Richerson's theory of cultural group selection.[71] MacDonald argues that evolutionarily stable group strategies support rather than compete with genetic interests, and illustrates his point with case studies of Catholicism and Judaism in the Middle Ages.[72] These case studies are extensive applications of the theory of cultural group strategies that find no overall conflict between these religious meme complexes and the genetic interests of the majority of Medieval Catholics or Orthodox Jews. MacDonald's thesis is that persistent cultural group strategies such as the major religions are not pathogenic mind viruses as in Dawkins' metaphor, but more like the mutualistic gut flora that aid in digestion. D. S. Wilson confirms this thesis in a comparative study of several religions—Calvinism, the water temple system of Bali, Judaism, and the early Christian Church.[73] He also reports on research in progress on 25 randomly-chosen religions from around the world. Wilson's conclusion is that religion is functional at the group level, that it is 'organismic' in that it organizes adherents into 'adaptive units'.

In his comparison of minority ethnic traditions, Spicer found that Judaism is the most tenacious of all, despite a dispersion lasting for two thousand years.[74] In human terms, this period of time approaches evolutionary dimensions, and sheds light on whether the memes constituting traditional Judaism have been mutualistic with the Jewish people's genetic continuity, or parasitic upon it. R. Paul distinguishes Judaism's cultural and genetic components, and finds strong mutuality:

> [D]espite the apparent coequality of the two parallel tracks of transgenerational inheritance (genetic and cultural), the ultimate goal of religious practice in biblical Judaism, including the transmission of nongenetic, symbolic replicators, is still the genetic success of the House of Israel as a biologically defined descent group. . . . [T]here is the clear promise that, if Israel keeps up its side of the covenant with God by obeying and transmitting his commandments, God will ensure the success of the people as a whole across generations. . . . [C]ultural reproduction [thus] remains a practice in the service of, and yoked to, the still higher aim of biological reproductive success for the Israelite nation as a biological descent group.[75]

Paul then compares Judaism with Christianity, concluding that the latter is a self-serving meme complex, applicable to any assemblage of descent groups. 'With this stroke it became possible for the symbolic replicators, the "customs, ceremonies, and dogmas" of the Judeo-Christian religion, to propagate through populations unrelated by kinship or ethnicity, fictive or otherwise, to the Israelite lineage. . . .'[76] The decoupling of cultural and biological reproduction allows greater scope for meme-gene mutualism to break down, though as MacDonald[77] and Akenson[78] argue, that decoupling has not been consistent or complete. Christianity has at times been constructed as an ethnic group strategy, and for the majority of its existence has been an instrument of social cohesion and regulation that benefited local communities. Nevertheless, the ability of Christian doctrine to spread across lineages has caused it to multiply much faster than ethnically-tied Judaism, as Paul argues. 'Liberated from any biological grounding for membership in the group or for receipt of the teachings, Christianity is able to conceptualise a far stricter dichotomy between spirit and flesh than did Judaism, assigning guilt, sin, and corruption to the flesh and finding a possibility for purity only in a spirit that would be better off freed from its carnal vehicle.'[79]

Meme-gene mutualism but not conflict is stable over evolutionary time. An illustrative case is the phylogeny of indoctrinability. Eibl-Eibesfeldt argues that humans are genetically predisposed to be indoctrinated to identify with clan and tribe, which in the Environment of Evolutionary Adaptedness were, for most members, stores of genetic interest relative to neighbouring tribes.[80] P. Wiessner concurs, but observes that greater effort is required to indoctrinate the young to develop group attachments outside the family and clan, such as with the tribe.[81] Significant effort is expended on initiation rituals that bond teenagers to peers from other clans. In highland New Guinea ancient initiation traditions are augmented by more recently developed cults, ritual practices that are purchased from tribes admired for qualities believed to be fostered by the cult.[82] Tribal members monitor and discipline each other with a rigour that discourages free riding on the tribal solidarity of others.[83] Thus it appears that initiation rituals and cults, in Boyd and Richerson's terminology cultural group strategies, do not harm but enhance the genetic interests of the initiates. Evolutionarily stable cultural group strategies do not reduce the inclusive fitness of participants. It is possible for a group to adopt a strategy that is evolutionarily unstable, such as mass suicide or celibacy to achieve some spiritual goal. But such strategies are rare. If common they would become rare within a generation because they reduce the number of participants. If such a group strategy somehow continued for many generations, it would be self-eliminating by reducing the frequency of the participants' distinctive genes that made them vulnerable to such a destructive ideology in the first place. Groups that adopt fitness-depressing memes are rapidly eliminated and are replaced by groups practising adaptive cultural strategies. This is a central finding of the theory of cultural group strategies.[84]

Hirschfeld's findings are consistent with Boyd and Richerson's theory. Arguably an innate mental module for distinguishing descent groups from other social categories could be selected when it directed cultural group strategies to groups with elevated kinship. But on this reading cultural group strategies are not themselves ultimate interests but aspects of the environment, albeit man made. They affect genes as agents of selection, and although they replicate themselves to a certain extent, they are not germ-line replicators. Can selection pressures be interests? Species have ultimate interests separate from the selection pressures that shaped them. Indeed those same pressures are often invidious. They are dangers and miseries we try to escape. In principle a cultural group strategy could be viable as a reproducing meme complex that treated humans as expendable fodder. An economic or cultural structure could be maintained at the cost of its inhabitants' genetic interests, so long as 'new blood' was always available. Families and ethnies could be systematically incorporated, indoctrinated, disciplined, and prevented from reproducing, assuming a steady stream of new sacrificial workers was provided from outside the system. An example is extreme religious sects that prevent procreation. Another is modern Western societies with below-replacement birth rates combined with high levels of immigration from genetically distant populations. But I cannot see how it follows that such a meme complex would ever constitute an *interest* for its inhabitants, let alone their ultimate interest. From the evolutionary standpoint, that is, one that adopts reproductive fitness as the criterion of adaptiveness, the inhabitants of such a meme complex are *victims*, not beneficiaries. How can a mortal risk constitute an interest? This dubiousness applies as much to natural as to constructed environments. To repeat, organisms have genetic interests independent of the physical or cultural environment in which they find themselves. Retaining a set of (constructed) environmental conditions in no way compensates for loss of genetic interests occasioned by that environment.

Culturally defined groups that persist for many generations cannot escape the calculus of genetic fitness, as implied by E. Sober and D. S. Wilson's analysis of multilevel selection.[85] They note that: 'Inclusive fitness theory, evolutionary game theory, and selfish gene theory . . . are not regarded as competing theories . . .' This is most apparent in the case of ethnic nepotism theory where the ethny is conceptualized as a super family. My analysis in this book amounts to a multilevel selection theory (though prescriptive rather than descriptive) in which individuals strategize to invest in concentric circles of kinship. But I construct this analysis from the 'bottom up' within the rules of radical neo-Darwinism (corresponding to what Sober and Wilson describe as inclusive fitness and selfish gene theory). Commenting on Sober and Wilson's multi-level analysis, H. K. Reeve argues that the different levels of selection are perfectly translatable into inclusive fitness. '[T]he "new" group selection models, in which subgroups of the population, rather than individuals, can be seen as vehicles of selection, are not mathematically different from the broad-sense individual selection (e.g. inclusive fitness) models at all.'[86] This is not a completely accurate assessment, since as D. S. Wilson points out, competition between subdivisions of a metapopulation can cause behaviours to evolve that would not have evolved under strictly individual selection.[87] However, this does not invalidate our definition of genetic interests. Since an individual's inclusive fitness is his effect on the frequency of his distinctive genes (see this volume, pp. 120–121), it follows that genetic interest remains regardless of shifts between levels or vehicles of selection that structure the lineage. (I discuss group selection at greater length in Chapter 5, pp. 127–130. D. S. Wilson's view that some cooperative groups constitute units of selection and even super organisms might mean that genetic interests can exist at the group level, but that is properly the subject for another discussion.)

A final note on memes has the flavour of science fiction, but one that is instructive as a mind experiment. What if the genes' function were performed by other replicators? The brilliant computer pioneer Alan Turing raised the possibility that the human mind is a sort of software that can be 'run' on hardware other than the brain.[88] The digitization of the human mind has been predicted by Ray Kurzweil, a leading computer scientist.[89] If this is possible, computer programs might become more than analogous to the mind and subjectively human consciousnesses might reside in the micro-circuits of computer. A philosopher of consciousness, John Searle, rejects Kurzweil's prediction as impossible.[90] His argument is that simulating aspects of brain function cannot replicate the subjective aspects of consciousness. In that case the notion of a virtual human is an oxymoron—'they' would be 'us' no longer.

Parenthetically, Searle's argument, even if true, does not prevent computers from having ultimate interests; but they would not be human interests. Searle is concerned only with consciousness, not with issues of evolution or survival. Organisms lacking consciousness still have genetic interests, as argued by Alexander. From the evolutionary perspective it is interesting to wonder whether computers will someday become organisms by gaining the capacity to reproduce, otherwise behave adaptively, and have these behaviours modified down the generations by natural or artificial selection.

If Kurzweil is right and a human mind is someday digitized, what would be its ultimate interest? That interest would comprise the memes coding for conscious and unconscious thought processes as well as the computer hardware housing them. If digitization of mind was constrained to simulate the epigenetic (developmental) rules contained within the genes then the functional information contained within a subset of genes—those coding for the nervous system and related motivational systems—would be continued in the digital domain. Our ultimate interest would then be the accurate replication of this software and the maintenance of suitable hardware for storing and powering it. Also, if virtual humans did have reproductive ultimate interests, they would necessarily be mortal and still evolving. As implied by the discussion of Hamilton's inclusive fitness theory in Chapter 5 below, the only organism that lacks ultimate interests is one that has ceased evolving and is thus no longer dependent on transgenerational replication. If virtual existence allowed a consciousness to replicate itself perfectly, dispense with sexual reproduction and achieve immortality, its personal survival would become its only ultimate interest. In the meantime we mortals are burdened with our evolutionary heritage and with being dissolving links in evolving chains of life.

To conclude the discussion of memes and of culture in general, it is clear that these can be interests when they function as adaptive tools. Some might even be ultimate interests in the sense that they are replicators and help shape traits. However, memes are subject to genetic veto. Memes and anything else that reduce an individual's inclusive fitness are not interests but liabilities.

One final contender for the status of ultimate interest is 'communicating genes' as postulated by R. Buck and B. Ginsburg.[91] Their idea is that genes do not stand alone, but always act in concert with other genes. No gene is an island. Buck and Ginsburg are critical of what they contend is the sociobiological assumption of genetic atomism, that genes are selected for their individual effects rather than their joint actions with other genes. Those joint actions are due to communication between genes. Selection occurs 'at the level of the communicative relationships of genes', starting with dyads of genes.[92] Gene dyads can be in the same cell, in different cells, and in different organisms. It is these dyads that are the basic unit of selection, not individual genes, Buck and Ginsburg believe. If so, communicating gene dyads, not single genes, are ultimate interest.

This presents no obvious problem for the concept of genetic interests. Different individuals and groups will still possess different gene frequencies, and those differences will mean differences in genetic interests. What does present a difficulty is Buck and Ginsburg's rejection of the logic of inclusive fitness theory. This appears to be a difficulty for their theory, not for mainstream neo-Darwinism.

Buck and Ginsburg approve of the view that prosocial motivation is directed at the survival of the species. They reject the argument that species-directed altruism is an extreme form of group selection that must fall victim to freeriders, with the sentence: 'We reply that such disruptive tendencies are countered from the outset—literally from the beginning of life—by prosocial tendencies based upon dyad-level selection.'[93] Yet the free rider argument is taken very seriously by most evolutionary theorists. It was sufficient to undermine the 'good of the species' thinking that had prevailed in biology up to the publication of Wynne-Edwards's 1963 monograph *Animal Dispersion in Relation to Social Behaviour*. Wynne-Edwards argued that birds limit their reproduction to preserve resources for the species as a whole. This view was criticized by J. Maynard Smith and G. C. Williams based on the argument that birds that did not limit clutch size would rapidly outbreed altruists who limited theirs, resulting in replacement of the genes (or gene dyads) responsible for the altruistic behaviour.[94] Buck and Ginsburg's failure to discuss these critical theoretical issues is indicative of a general absence of consideration of selection and competition in their theory. Do not communicating gene dyads compete with one another? Or is the whole gene pool one big gene dyad? The failure to deal adequately with competition, a core element of Darwinism, raises the problem of how to apply their theory to the real world. The problem is, I think, insurmountable.

Despite a thriving pool of pretenders, genes appear safely enthroned as humans' (and all other species') ultimate interest.

### *(4h) Objection from inequality in ethnic genetic interests*

An objection to the concept of ethnic genetic interests might be that since some ethnies are too small or undistinguished to carry significant genetic interests for the rest of humanity, the concept does not apply to them. This criticism is based on some erroneous assumptions. The simplest is that genetic interests are defined by reference to a larger group. In fact what outsiders think about someone's group is irrelevant to that group's genetic worth to itself. Genetic interest is an objective property of kin and ethnic groups. Even if all humanity held some ethny in contempt, and the members of that population had been persuaded that they were in fact worthless, that group would still bear a precious genetic interest for its members. As argued in Chapter 2, genetic interests are concentrated in families and ethnic groups, so that even relatively small ethnies contain a large fraction of their member's total genetic interests.

Neither do group differences detract from any group's genetic interests. For example, from a biological perspective, personality and intelligence are strategies, not ends in themselves. A pronounced group identity usually results in group pride, the belief that one's group is in some way distinguished. Notions of ethnic superiority are used as ideological weapons in inter-group competition, to bolster the ingroup's confidence and break the spirit of other groups, in effect, claiming precedence for the assertive party. Chauvinism thus has its functions, and might be considered morally superior to arrows and hatchets as a means for conducting intergroup competition, but it should never be confused with truth. And it is simply untrue, from a modern biological perspective, that the value of a group's fitness can be graded according to differences in phenotypic characteristics. In fact, no group difference—whether in intelligence, cultural achievements, athletic performance, or health—is relevant to evaluating a group's value to itself, which is innate to all populations (indeed, to all evolved species).

It is not rational—though perhaps it is adaptive—to claim precedence for one's ethny on the grounds of superiority. And it is certainly irrational as well as imprudent to grant another group precedence over one's own, whatever claims are made about relative capacities. In principle some other ethny's genetic interests might rationally be accorded special weight in a way that was adaptive for all grades. That is when some special characteristic of a group is adaptive (or maladaptive) for humanity as a whole. In such a case, it would be prudent for all parties if that special group had some kind of precedence (or restriction). In the case of a group bearing some unique benefit for mankind, other ethnies might guarantee its survival and even help it grow. However, several factors militate against the prudence of granting precedence to any group. If survival is sufficient to preserve an ethny's special quality, then no precedence is needed beyond the general rule of preserving all groups. If elevated fitness is required, how should groups charged with granting precedence be sure that this will benefit them as opposed to enslaving or replacing them? How can they be sure that the claim to special status is not itself a competitive move? Which disinterested party could be trusted to anoint a group with special status?—surely not an ethny's own spokesman! If a group were shown to possess some uniquely precious adaptation, that adaptation would need to be genetic, not cultural and thus readily transmittable, for it to justify precedence for the group's *genetic* interests. Even if a characteristic is genetic, it can be spread over a few generations, through intermarriage or reproductive technologies, to any receptive group without the insult of being replaced wholesale. It is inconceivable that group differences can justify one group's replacement of another in the name of the general welfare. This is a special case of a general flaw in the 'humanist strategy' that I discuss towards the end of Chapter 6.

*Notes*

1 Hamilton (1996, p. 2).
2 Alexander (1979, pp. 80–81).
3 Alexander (1995/1985, pp. 182–3).
4 Salter (in press-b).
5 Alesina et al. (1999); Faist (1995); Sanderson and Vanhanen (in press).
6 Blok (2002).
7 Hamilton (1971, p. 73).
8 Rushton (1995).
9 Herrnstein and Murray (1994).
10 Lynn and Vanhanen (2002).
11 E.g. Buchanan (1998; 2002).
12 E.g. P. Brimelow (1995) and journals such as *The Social Contract*, published in the United States, refer to the demographic harm done by mass immigration to Western societies.
13 Lynn (1996; 2001); Vining (1986).
14 Kaplan and Lancaster (1999).
15 Menzel and D'Aluisio (2000).
16 *International Herald Tribune* 20.11.2000, p.22.
17 Quoted in Menzen and D'Aluisio (2000, p. 25).

18 Dawkins (1989, p. 1).
19 Hamilton (1971/1996, p. 219).
20 Dawkins (1989, p. 8).
21 Dawkins (1976; 1982a).
22 Dawkins (1982a, p. 83).
23 As noted in section 4a, phenotypic and genetic 'striving' correspond most closely in the environment of evolutionary adaptedness, with mismatches expected if environmental change outpaces genetic selection or genetic change drifts from environmental fit.
24 Hamilton (1964/1996, p. 38).
25 Wolf et al. (2000).
26 Eibl-Eibesfeldt (1970).
27 Hamilton (1996, pp. 21–6).
28 Dawkins (1978).
29 Rushton (1995).
30 Freedman and Freedman (1969).
31 Hamilton (1964).
32 Dawkins (1978, p. 191).
33 It is sometimes argued that helping behaviour that aids kin is not altruistic because it enhances the helper's inclusive fitness. On this reading, only behaviour that reduces the actor's inclusive fitness can be called altruistic. This interpretation has several weaknesses. First, it ignores the historical development of the altruism concept and related theory. Secondly, it tends to confuse analysis by replacing a readily-observed criterion (helping behaviour that carries a net cost to the helper, with motivational correlates) with a difficult-to-observe, analytic, one (inclusive fitness). Thirdly, and most importantly, it confuses the actions of genes with the behaviour of phenotypes, as if genetic 'selfishness', to use Dawkins's metaphor, means that mother love and heroic self sacrifice are themselves selfishly motivated. Confusion over the proper use of the altruism concept is intensifying despite excellent clarifying discussions by D. S. Wilson (1992) and Smuts (1999).
It is often asserted that an act is not altruistic if it serves reciprocal relationships (tit for tat) or aids kin. This interpretation is based on the definition of altruism as social behaviour that reduces the actor's inclusive fitness. Thus altruism is ruled out as an explanation of a behaviour when nepotism is the motive. According to this view, a honeybee's self-sacrifice in defence of the hive might appear to be altruistic, but in reality it is selfish because it serves to increase the bee's inclusive fitness by helping close relatives. This is a valid interpretation if one accepts the definition of altruism as necessarily maladaptive. But it is difficult to imagine a case for which such a definition aids understanding. The definition breaks with the history of evolutionary theories about altruism by confusing individual with inclusive fitness. This distinction is critical to the theory, as indicated in E. O. Wilson (1975, pp. 3–4; see also p. 117): '[T]he central theoretical problem of sociobiology [is]: how can altruism, which by definition reduces personal fitness, possibly evolve by natural selection? The answer is kinship: if the genes causing the altruism are shared by two organisms because of common descent, and if the altruistic act by one organism increases the joint contribution of these genes to the next generation, the propensity to altruism will spread through the gene pool.'

The problem posed by altruism was originally recognized by Darwin, who wondered how his theory of individual selection could account for self-sacrificing behaviour. Darwin entertained the possibility of group selection as the answer to this puzzle, but mainstream sociobiology has gone with the smallest unit of selection, individual genes. The theoretical breakthrough came from the study of social insects, in which individual altruism takes the extreme form of worker ants and wasps failing to reproduce individually, instead helping their queen to do so (Hamilton 1964). The insight, formalized as inclusive fitness theory, is that the genes coding for altruism need not go extinct if that altruism is directed towards kin, thus aiding the reproduction of copies of the genes shared by the kin. This idea, that genetic payoff allows altruistic behaviour, was later extended by Trivers (1971) to try to explain friendships and alliances with a theory he dubbed reciprocal altruism theory. Notice that Trivers did not call his theory 'reciprocal selfishness'. He used altruism in the behavioural sense, while maintaining that the selfishness metaphor is appropriate for genes. Retaining the behavioural definition of altruism has the added advantage of concurring with subjective experience. On this score, mothering deserves to be described as altruistic because it is an affiliative, affectionate helping behaviour. It is not selfish even though it most decidedly increases the mother's inclusive fitness.
Parental care does not qualify as altruism based on an otherwise useful definition offered by D. S. Wilson (2002, p. 235, fn. 2) based on multi-selection theory. 'A behavior is selfish when it increases the fitness of the actor, relative to other members of the group. A behavior is altruistic when it increases the fitness of the group, relative to other groups, and decreases the relative fitness of the actor within the group.' The definition generates inconsistencies because it relies on the criterion of fitness rather than motivation. Consider the case of an individual member of a clan or ethny embedded in a larger society, and consider an act that benefits the clan or ethny, such as risking his life in its defence, while lowering the actor's fitness relative to other clan/ethny members. According to Wilson's definition the individual is being simultaneously altruistic (towards the clan/ethny) and selfish (towards the larger society).
Resolution of the problem of defining altruism lies with Eibl-Eibesfeldt's distinction between behavioural (including psychological) altruism and selfishness. 'Much of the confusion derives from mixing up selective consequences and individual intention. All behaviors which contribute to fitness can be said to have egotistic consequences, since they promote the actor's genes. But the intention of the individual may very well be called altruistic, if it is spontaneous and motivated by empathy and aids another individual even at high risk' (Eibl-Eibesfeldt 1989, p. 99).

34 Hamilton (1996, p. 19).
35 *Ibid.*
36 Lewontin (1972).
37 P. Frost (personal communication 23.11.1998); Karow (2000); for another estimate see Holden (1998).
38 E. O. Wilson (1999, pp. 48–50).
39 Dawkins (1989, p. 10).
40 E. O. Wilson (1984).
41 Dawkins (1978, p. 193).
42 Trivers (1974).
43 Wright (1943).
44 Dawkins (1976, Chapter 11).

45 *Ibid.*, pp. 207; 212–15.
46 *Ibid.*, p. 207.
47 Dawkins (1978, p. 68).
48 Bateson (1978).
49 And see Lynch (1996).
50 Dawkins (1978, p. 68).
51 Alexander (1979, p. 79).
52 *Ibid.*, p. 80.
53 Dawkins (1982a).
54 Thiessen (1997).
55 D. S. Wilson (2002).
56 McNeill (1979, p. 34).
57 *Ibid.*, pp.32–3.
58 MacDonald (1995).
59 Irons (2001).
60 Blackmore (1999, p. 159).
61 E.g. Segal (1999).
62 Boyd and Richerson (1985; 1987; 1992); Richerson and Boyd (1998/2002).
63 Price (2000, p. 10).
64 Richerson and Boyd (in press).
65 Tajfel (1981).
66 Sherif (1966); and see Rabbie (1992).
67 And see Triandis (1990).
68 Boyd and Richerson (1985).
69 Hirschfeld (1996).
70 E.g. Golby et al. (2001).
71 MacDonald (1998, Chapter 1); Boyd and Richerson (1985; 1987).
72 MacDonald (1994; 1995).
73 D. S. Wilson (2002).
74 Spicer (1971).
75 Paul (1998, p. 389).
76 *Ibid.*, p. 390.
77 MacDonald (1995).
78 Akenson (1992).
79 Paul (1998, p. 391).
80 Eibl-Eibesfeldt (1998/2002).
81 Wiessner (1998/2002).
82 *Ibid.*
83 Boyd and Richerson (1992); Eibl-Eibesfeldt (1982); Frank (1995).
84 Soltis et al. (1995).
85 Sober and Wilson (1998, p. 98); and see D. S. Wilson (2002).
86 Reeve (2000, p. 65).
87 D. S. Wilson (2002, Chapter 1).
88 Turing (1959).
89 Kurzweil (1999).
90 Searle (1999).
91 Buck and Ginsburg (1991; 2000).
92 Buck and Ginsburg (2000, p. 4).

93 *Ibid.*, p. 13.
94 Maynard Smith (1964); Williams (1966).

# Part II: Strategies

# 5. Can Ethnic Altruism be Adaptive? Hamilton's Rule, Free Riders, and the Distribution of Altruism

*Summary*
Aid to co-ethnics that reduces the giver's individual fitness (ethnic altruism) risks being maladaptive, or 'evolutionarily unstable'. Hamilton formulated a criterion for adaptive altruism, referred to as Hamilton's Rule. This is based on his theory of inclusive fitness theory, which he developed to explain altruism. The theory states that altruism can be adaptive when it tends to preserve or increase the altruist's genetic representation in the next generation. However, adaptive altruism must overcome the risks of free riders and maldistribution of resources. Despite these risks, Hamilton argued that altruism directed towards the tribe can, in principle, be adaptive. The large aggregate kinship contained by ethnic groups supports Hamilton's view.

## *Introduction*

In Chapters 2 and 3 we saw that ethnies carry large stores of their members' distinctive genes, especially when contrasted with ethnies of different racial background. It is therefore plausible to assume that it would be adaptive to direct a great deal of altruism towards one's ethny. After all, if it is adaptive for a parent to make sacrifices for a family containing a total genetic interest of a few children, it is easy to conclude that efforts to preserve a population carrying the equivalent of thousands or millions of children must be at least as adaptive. The adaptiveness of ethnic patriotism would seem to follow directly from a rule formulated by W. D. Hamilton for deciding whether altruism is adaptive (see below in this Chapter).

In fact, this common sense proposition is controversial, when it is discussed at all. Dawkins dismisses the notion that racial similarity denotes any significant degree of kinship.[1] Commentators responded with equal brusqueness to Rushton's argument that competition between ethnies could amount to group selection of similar genes.[2] The likelihood is hardly discussed, rare exceptions coming out against the idea that ethnic altruism is[3] or even *could* be adaptive.[4]

The issue has practical implications. A potential interest cannot reward behaviour unless it is attainable by human agency. The telescope observation of a fabulous field of diamonds on Mars could yield no advantage until it became feasible to mine and transport the stones. The same might apply to genetic similarity distributed across a large population. If there are no practical ways to defend group genetic interests then those interests must play a diminishing role in shaping political decisions by rationally informed actors. The better informed individuals were about ethnic strategies, the less likely they would be to show ethnic solidarity. Ethnic loyalty would properly be called irrational. Individuals influenced by ethnocentric values would be squandering resources. If on the other hand there *are* effective strategies for advancing ethnic genetic interests, as I shall presently argue, then those interests can be rationally advanced via ethnic solidarity.

The point at issue is more subtle than deciding whether it is adaptive for an ethny as a whole to avoid aggression and retain resources such as a territory, although these are basic to strategies for defending ethnic genetic interests (see Chapter 6). The crucial issue for individual actors is whether ethnic altruism is adaptive; can they sacrifice their individual reproduction for the sake of their ethnies without losing genetic interests, that is, without contributing to the elimination of their distinctive genes from the gene pool?

In this chapter I apply Hamilton's Rule to the data on ethnic kinship reported in Chapter 3, before discussing two possible criticisms of the view that altruism directed towards ethnies can be adaptive—based on free riders and suboptimal distribution of altruism. Based on a mathematical argument presented by Hamilton in 1975, I conclude that there is no theoretical reason to doubt the feasibility of ethnic altruism, though free riders and suboptimal distribution of altruism, together with other strategic considerations, do restrict the types of ethnic altruism that are adaptive. In the next chapter I discuss some of the strategic constraints on adaptive ethnic altruism.

### *Inclusive fitness theory*

Before proceeding, I shall now introduce some of the concepts concerning inclusive fitness, in the process distinguishing it from genetic interests. Evolutionary biologists speak of 'investing' in a person or group, meaning an act of altruism—unreciprocated giving or assistance. Like so much else in evolutionary biology, metaphors are meant to be taken in only one way. Outside biology, investing is a deliberate activity towards a goal. It is teleological. But conscious choice is not a necessary part of biological investment; indeed, it has been absent for much of evolutionary history. Parent birds invest heavily in their hatchlings by feeding them, but the motivating mental states are emotions and fixed action patterns released by nestling markings and behaviours, not conscious reflection. Thus while 'investment' and 'strategy' always describe behaviour, they do not necessarily describe intentions.

Another piece of theory relevant to present purposes is the distinction between units and vehicles of selection, already touched on in Chapters 1 and 4. According to Dawkins a unit of selection is necessarily a replicator, something that is reproduced each generation and thus whose number of copies can rise or fall depending on its effect on the vehicle carrying it.[5] In sexually reproducing species individuals and groups are not replicators because mixing and reassortment of parental genes prevents offspring from being identical. That leaves the gene as the least controversial unit of selection, the reason for the title of this book. A vehicle is any 'relatively discrete entity' that carries replicators and is influenced by them so as to influence their reproductive success.[6] Dawkins offers the metaphor of vehicles as survival machines programmed by replicators. The least controversial vehicle is the individual organism, the most controversial is the large group, such as the ethny. In this book I argue that ethnic groups are relatively distinct entities that carry concentrations of their members' distinctive genes and are therefore potential vehicles, even if they have not been so in the past.

Another string of relevant concepts involves kin recognition and free riders. It is critical that individuals successfully distinguish kin from non-kin if they are to avoid investing in free riders, individuals who benefit from someone else's altruism without yielding a fitness benefit to the altruist. If investment in free riders is not avoided, the genes causing the altruism come under selection pressure. A variety of 'kin-recognition mechanisms' has evolved, including mothers' individualized bonding with their new-born babies. Other mechanisms include cohabitation, use of kinship terms, perhaps similarity in body odour and other characteristics, and symbolic knowledge about relatedness based on observation and reports, aided by analogies such as 'shared blood'. Kin recognition mechanisms help cue nurturant and other helping behaviours, especially from parents, though behavioural cues such as suckling (which releases the bonding hormone oxytocin) and the care-eliciting baby schema also play important roles.[7] Altruism between close kin is the most prevalent form of an inclusive fitness mechanism because kin selection is the most readily 'evolutionarily stable' strategy. 'Stable' in this sense means sustaining the genes responsible for a behaviour across many generations. An evolutionarily stable strategy, such as maternal nurture, maintains or increases the frequency of the genes coding for maternalism in the gene pool. Indiscriminate altruism such as foregoing reproduction to aid nonkin to reproduce, will weed out the genes that code for such behaviour, if maintained over many generations.

To briefly recapitulate some of the theory discussed in the first two chapters, the core principles in this book derive from Hamiltonian inclusive fitness theory.[8] The theory is an attempt to explain the puzzle of altruism—how can self-sacrificial behaviour survive in the face of natural selection? Should not all individuals be perfectly selfish, caring only for their own survival and reproduction? The last phenomenon hints at the answer found by Hamilton. Maternalism, the original form of altruism in evolutionary history,[9] raised the survival chances of offspring and hence of their genes, including the genes coding for maternal care. Maternalism spread to other forms of kin altruism—including paternal and sibling nurture, through the process of kin selection. As noted in Chapter 2, Hamilton's initial explanation of kin altruism turns out to be a special case of inclusive fitness theory. The latter sets the general conditions under which a gene coding for altruism increases its overall representation in the gene pool when it causes its phenotype to aid the reproduction of other phenotypes that share copies of itself. The latter need not be genealogical kin. This selective help between phenotypes bearing a gene for altruism results in altruists having, overall, more offspring so that the gene spreads throughout the gene pool.

### *Genetic interest is not inclusive fitness*

Inclusive fitness is widely misunderstood to mean genetic interest, though the two concepts are distinct. It is often stated that inclusive fitness is the number of copies of ego's distinctive genes existing in offspring and collateral kin. That definition actually describes familial genetic interests. Inclusive fitness is the effect of an individual's behaviour on the reproduction of his distinctive genes in himself and others (usually kin and fellow ethnics). Thus fitness is a behavioural effect across some unit of time, not a static number of genes. Fitness is a type of speed. In terms of discrete generations, fitness is prospective, the number of offspring caused by some behaviour. In continuous time fitness is the instantaneous rate of making copies of genes. Fitness is the effect of investing in copies of one's genes. It is not the aggregate count of those genes—the genetic interest. Behaviour is adaptive when it has the effect of preserving or increasing the actor's genetic interest, which is the sum of copies of his own genes (actually germ-line alleles) in the population. Such behaviour is said to increase the actor's inclusive fitness.

Genetic interests can be thought of as a type of capital, and fitness the result of investing in that capital. It is the type of capital from which a profit is only realized through constant investment, like a farm or a business. Left to itself farmland does not disappear, but it produces no yield. It can degrade in value due to weeds and erosion. In a lawless land a neighbour might claim parts of a farm whose fences have not been kept in good order, so that when the farmer finally takes up the plough he has less capital in which to invest. Capital can be put to work or left idle. It can be protected or left vulnerable to filching. It can be forgotten, due to poor bookkeeping or laziness. It can be squandered in bad investments. Strategies for investing in it carry different degrees of risk. Ethnic genetic interest is a form of capital that is a collective good for every member of the ethny.

To continue the analogy, inclusive fitness is the farm's yield, not the farm itself. The members of an ethnic group who cease striving to preserve their ethnic genetic interest have their inclusive fitness—their adaptiveness—reduced to the individual fitness of children and close family. They are, in effect, leaving their ethnic genetic capital to chance—the vagaries of nature and the good-will of competing groups.

Inclusive fitness has some counter-intuitive properties. An individual can have a large ethnic genetic interest but, unless he or she is investing in that interest, little or no inclusive fitness. The effort needed to bring all of one's ethnic group within the compass of one's inclusive fitness can be slight. Mathematically, inclusive fitness can rise or fall by millions depending on whether a person takes any heed of his ethnic genetic interests. Imagine a wealthy citizen whose inclusive fitness is no larger than the fitness of her children, because she contributes nothing to ethnic fitness. One day she has a change of heart, and makes a large donation (though small change for her) that bolsters the community. With that single act the woman has increased her inclusive fitness by several orders of magnitude. Ethnic duties can thus be light, constituting nothing more than vigilance. Wider inclusive fitness is maintained merely by looking up from one's private world, assessing the situation, and making a contribution (or no contribution) as needed. Since one has interacted with that larger gene pool, its fitness becomes part of one's inclusive fitness. If it prospers and continues, the vigilant member also achieves genetic continuity. If the member's vigilance or strategic assessment fails and the ethny declines in number (meaning a decline in genetic interests), that is a large loss of inclusive fitness for the member, and he can be said to have depressed his inclusive fitness. Most counter intuitively of all, that decline would not count as a loss of inclusive fitness for someone who cared only for the private realm (unless we count lack of vigilance or patriotism as a behaviour, a sort of negative interaction). That is one reason why gene frequencies are more appropriately nominated as genetic interest, and the concept of inclusive fitness reserved to mean the effect that an individual has on those frequencies.

Because this is an important distinction, let me present a different chain of examples, comparing a woman's fitness and genetic interest across her lifespan. A girl is born with zero inclusive fitness because she is incapable of aiding others' reproduction. However, at birth the same child possesses a genetic interest described in Chapter 2, consisting of copies of its distinctive genes in family and ethny. Barring catastrophes, this interest will not change much during the entire lifespan. During her development the girl's fitness might be somewhat positive if she helps relatives reproduce, for example by babysitting or doing chores around the house. Her fitness could also be negative, if she hinders her parents' bearing another child. Her genetic interest can also fluctuate for the same reasons, but around a high positive value, since it includes the copies of her genes contained in her extended family and ethny. During her reproductive career she raises her fitness each time she bears a child and each time she cares for her children (or other kin). Similarly, her genetic interest increases with each child. The instant a mother ceases caring for her family, her fitness drops to zero, since she is no longer making a difference. But her genetic interest does not change, because no child has died or become infertile. Actually her fitness is unlikely to fall so quickly to zero because parents usually continue to interact with their children after they leave home, giving advice and emotional support.

The comparison so far indicates that fitness and genetic interests are not so different, except that inclusive fitness is more variable. But let us not underestimate that variability. Consider two brothers who sire the same number of children who, in their turn, produce the same number of grandchildren. These brothers' familial genetic interests are equal. But their inclusive fitnesses could be radically different, if one brother invests nothing in his family, and the other one does. The fitness of an individual who ceases reproducing and aiding others is the same as if he or she were dead. Behaviour also affects genetic interests, in the way described by inclusive fitness theory, but the genetic interest remains when the behaviour ceases. In principle past generations have genetic interests.

Another way that genetic interests differ from inclusive fitness is that the latter involves only genes coding for altruism, not distinctive genes as such.[10] An individual's genetic interests are wider than this, embracing all of his or her distinctive genes.[11] Is another theory needed to account for the inclusive fitness of 'non-altruistic' genes? Thankfully, no. Hamilton's inclusive fitness theory is the critical analytic tool for mapping concentrations of all an individual's distinctive genes because it is based on coefficients of relatedness that apply to all genes. Individuals who show kin altruism due to the action of 'altruistic' genes thereby boost the inclusive fitness not only of those 'altruistic' genes but of all their distinctive genes, since these are replicated in kin in the same proportions. Strategies for defending broad genetic interests thus safely piggy-back on inclusive fitness theory through the minefield of free riders. It follows that adaptive ethnic nepotism not only selects for ethnic altruism, but for all gene-based characteristics distinctive to the group. This applies not only to traits with high heritability, such as cognitive profile, personality, physiognomy and overall morphology, but to cultural traits passed on through socialization within the family, such as religion, cosmology, and political culture. This is not to deny the existence of other selective pressures operating on these genetic and cultural traits, such that they can rise and fall in frequency even while the ethny retains stable numbers.[12] And cultural change can select for different gene frequencies.[13] However, it remains plausible that adaptive ethnic nepotism will contribute not only to preserving ethnic genetic interests but to preserving group characteristics in general.

Eibl-Eibesfeldt observes that the phylogenetically most ancient form of kin investment goes from mother to offspring, cued by innate kin-recognition and bonding mechanisms.[14] Dawkins reminds us that altruism is programmed in simple behavioural rules in which the actor is blind to any genetic effects. 'The kin altruism gene does not program individuals to take intelligent action on its behalf . . .'[15] This has been true for most of evolutionary history but in recent decades one species has become aware of the genetic dimension of kinship and can, in principle at least, devise strategies for deliberately helping kin and other individuals bearing copies of distinctive genes. Culture and tradition were important to human adaptation well before the advent of the science of genetics. But now, in an age of growing genetic knowledge, culture might be fashioned to conserve genetic interests, especially in ethnic groups, if such a strategy is at all possible.

### *Hamilton's Rule for Adaptive Altruism*

Hamilton's theory of inclusive fitness allows us to calculate the number of co-ethnics that must benefit if an altruistic act is to be adaptive. Hamilton formulated a rule for calculating when an act of altruism is adaptive, or 'evolutionarily stable', such that the altruist's genes are not reduced in frequency in the population. The rule is not complex, and is included in some undergraduate textbooks on evolution.[16] Hamilton formulated his rule in terms of relatedness $r$, for which $2f$ is substituted here to retain compatibility with the data of population genetics. Hamilton's rule states that altruism is only evolutionary stable when

$$b/c > 1/2f \qquad \ldots 5.1$$

where

$f$ = the average coefficient of kinship between the altruist and the recipients of the altruism;

$b$ = the sum of fitness benefits to all individuals affected by the altruistic behaviour;

$c$ = the fitness cost to the altruist.

Hamilton characterized his rule thus:

> To put the matter more vividly, an animal acting on this principle would be sacrificing its life adaptively if it could thereby save more than two brothers, but not for less.[17]

Hamilton's rule eases the condition for adaptive ethnic altruism in the case of repeated altruistic acts of small cost to the giver, the more so when the benefit to the receiver is a multiple of that cost. For example, small change given to street beggars can have much greater value to the receiver than to the giver. Also, someone with discretionary control over hiring or awarding contracts can dispense large benefits at little or no personal cost. Altruism would also appear to be highly adaptive when it benefits a large number of fellow ethnics, even when costly for the altruist. Based on the example discussed in Chapter 3, an act of charity or heroism by an Englishman that prevented 10,000 Danes from replacing 10,000 English would be adaptive even if the act cost the altruist his or her life and with it all prospects of raising a family (at least a family of less than 167 children), since this would save the equivalent of 167 of the altruist's children. Preventing replacement by 10,000 Bantu would warrant a much larger sacrifice because the genetic benefit is about 65 times larger; random Englishmen are almost as related as parent and child compared to the relationship between Englishmen and Bantu. Adaptive altruism need not consist of a single act, but a series of acts that impose accumulating cost on the altruist while reaping accumulating fitness benefits. Thousands of discriminations large and small, spread over a lifetime, could amount to a sizeable fitness gain compared to a lifetime of ethnically neutral conduct.

*Free riders*

Rushton argues that ethnic altruism is maladaptive due to free riders.[18] He points out that genetic competitors also exist within one's ethny, so that many acts of altruism directed towards co-ethnics will in fact assist competitors. These individuals will tend to outbreed the altruist, replacing his genes, making the strategy of ethnic nepotism maladaptive. Rushton argues that favouritism based on individual similarity is better proofed against free riders. However, the risk of free riders would seem to be accounted for by the 'average coefficient of relatedness' in Hamilton's Rule. It does not matter if the relatedness (or kinship) of recipients of altruism varies, so long as the average relatedness between nepotist and benefactors is higher than that between nepotist and competing groups which do not receive nepotism. These are sufficient conditions to allow Hamilton's Rule to apply.

Hamilton's 1975 model of a genetic basis for tribal altruism shows that it is theoretically possible to defend ethnic genetic interests in an adaptive manner, even when the altruism entails self sacrifice. He argued mathematically that an act of altruism directed towards the tribe was adaptive if it protected the aggregate of distant relatives in the tribe. In sexually-reproducing species a population's genetic isolation leads to rising levels of interrelatedness of its members and thus makes greater altruism adaptive. Low levels of immigration between tribes allow growing relatedness of tribal members, which in turn permits selection of altruistic acts directed at tribal members, but only if these acts 'actually aid group fitness in some way—reduce its chance of sudden extinction . . . , or increase its rate of emission of migrants'.[19] In a world where cooperation brings so many benefits, altruism is 'precious stuff'.[20] Closely related individuals are less likely to free ride and more likely to invest in and thus strengthen the group as a whole, improving the fitness of its members. In an earlier paper, Hamilton stated that altruism of the heroic kind could become adaptive after many generations of inbreeding within a population. Members should be more willing to invest in the group the more dense the ties of kinship within it. The theoretical basis was provided by Wright's concept of 'isolation by distance' in which relatedness falls away in a normal distribution from place of birth.[21] R. A. Fisher, a founder of the neo-Darwinian synthesis of Mendelian genetics and natural selection theory, also discussed the evolutionary origins of heroism and suggested that it benefited group 'prosperity'.[22] The precursor to all of this was Darwin's argument that moral behaviour, including patriotism, benefited the group as a whole even if it worked against the individual fitness of the actor.[23]

In one model that Hamilton considered realistic, the coefficient of relatedness between random members could rise as high as 0.5, even with some intake of immigrants. He notes that this is the same relatedness found within nuclear families, and 'we therefore expect the degree of amicability that is normally expressed between siblings'.[24] In such a population altruism between family members should also be stronger than usual. Most strikingly, Hamilton stated that 'we expect less nepotistic discrimination and more genuine communism of behaviour' in such a population.[25] However, at the group boundary relatedness drops sharply with a likely rise in intergroup hostility, such that 'a minor benefit from taking the life of an outsider would make the act adaptive'.[26]

Hamilton does not discuss overlap of kinship, where some inter-group pairs have closer relatedness than some within-group pairs. Should this occur there will be a conflict of genetic interests between fellow ethnics. Kinship overlap is not a feature of Hamilton's town model, perhaps because the towns are discreet territorial units, unlike the interpenetrated fuzzy boundaries between real populations. Another reason could be that kinship overlap is statistically unimportant when intra-ethnic kinship reaches the high levels predicted by Hamilton. As we saw in Chapter 3, Hamilton's prediction is confirmed by global genetic assays, though I could find no analysis of overlap in the literature of population genetics. Any overlap will be greatest between closely related ethnies, and least between ethnies drawn from different geographic races. Kinship overlap has strategic and ethical consequences, discussed in subsequent chapters.

Hamilton also remarked the disadvantages of high levels of immigration. While complete isolation risks inbreeding depression in small groups, high levels of immigration tend to reduce group cooperation by increasing the benefits of free riding. In competitive environments, groups that cannot cooperate risk being outcompeted and encroached upon by more cooperative groups. Thus it is critical to regulate immigration to maintain a workable level of group solidarity, either by keeping intake low or by selecting altruists as immigrants.[27]

E. O. Wilson has presented an argument similar to Hamilton's.[28] He notes that group selectionist theories for the evolution of noble traits such as 'team play, altruism, patriotism, bravery on the field of battle' originate with Darwin's *The Descent of Man*. These theories have been developed with increasing depth by A. Keith, A. E. Bigelow, R. D. Alexander, and most recently by E. Sober and D. S. Wilson.[29] These theories assume a prehistory of endemic armed conflict between groups, and except for the last assume that these groups were comprised of closely related individuals—that is in which the group represented an extended kin group and thus a concentrated store of members' distinctive genes. The last theory is consistent with this assumption but proposes a more general mechanism for the evolution of group altruism.

The most obvious form of free riding afflicts an altruist who invests in an unrelated individual or group that does not reciprocate. But free riding can occur within families and within ethnies due to conflicting selection pressures between different units of selection, such as individuals and groups. In the following quotation E. O. Wilson expresses the conditionality of conflict between different units of selection representing different pools of genetic interest: '[W]hat is good for the individual can be destructive to the family; what preserves the family can be harsh on both the individual and the tribe to which its family belongs; what promotes the tribe can weaken the family and destroy the individual; and so on upward through the permutations of levels of organization.'[30] Thus, as argued in the previous chapter, free riders pose a real threat to altruists' genetic interests. However, there are circumstances in which sets of interests coincide, such as external threat to the group or in competitive relations between ethnic groups within a multi-ethnic society. Since the rise of the first empires several thousand years ago, whole ethnies have been subordinated by invaders or by groups that were more competitive politically or economically. Arguably this has tended to reduce the fitness for all members of the subordinate ethny, who were enslaved or deprived of access to resources including land and females. The most drastic outcome of tribal or national defeat is large scale killing. When the action of defending the ethny simultaneously serves kin and ethnic genetic interests, as has so often been the case, it is adaptive for individuals to make substantial sacrifices in defence of the ethny. As Wilson implies, it can also be adaptive to sacrifice individual and family interests for the tribe.

E. O. Wilson adds a territorial dimension to his argument, suggesting that displacing other groups from their land was a central feature of ethnic competition from its conception in the tribal past. At the same time he suggests that genocide and ethnic cleansing were primordial human strategies. (The centrality of territory as an asset in ethnic group strategies is discussed in Chapter 6, section e.)

> A band might then dispose of a neighboring band, appropriate its territory, and increase its own genetic representation in the metapopulation, retaining the tribal memory of this successful episode, repeating it, increasing the geographical range of its occurrence, and quickly spreading its influence still further in the metapopulation. Such primitive cultural capacity would be permitted by the possession of certain genes.[31]

Group selectionist theories have been criticized for not accounting for the action of free riders. Three points need to be made here. First, the free rider problem among humans was probably solved long ago in small-scale societies by monitoring and punishment, as reported in contemporary band societies.[32] A recent experimental study indicates that humans monitor group members' contributions to public goods and punish cheats.[33] Small-scale societies are notorious for the intense mutual monitoring and informal social controls placed on members. Monitoring is effective because in these societies everyone knows each other's business; resources and social manoeuvres are difficult to disguise. Individuals will go out of their way to punish those who attempt to free ride on collective goods.

Secondly, the group selection under consideration here is actually extended kin selection, since tribes and ethnies are extended kin groups. As Hamilton pointed out, rising intra-group kinship reduces the risk from free riders somewhat by making free riding attempts less likely and by mitigating the maladaptiveness of free riding when it does occur. The theory of kin selection and attendant conditions are well known and generally accepted in the relevant biological disciplines.

A third point about group selection is that it is not necessary to the concept of genetic group interests. The idea of genetic group interests is logically distinct from the view that tribes have evolved through group selection. As discussed earlier in this chapter, Hamilton's 1975 'town' paper shows how concentrations of kinship can build up within territories even with a steady trickle of migration. That is why individual altruism directed towards the group can be adaptive from a strictly gene-selectionist perspective, by virtue of such altruism protecting the individual's genetic interests contained within the group. But none of this theorizing is necessary to accept the existence of ethnic genetic interests. The existence of tribal genetic interests, and its modern ethnic counterpart, does not require taking a position on the group selection debate. All that matters is that these populations are descent groups—whatever the cause—that hold greater concentrations of members' genes than do other populations. Neither does the existence of tribal genetic interests entail humans having special psychological adaptations for defending those interests. If we are adapted to defend tribal genetic interests, the explanation is more likely to reside in culturally elaborated kin selection mechanisms.

It might be argued that group selection theory must be embraced by anyone who wants to argue that ethnic nepotism can be evolutionarily stable. In other words, even accepting that individuals have many copies of their distinctive genes in their ethnies, any attempt to defend that interest must be evolutionarily unstable (that is, maladaptive) *unless* one accepts the possibility of group selection. The critic would then conclude that anyone who rejects group selection theory must also reject the concept of ethnic genetic interests as applicable to the real world. There are two answers to this, applicable to two types of critics, those who support group selection and those who support only kin selection and compatible mechanisms such as reciprocity. The first type concedes that group selection theory has found a way around the free rider problem, and so should be amenable to the possibility of adaptive ethnic nepotism. The second type of critic, those who accept kin selection, will not object to the possibility of adaptive ethnic nepotism so long as it meets the well-known conditions for kin selection theory, namely Hamilton's Rule for adaptive altruism and the control of free riders. Both conditions are explicitly addressed in this chapter, as well as conditions special to adaptive ethnic altruism in Chapter 6.

Another response to an attempt to tie ethnic genetic interests to group selection would be to point out that, according to one definition, kin selection is a type of group selection. This would be possible if there were significant genetic variance between groups, as is usually the case between ethnies. It follows that one type of group selection is really kin selection. More generally, some of the disagreement in the debate over group selection is terminological. For example, D. S. Wilson apparently equates kin selection with inclusive fitness processes, while Hamilton and others have argued that the former is but one type of the latter (see discussion this volume, pp. 41–43).[34]

Dawkins criticizes aspects of that part of Hamilton's analysis where he describes strategies for defending tribal genetic interests. Dawkins offers no criticism of Hamilton's model of relatedness within and between tribes, but disagrees with Hamilton's ideas about consequent adaptive strategies. Just because within-tribe relatedness rises to that of full siblings, Dawkins notes, it does not follow that members of the tribe will show sibling-like altruism towards each other. True, 'random town members will be more altruistic towards each other than they are to recent immigrants from other towns, for the latter will be noticeably less closely related to them'.[35] But this within-tribe altruism will only appear strong in comparison to the xenophobia shown to immigrants, he thinks. In comparison to sibling altruism random town members' interactions may appear selfish or indifferent. Dawkins's basis for this claim is Hamilton's reckoning that true sibling relatedness in such a semi-isolated tribe would rise well above the 0.5 level, so that it would remain adaptive for sibling altruism to be stronger than random tribe member altruism.

Dawkins's criticism is useful for the way it qualifies Hamilton's conclusion that altruism between random tribal members would rise to that of true siblings, and for emphasizing the genetic rationale for continued within-tribe competition, even with elevated levels of relatedness between random members. But he fails to report all of Hamilton's model and in consequence misses its full subtlety. The model does not banish nepotism but confines itself to predicting that there will be less of it 'and more genuine communism of behaviour'. It is a matter of degree, not absolutes as Dawkins interprets it. The gap will close significantly between nepotistic and competitive behaviour within the tribe, but not disappear. This follows from Hamilton's model of rising relatedness. Between random members it rises to 0.5, but between siblings it rises above 0.5, though he does not specify just how high (but see Chapter 8, p. 261). It certainly does not rise to 1.0, which it would need to do in order to maintain the 0.5 gap between sibling and random relatedness in a panmictic population (in which random relatedness is zero). Dawkins also misses the significance of competition between tribes. In Hamilton's model, greater altruism between random tribal members is justified when this prevents harm to the tribe, which would include replacement by a different tribe. Also, the risk of free riders is reduced by high intra-tribal relatedness. So 'selfish or indifferent' interactions between random tribal members should decline in frequency as within-tribe relatedness rises.

Hamilton's 1975 model thus indicates, at least theoretically, that patriotic sacrifice and lesser forms of altruism directed toward the ethny can be safe from free riders. Free riding is still a risk, but it does not inevitably render group altruism maladaptive. Indeed, the large store of distinctive genes contained by ethnic super families means that most common forms of ethnic nepotism, especially when they serve group interests against competition from genetically distant ethnies, stand a good chance of serving the actor's genetic interests.

*Suboptimal fitness investment portfolios*

Individuals can, in principle, devote too much or too little of their time, resources, and personal security to the collective welfare of their ethnies, failing to accumulate a balanced portfolio of investments in reproduction. 'Portfolio' in this sense means the pattern of an individual's allocation of lifetime resources (work, time, money) across different genetic interests. It would be maladaptive to sacrifice personal reproduction unless doing so helped alleviate a real threat to the ethny. A patriotically-minded individual might give his or her life in an unnecessary war or suspend economic or status competition with co-ethnics when this made no difference to group security or prosperity. At the other extreme, an individualistically-minded person might not show much ethnic altruism at all, or devote too much to family even in times of national emergency. Continuing to amass resources for one's family in competition with fellow ethnics can be maladaptive when it reduces the group's ability to survive or prosper.

The allocation of altruism over the life span is affected by the cultures in which we live. In the past ethnocentric culture has usually been adaptive. Indoctrination is a powerful strategy for encouraging ethnocentric thinking, one that allows leaders to mobilize the community for defence.[36] Ethnocentric cultures indoctrinate members to code social information with emotional tags systematically connected to perceived group interests. Ideas, policies, and cultures that are perceived to be detrimental to group interests are tagged with negative emotions, such as anger, contempt, and disgust. Positive emotions such as happiness, love, and respect are reserved for things that benefit the group. The result is a generally higher level of group activism and greater group cohesion in pursuit of perceived group goals, especially competition with other groups. Rituals that rehearse ingroup solidarity and external threat are also able to engender ethnocentrism, as do folk histories of past glories and defeats. These 'oppositional symbols' keep the community at a heightened state of readiness for war and other forms of contest.[37]

Institutions that indoctrinate belligerence into a large proportion of the population can be maladaptive. J. van der Dennen has compiled anthropological evidence showing that even the most belligerent tribes also have mechanisms for making and sustaining peace, including feasts, and exchange of brides and gifts.[38] Tribes that cannot make alliances and conduct economic exchange can find themselves isolated and vulnerable. An overly warlike stance is also maladaptive for many group members who can sacrifice individual resources and even lives for a war from which they or their kin stand to gain little.

The existence of a genetic interest will, under the right circumstances, select for strategies to protect it. Defending tribal interests has probably shaped aspects of tribal politics, selecting for cultural mechanisms underlying solidarity and defence. As discussed in the next chapter, some theorists believe that we are equipped with psychological predispositions specially adapted to defend the tribe. But if environmental conditions change too rapidly for selection to keep up, genetic interests can exist undefended. We would then be unwise to rely on the 'direction of striving' to identify our interests, and we would need to fall back on general principles developed from the study of life. I shall argue in Chapter 6 that this is the situation regarding ethnic genetic interests in modern societies, and to some extent with familial genetic interests as well.

So it is certainly possible for altruism to be distributed maladaptively, due to both idiosyncratic and culture-wide choices. But there is no reason to believe that it is impossible to choose an adaptive investment portfolio. Even if there were some universal epistemological problem that prevented the systematic choice of an optimal (or at least comparable) portfolio balance, then everyone would suffer and noone would suffer loss of relative fitness. Thus the appropriate question is not whether an adaptive mix of altruism is possible, but rather when it is adaptive for a group member to give his or her all for the ethny, and when to hold back and preserve the family (more generally, to preserve the potential for personal reproduction)? As I hope the examples described above indicate, the answer depends on circumstances. There is no fixed rule for distributing altruism, except that the extremes of completely failing to invest in family or ethny are imprudent. Failure to invest in family, whether in one's offspring or other kin, risks loss of relative inclusive fitness within the ethny. Failure to invest in the ethny risks loss of relative inclusive fitness between populations. In the next chapter I consider some different threats to fitness and the strategies, especially investment portfolios, likely to meet them.

*Notes*

1 Dawkins (1981); Miele (2002, p. 35).
2 Rushton (1989b, p. 518).
3 E.g. Rubin (2000); Masters (in press).
4 Silverman and Case (1998/2002); Dawkins (1981).
5 Dawkins (1982b).
6 Dawkins (1982a, p. 302).
7 Eibl-Eibesfeldt (1989).
8 Hamilton (1964; 1975).
9 Eibl-Eibesfeldt (1970).
10 Dawkins (1978, pp. 192–3).
11 For this reason the ultimate stake in competition between identical twins is difficult to identify, because whoever wins, the same genes benefit in fitness. The ultimate stake in competition between twins, say for resources from parents, is the inclusive fitness of the pair, not of either twin. This is because, statistically speaking, twin A passes on the same genes as twin B. This reduces the advantage of winning such a contest. Competition is maladaptive if the twins are likely to pass on more of their genes if they cooperate than if they compete. Segal and Hershberger (1999) present experimental evidence of greater cooperation between identical than nonidentical twins.
12 Cavalli-Sforza and Feldman (1981).
13 Boyd and Richerson (1985); Durham (1991).
14 Eibl-Eibesfeldt (1970).
15 Dawkins (1979, p. 192).
16 An excellent example is the textbook by Boyd and Silk, which offers a clear exposition of the rule (1997, pp. 258–63).
17 Hamilton (1963/1996, p. 7).
18 Rushton (1989).
19 Hamilton (1975, pp. 141–2).
20 *Ibid.*, p. 142.
21 Hamilton (1971, p. 74).
22 Fisher (1930/1999, p. 246) believed that heroism in battle, in 'useful border expeditions', preserved the 'prosperity of the group' and thus made the 'sacrifice of individual lives occasionally advantageous'. But much more important was sexual selection of heroic qualities. The hero benefits all his kin by the prestige conferred on them due to his heroic sacrifice. But Fisher does not explain the selective logic of conferring prestige on heroism. A possible answer is that those who do the conferring—nonkin tribesmen—benefit from the heroism and reward it with delayed reciprocity to the hero's family. Granting a precious resource to nonkin would otherwise be evolutionarily unstable.
23 Darwin (1871).
24 Hamilton (1975, p. 143).
25 *Ibid.*, p. 144.
26 Ibid.
27 Hamilton (1975, pp. 141–2). Hamilton did not suggest that altruism was the only criterion for selecting immigrants. It would hardly be adaptive for a group to allow immigration on such a large scale, even if they were altruists, that relative fitness was significantly reduced.

28 E. O. Wilson (1975, p. 573).
29 Keith (1968/1947); Bigelow (1969); Alexander (1971); Sober and D. S. Wilson (1998).
30 E. O. Wilson (1975, p. 4).
31 *Ibid.*, p. 573.
32 Campbell (1983); Eibl-Eibesfeldt (1982); Boyd and Richerson (1992); D. S. Wilson (2002, p. 19).
33 Fehr and Gachter (2002).
34 In answering the question, 'What constitutes fitness?', D. S. Wilson (2002, pp. 37–40) does not mention inclusive fitness. Yet he states that an organism is fit when its reproductive success is higher than that of conspecifics. Inclusive fitness is the broadest measure of reproductive success, and so is compatible with D. S. Wilson's theory of group selection.
35 Dawkins (1979, p. 193).
36 Eibl-Eibesfeldt (1982; 1998/2002).
37 Spicer (1971).
38 van der Dennen (1995, Chapter 7).

# 6. Allocating Investment between Family, Ethny, & Humanity: Optimal and Actual Fitness Portfolios

*Summary*

When resources are limited, the main challenge for an adaptively minded individual is to strike a balance between investing in kin, ethny, and the whole species. The balance point shifts radically with contingency. Investment in any group is adaptive only when kept within the bounds set by Hamilton's Rule for adaptive altruism. This involves a number of criteria.

(a) Confidence of relatedness;
(b) The genetic distance between ingroup and competing groups;
(c) Group size;
(d) The salience of intergroup versus interindividual competition;
(e) Costs and benefits.

A further variable is not raised by Hamilton:

(f) The availability of collective goods. Territory is a collective good fundamental for harmonizing familial and ethnic genetic interests and securing long-term genetic continuity.

I discuss strategies for meeting each criterion. Comparing optimal and actual strategies, the match is weakest where modern society differs most profoundly from the Environment of Evolutionary Adaptedness. In the evolutionarily novel environment of the urban centralized state humans are imperfectly adapted to recognize and defend their genetic interest, and must rely on cultural strategies to substitute for innate mechanisms. Ideologies act as fitness portfolios, affecting believers' allocation of life resources across family, ethny, and humanity in general. Individualism, nationalism, and humanism are each adaptive under different circumstances, but in general the most prudent apportionment is family > ethny > humanity.

In the previous chapter I argued that it is possible to invest in one's ethny in an adaptive way despite the dual risks of free riders and suboptimal distribution across genetic interests in family, ethny and humanity. I based this argument largely on Hamilton's Rule for adaptive altruism, as well as his model of adaptive altruism towards the tribe.[1] Hamilton maintained that there 'should be restraint in the struggle within groups and within local areas in the interests of maintaining strength for the intergroup struggle', and went on to argue mathematically that such restraint could take the form of altruism between random pairs of an ethny, and still be adaptive.[2] This means that it is generally (though not always) adaptive to contribute to one's ethny rather than to the species as a whole, and that it can even be adaptive to direct some investment away from the family to the ethny.

Hamilton's analysis must be supplemented if it is to generate strategies for investing in modern ethnies. Modern ethnies are larger in population and territorial extent and are therefore more anonymous than were ancient bands and tribes. Modern ethnies are often distributed across two or more continents. Because of the concentric character of ethnicity, large ethnies usually comprise many cultures, languages, and religions. Even a territorially compact, culturally homogeneous modern ethny such as the Japanese or Bavarians are much larger with a far more complex division of labour than archaic human societies. These novel features mean that individuals cannot be familiar with more than a small fraction of the population. Marked differences of wealth and lifestyle arise. Finally, while the nuclear family has remained a vehicle for defending genetic interests comparable in importance to primordial times, the clan has declined in importance.

Hamilton did not consider these changes. But knowledge from sociology and politics allows us to apply Hamilton's basic principles to modern conditions, to formulate optimal strategies for distributing labours of love between family, ethny, and humankind.

As argued in Chapter 5, the risk of free riders is a real threat to investing in one's ethny, especially when the investment is so large that it reduces personal reproduction or investment in kin. The theme of free riders thus recurs throughout the following discussion. Also, when resources are limited, as they always are, a challenge for the individual seeking to defend his or her genetic interests is to strike a balance between investing in kin, ethny, and unrelated individuals (humanity). The balance point shifts radically with contingency. People do sometimes invest too much in their ethnic groups, though they often invest too little. Over-investment is often due to being tricked by fictional-kinship ideologies, as suggested by Masters,[3] though less remarked is the under-investment due to fictional *non-kinship*, which is nowadays prevalent in western societies.

Below I argue that there are several interrelated criteria for adaptively apportioning altruism between family and ethny. The criteria apply to individual decision makers. They are factors affecting people's ability to defend their total genetic interests, whether located in the family, ethny or species. Investment in the group is kept within the bounds set by Hamilton's Rule for adaptive altruism. This involves a number of variables:

(a) confidence of relatedness;
(b) the genetic distance between ingroup and competing groups;
(c) group size;
(d) the salience of intergroup versus inter-individual competition;

(e) costs and benefits.

A further variable is not explored by Hamilton:

(f) the availability of means by which the individual can contribute to group competition: collective goods including territory.

In following sections under these headings I discuss strategies for meeting each criterion. Each time I begin by postulating optimal investment strategies, denoted by bold type. I then compare these hypothetical strategies with real-world behaviour.

The postulated optimal investment principles are based strictly on these criteria without regard for ethical sensibilities, which are discussed in the third section of the book. One formulation or another will offend those who give priority to the individual or family or ethny or humanity as a whole. It will, of course, offend those who deny the objective existence of race and ethnicity as genetic categories.

At times sociobiologically oriented analysis is bound to jar with our subjective moral sense, especially when the latter is informed by outdated biology or by no biology at all. The same is true of any ethic that allows for winners and losers in some circumstances, even if it also allows for mutual benefit through cooperation. Since genetic interests are highly valued within neo-Darwinism, a strategy can only be criticized as imprudent within that frame on the basis that it is maladaptive, that is when it reduces inclusive fitness. Recall that the Darwinian thinker cannot overemphasize the importance of reproductive fitness as an interest. Morality itself is a product of the evolutionary process, an adaptation to living in groups.[4] Genetic interests are not only more important than other interests; they have absolute priority or to use another metaphor, infinite weighting. E. O. Wilson notes that philosophers usually recoil from the idea that 'in evolutionary time the individual organism counts for almost nothing'.[5] Yet contemporary Western morality is highly individualistic and often rejects or is oblivious to genetic continuity as an imperative. Hamilton also realized how shocking the evolutionary perspective can be. 'The evolutionary process certainly has no regard for humanitarian principles. Pain is itself evolved to teach the animal to avoid harmful stimuli. In the immediate situation pain warns of threat to fitness.'[6] He also echoed T. H. Huxley's view that policy cannot be justified merely on the basis of what happens in nature; usually civilized existence is the result of striving to *avoid* nature 'red in tooth and claw'.[7] But neither is civilization meant to be maladaptive. On the contrary, it is the culmination of a long process of cultural adaptation, a way of surviving and prospering. It is unreasonable to expect people to tolerate aspects of civilized life that harm them. After all, civilization's main attraction is that it benefits its citizens. In Chapter 9 I discuss the ethics of defending genetic interests.

### *(6a) Confidence of relatedness and the problem of free riders*

Without reliable recognition of co-ethnics, individuals risk investing in free riders, unrelated individuals who use the investment to increase their reproduction at the expense of the altruist. The altruist risks losing relative fitness and often incentive to invest in the group. As argued in Chapter 5, free riders are not a problem for the concept of ethnic genetic interests. But they can be an obstacle to the realization of those interests. Theoretically, if the free rider problem is not solved, then over generations altruistic motivation to invest in the group is selected out, and perhaps the group gives way to another more cohesive ethny that has found a way to make ethnic nepotism pay off genetically.

*Collective goods and free riders.* The free rider problem was first recognized by economic and political theorists as a problem of public goods.[8] A public good is some benefit possessing two properties, jointness of supply and nonexcludability. Jointness of supply means that the good is not diminished by consumption, so that any number can benefit. Nonexcludability means that no member of the group can be prevented from consuming the good. The classic example of a public good is a lighthouse, whose warning beams can equally assist any number of boats and which no boat in the vicinity can be excluded from utilizing. In cases where a benefit is restricted to a particular group or society, such as a nation's school or welfare system, 'collective goods' is a more appropriately term, and will be favoured below.

Mancur Olson introduced to a large audience the issue of how best to provide collective goods in the face of free riders in his 1965 book, *The Logic of Collective Action.* As a solution Olson argued that enlightened self interest should encourage individuals voluntarily to contribute to collective goods. Recent experimental studies indicate that Olson was right in circumstances of social transparency, where everyone's contribution to collective goods is known. Cheaters, those who take from the collective good but do not contribute at the same level as other participants, are punished by the group, the effect being to deter and compensate for free riding.[9] Unfortunately, individual behaviour is not transparent in modern mass societies. From an evolutionary perspective many collective goods in modern societies represent an opportunity for enlightened free riding. Hardin plausibly maintains that in modern societies only sanctions imposed by the state can prevent free riding and the 'tragedy of the commons'.[10] Unless the state systematically directs the ethnic altruism of contributors to co-ethnics or maintains reciprocity between ethnies, individual investment is most prudently directed not towards the state but towards individuals and groups known to be kin, such as the family and the local ethnic community. In mass anonymous societies where the clan is scattered and high residential mobility prevents the formation of local ethnic communities, **this favours investing preferentially in the family and, less so, in co-ethnics bearing hard-to-fake ethnic markers such as dialect and racial features** (Optimal Strategy).

*Ethnic identity.* **Within multi-ethnic societies certainty of relatedness will be raised by investing in personal or organized efforts to identify co-ethnics and direct altruism only to them** (Optimal Strategy). However this strategy imposes the cost of such efforts, which is only justified if a significant fitness payoff would result. There is the separate issue of whether identifying and discriminating in favour of co-ethnics with sufficient reliability on a large scale is technically feasible. Such techniques would be of great value. In the absence of such techniques, the resulting uncertainty **favours investing in family and, less so, in reliably marked co-ethnics** (Optimal Strategy).

It can be in an ethnic group's interest to participate in a multi-ethnic society. Aggregating two or more ethnies into a larger population provides the benefits of size—stronger mutual defence and a larger more diverse market. Reciprocity is an attractive strategy for conducting inter-ethnic relations in a world made small by telecommunications and mass transportation. Mutual benefit can in principle be gained from a division of labour between groups with different talents, whatever their genetic distance. By controlling conflict, all can profit from trade and cultural exchange. The protection of a strong and fair legal system and the availability of social and economic mobility can minimize destructive ethnic conflict.[11] In addition to legal constraints, ethnic conflict can be further reduced by the blurring of group identities as they enculturate to a common standard of language and lifestyle. But there are risks in eliminating borders between ethnies. Loss of identity also translates into lost ability to strategize as a group, should the multi-ethnic experience sour. Ethnic polarization and conflict reduce the quality of life, and it is maladaptive for groups that have lower growth rates, due to immigration or reproduction. As argued in Chapter 3, replacement of an ethny is a major blow to the genetic interests of its members, especially when the replacing group is genetically distant. The loss will be much less when the two groups are kindred ethnies.

**Multi-ethnic societies are best entered when at least one of the following conditions is met: (1) when one's ethny is sure of maintaining parity or higher in relative numbers; or (2) when one's ethny is endogamous and sure of maintaining high status and resources whatever its proportion of the population; or (3) when ethnic partners in the multi-ethnic enterprise are**

**genetically close; or (4) when one's ethny is guaranteed sole use of some delineated and viable territory as a fall-back should its representation decline in the mixed part of the multi-ethnic society** (Optimal Strategy). I expand on territory as a strategy in section 6f, and on the relative advantages of ethnic nations and multicultural states in the next chapter.

*Social controls.* Government administered social controls that prevented ethnic free riding would reduce the need for individual citizens to distinguish ethnies and make it more adaptive to pay taxes and otherwise support the state. Free rider controls would pay off if they were less costly than individual identification or wholesale separation (see below). Controls would be attractive to majorities faced with rapidly rising minority populations, and to minorities faced with aggressive majorities. Such social controls would aim to ensure reciprocity, for example by ensuring that no group drew more welfare and other collective goods than it contributed. They would also preserve ethnies' relative fitness by, for example, counteracting higher-than-average reproduction by any ethnic group and preventing large-scale immigration of an ethnic mix that differs from that of the host society.

This would seem to counter Dawkins's assertion that human societies can do nothing to prevent their extinction. He argues that any attempt to uphold a generally high level of altruism, to keep the society strong as a whole, will surely fail due to free-riders immigrating and breeding at the expense of the self-restrained altruists. 'Even while the group is going slowly and inexorably downhill, selfish individuals prosper in the short term at the expense of altruists. The citizens of Britain may or may not be blessed with foresight, but evolution is blind to the future.'[12] Yet immigration controls or other culturally-based group strategies can shape human evolution. One population being swamped or replaced by another is a form of evolution since both cause genetic change. Evolution is not blind, or need not be, when a country chooses who can enter, and that choice affects future gene frequencies within the society.

One fascinating suggested method for governmental control of reproduction was advanced by Hamilton. He suggested a method for controlling overall population growth while keeping the mutation load within tolerable bounds. While Hamilton was not discussing ethnicity, his strategy would also tend to restrict free riders who breed at the expense of those who control their reproduction. In his scheme every individual would have an automatic right to two children. Unexercised rights could be given away at the discretion of the holder. Hamilton thought that most transfers would occur within families.[13] The effect would be to limit the rate of growth of any ethnic group, though this would not prevent growth rate differences if some groups failed to use their quotas.

While no multi-ethnic state controls reproductive free riding, the principle is still worth stating. **The presence of effective social controls on ethnic free riders favours investment in the multicultural state in addition to family in multicultural societies** (Optimal Strategy).

*Aggressive social controls.* Social controls can be used aggressively by a group to increase its relative numbers or parasitise others. Aristocracies and such systems as South African apartheid are examples of macroparasitism.[14] In democracies aggressive parasitism is employed mainly by majorities, while passive parasitism is the strategy of poor ethnies who draw more from welfare and other public goods than they contribute. Aggressive use of social controls is adaptive if it aids continuity or advances relative fitness, but can become maladaptive if it undermines collective goods to such an extent that society as a whole suffers and every group's absolute size shrinks. One countermeasure that might be used by minorities with little prospect of becoming majorities would be to prevent other ethnic groups from forming political majorities. Minority coalitions might agitate for the immigration of groups unrelated to the existing largest group, boosting the coalition's influence at the expense of the majority. This would make it difficult for any group to dominate government and impose social controls.

*Separation.* An alternate strategy for overcoming uncertainty of relatedness is separation in which part or all of an ethny secedes from a multi-ethnic society. The strategy amounts to a decision to conduct ethnic relations between societies rather than within the one society. This approach entails the one-time cost of identifying co-ethnics during the separation process, but avoids the cost of pervasive, open-ended identification and discrimination needed to avoid free riders in mixed societies. The latter cost can be high, both directly and in the communal polarization and atomisation that can result. Civil ethnic conflict is a major cause of warfare in the modern world.[15] Even without war, institutional discrimination can undermine liberal institutions and culture and the economic and social benefits associated with them. Costs are not eliminated by multiculturalism, which encourages different ethnies to maintain their separate traditions within a single state. The associated state and corporate apparatuses designed to suppress spontaneous discrimination and separation and indoctrinate support for public goods come at a price paid disproportionately by the majority ethny (see Chapter 7, pp. 188–190).

Separation is a proven method for ending endemic ethnic strife[16] that facilitates the development of nondiscriminatory civil and national societies with higher levels of voluntary investment in collective goods.[17] Furthermore, since the state is the most powerful means for implementing collective policies, a nation state has the great advantage of increasing people's capacity to strategize on behalf of their ethnic genetic interests. From this perspective the nation state is an innovation—a cultural strategy—for identifying and defending ethnic genetic interests. Any dilution of homogeneity must reduce that capacity by reducing the efficiency of the state as a vehicle of majority ethnic interest. Conversely, it will often be in the interests of minorities to maintain separation of 'church and state'. Minorities stand to benefit from this separation, because they are typically more ethnically aware and mobilized than majorities. The question of the fitness value of the nation state is of such importance that I devote Chapter 7 to the subject.

In general, any action that reduces an ethny's capacity to organize and otherwise strategize to achieve collective goals is likely to reduce members' genetic interests in the long run. Such actions include changes to ritual and culture that de-emphasize ethnicity, for example through the promotion of immigration of culturally distant groups. The predictable effect is to confuse identity thus lowering mobilization. Lowering mobilization from high levels is not necessarily harmful, since over-mobilization tends to sever mutually beneficial trade and cultural relations and creates opportunities for elite exploitation, for example in promoting mutually destructive warfare. At the other end of the spectrum, sustained low levels of mobilization risk the loss of identity due to immigration and cultural change, and a confused identity reduces the ability to mobilize adaptively in the face of threats to the group.

**Secession leading to the formation of a nation state is thus one way to arrest an ethny's decline. When the benefits of separation would be high, the project warrants large investment. Ethnic mobilization should become less adaptive once separation has been achieved and the new state established.** By the same reasoning, **an aggressive strategy is to prevent a competing ethny from forming a nation state by confusing its identity, for example by promoting immigration of culturally distant ethnies, manipulating its culture to de-emphasize ethnicity, opposing secessionist ideas and legislation, and severing existing 'church-state' links** (Optimal Strategy).

*Actual behaviour regarding confidence of relatedness and the problem of free riders.* Some modern human behaviour patterns match fairly well with optimal strategies for ensuring confidence of relatedness, especially with regard to close kin. The overall fit is fair. Hamilton notes the abiding interest humans have in kinship. We invest time and energy in tracking our lineages. Humans do indeed invest most intensively in close genetic kin. The great majority of parents care for their biological children. Adopted children are generally less well cared for by step-parents.[18] For example, in the United States adopted children receive less education and even less food.[19] Moreover a recent study finds that more altruism is shown between full than half siblings in a Mormon community.[20] Male sexual jealousy helps prevent them from wasting paternal investment on offspring conceived with another man. Female romantic jealousy helps her to retain provisioning by her mate.[21]

The intense investment in close kin makes identifying children's parentage a critical issue for individual fitness, yet with rare exceptions no legal system currently takes genetic interests explicitly into account in paternity cases. An example is the Texan man who in 1999 discovered through DNA testing that he had fathered only one of his four children. His daughter was his but none of his three sons. In divorce proceedings the court would not consider the genetic evidence and refused to allow the man to stop paying child support for the boys. The same court cut off visiting rights to all the children, including his biological daughter.[22] Examples such as this show that genetic interests have not been systematically incorporated into family law.

Genetic interests do figure in the language of rights and duties used to debate the appropriate use of genetic fingerprinting. Those with the children's welfare in mind criticize fathers for trying to avoid responsibility for children with whom they have established a nurturing relationship. Undoubtedly it would be in the children's interests, proximate and ultimate, to have continuous paternal protection. What of cuckolded fathers? Of the 280,000 paternity tests conducted in the United States in 1999, 28 percent showed that the 'father' had not conceived the child in question.[23] Paternity law suits are becoming big business. Yet the arguments on their behalf do not canvas genetic interests. One frequent argument is that these 'fathers' should be allowed to halt child support because they were duped into believing the children were their own. This is not joined with the point that it is not in the father's genetic interest to invest in another man's child.

From the perspective of those who deny the social importance of human genetic diversity, there can be no genetic interest at stake in cases of cuckoldry. Since all humans are essentially clones it would be wrong of a man to claim a violation of his interests based on genetic data alone, since one child is as genetically precious as another, it is implied. Similarly a woman relying on genetic evidence alone would have no interest in claiming back a baby that had been accidentally given to another mother in the maternity ward. In this view there would need to be weightier matters than genetic relatedness, such as emotional trauma. The sociobiological view is that cuckoldry and substituted babies represent blows to genetic interests, because blood relatives carry copies of each other's distinctive genes, even in ethnically homogeneous populations. As Hamilton pointed out, even in a population of clones nepotism would be adaptive because it is a form of fitness insurance, a guard against mutations that arise in the germ line.[24] From this viewpoint it is little wonder that humans and other species have evolved a complex set of defences against mistaking paternity and to a lesser extent maternity (it is easier for a mother to keep track of her newborn baby than for a man to keep track of whose sperm fertilized his partner's ovum).

The law is catching up with genetic science. In 2001 the state of Ohio passed a law allowing a man to cease supporting children who are shown not to be his genetic offspring. New Jersey has introduced similar legislation. These laws protect litigants' genetic interests as if the principle had been explicitly recognized, but in fact the basis is a semi-analysed emotional reaction. The father's sense of grievance is being given precedence over the interests of children to receive continuous paternal care, although in most cases the parents have already separated so paternal care is disproportionately financial compared to that of live-in fathers.

What of identification of co-ethnics? Humans have universal characteristics that appear to be adaptations for identifying fellow tribal members. Humans are prone to develop ethnocentric attitudes and emotions about individuals and groups who differ from their own ethny in dress, dialect, and physical appearance. Experimental psychological methods have discovered a specialized mental ability that develops in children by age 3 that prepares them for distinguishing inherited differences from acquired ones.[25]

The data on altruism within primitive bands and tribes and the hostility between them fit fairly well with the optimal strategies formulated above and with Hamilton's 1975 model of intra- and inter-tribal altruism based on degrees of relatedness (see Chapter 2). Westermarck observed that altruism within primitive tribes was much higher than that within the Western societies that encountered them in the age of exploration.[26] Modern quantitative studies find evidence of the within-tribe competition predicted by E. O. Wilson,[27] but there is also evidence consistent with Hamilton's prediction of inter-tribal hostility. This is now known to be endemic in contemporary primitive cultures. Death rates from inter-tribal conflict, even among allegedly gentle cultures such as the Kalahari Bushmen, are well above the worst homicide rates in modern cities.[28] In the Indian subcontinent, a developing economic area, 90 percent of conflicts occur between ethnies.[29]

There are numerous examples of free riders being controlled, from informal expulsion and shunning in workgroups and primitive societies[30] to elaborate legislative and administrative structures, such as found in modern taxation systems. Informal social controls tend to be less effective in mass anonymous societies where activities can be masked in private dwellings and in the faceless crowd. C. Erasmus documented the cooperative behaviour of agricultural communities in three continents, including the Soviet Union and Maoist China. Successful cooperation is built around small groups, usually kin, who have intimate knowledge of one another and who expel or punish free riders. The resulting material inequality was fiercely opposed by communist regimes, but could not be defeated. If the group survived as a productive unit, so did control of free riders and the inequality that accompanied that control.[31]

Those who would break ranks with an ethnic group strategy are also controlled in some communities, especially in those classified by MacDonald as group evolutionary strategies, including the ancient Spartans, Medieval Catholicism, Orthodox Judaism, and the Hutterites of North America.[32] In mainstream Western societies majority ethnic group strategies have all but vanished and free riding is largely uncontrolled; indeed overt minority free riding is encouraged by establishment pluralist ideology in such policies as affirmative action, immigration, and encouragement of minority identity and mobilization.

An example of under-investment in ethny is the clannishness of the economically depressed areas of Southern Italy. Intense family loyalty correlates negatively with loyalty to the public realm, in what E. C. Banfield called a culture of 'amoral familism'.[33] The latter detracts from the creation of public institutions able to promote economic development that would, arguably, benefit the ethny as a whole and thus its constituent clans.

Ethnies from tribal times have adopted cultural group markers that distinguish them from other ethnies. Markers include language, dress and other physical culture, and scarification.[34] Traditions of ritual indoctrination are effective in expanding ingroup identification to encompass the tribe.[35] These cultural markers partly develop inadvertently[36] and are partly adopted as deliberate social technologies.[37] In modern multi-ethnic societies, enculturation to the majority language and customs blurs ethnic boundaries, raising the salience of racial markers. For example, in the United States assimilation is occurring more within the races than between them, leaving racial boundaries as major demarcation lines of group identity.[38] These lines mark steep genetic gradients. Ways are being sought to overcome racial discrimination, which is the form of discrimination most likely to be adaptive (see Chapter 3). Multicultural regimes deploy modern forms of ritual indoctrination to defeat inborn discriminatory responses to ethnic diversity, at least by majority ethnies. Indoctrination includes education programs (e.g. 'diversity education') and manipulation of messages in the press and films.[39] Recent research indicates that it is indeed possible to develop techniques to break down or neutralize ethnocentric responses to diversity.[40]

In multicultural societies where ethnic boundaries are evident, as between the races and traditional linguistic groups, individuals often follow optimal strategies by reducing contributions to public goods.[41] Members of other ethnies are often treated as potential free riders. As a result multiculturalism has a depressing effect on public altruism in most societies. World surveys of ethnic diversity and welfare find a robust inverse relationship between the generosity of redistributive welfare and ethnic diversity.[42] Relatively homogeneous societies invest more in public goods, indicating a higher level of public altruism. For example, the degree of ethnic homogeneity correlates with the government's share of gross domestic product as well as the average wealth of citizens.[43] Case studies of the United States, Africa, and South-East Asia find that multi-ethnic societies are less charitable and less able to cooperate to develop public infrastructure.[44] Moscow beggars receive more gifts from fellow ethnics than from other ethnies.[45] A recent multi-city study of municipal spending on public goods in the United States found that ethnically or racially diverse cities spend a smaller proportion of their budgets and less per capita on public services than do the more homogeneous cities. Those public services included education, roads, sewers, libraries, rubbish removal, and welfare.[46] A major cause of parsimonious welfare in the United States is the racial gap between predominantly white taxpayers and disproportionately black welfare recipients, contributing to taxpayer motivation to vote against generous welfare.[47]

Thus there is ample evidence that in all societies individuals continue to identify ethnic differences and display some degree of ethnic nepotism. Moreover, these cases of ethnic nepotism occur in societies in which the mainstream educational bureaucracies and mass media do not provide substitutes for tribal indoctrination, indeed, which proselytise against ethnic discrimination. Ethnic nepotism appears to be an innate, or at least conservative, component of ethnocentrism, and is probably adaptive in many situations. As Hamilton argued:

> [S]ome things which are often treated as purely cultural in man—say racial discrimination—have deep roots in our animal past and thus are quite likely to rest on direct genetic foundations. To be more specific, it is suggested that the ease and accuracy with which an idea like xenophobia strikes the next replica of itself on the template of human memory may depend on the preparation made for it there by selection—selection acting, ultimately, at the level of replicating molecules.[48]

Immigration policy for most societies in most ages has consisted of a blanket ban. Apart from isolated individuals, immigration has been resisted. It is only in the modern era that immigration has become a flexible policy for pursuing various goals, including economic, diplomatic, and humanitarian ones. Immigration policy in the United States has been influenced by perceived ethnic interests. The majority sought to control the increase of minorities by prohibiting immigration from East Asia in the late nineteenth century and finally by imposing the 1924 quota system in the face of large-scale immigration from Eastern and Southern Europe. This was a blow to the ethnic interests of those minority groups because their co-ethnics were denied the security and economic opportunities offered by the United States, and because their group influence within that country was demographically capped. Little wonder that some minorities lobbied against this legislation. The majority aim was to maintain the ethnic proportions as they existed in the 1890s, thereby retaining the country's Northwestern European ethnic identity, while the minority aim was to keep the door open to further immigration of co-ethnics, in solidarity with family and ethny. The quota legislation was finally overturned by a Democratic congress in 1965 during the Civil Rights era when that party had become the main vehicle for minority aspirations. Since 1924 the pendulum has swung from a prolonged period of declining minority representation and rising assimilation, to rapid expansion of non-European minorities, mainly due to immigration.

As noted in Chapter 3, wherever multicultural regimes have come to power in wealthy societies, the native population is set on a path to minority status. Included in the multicultural armoury are scientific methods for demobilizing ethnic majorities, for example by emphasizing the putative benefits of immigration and obscuring the costs, and by breaking the correspondence between national and ethnic identity, in order to make the latter more 'inclusive'. The social sciences have long been deployed to facilitate mass immigration, multiculturalism and thus, in effect, the partial replacement of native born populations. The 'Americanization' movement in the early twentieth century aimed at assimilating minorities to the established cultural norms. More sophisticated techniques are now evident. For example, psychologists have tested the efficacy of various techniques for counteracting hostility to immigrants among native-born Americans and Canadians. One method is to alter the perception of immigrants as competitors. Esses et al. use 'manipulations of the inclusiveness of national identity'.[49] There is evidence of the medicalization of multicultural social control techniques, including the treatment of ethnocentrism as a pathological condition. Perhaps the clearest example is the series of studies conducted by Adorno and colleagues at the Frankfurt School for Social Reserch in the late 1940s of the 'authoritarian personality' which pathologized patriotic and ethnocentric attitudes among Western majority populations.[50] In all such ideologically-dedicated research the majority ethnocentrist's views are not taken seriously, except as a threat to minorities. The Esses et al. paper, for example, never considers the possibility that immigrants do in fact compete with the native born. Neither does it countenance the possibility that ethnic 'prejudice' can have adaptive functions. No weight is given to non-economic interests. Reproductive interests are ignored along with values of sentiment. The research was funded by pillars of the establishment—the Social Sciences and Humanities Research Council of Canada and the National Institutes of Mental Health. Multiculturalism as presently constituted appears to be an unstable evolutionarily strategy for majority ethnic groups, but a more stable one (indeed, a boon) for immigrant ethnies.

While there is an overall match between real behaviour and optimal strategies in recognizing and favouring kin, the match is weaker for ethnicity, especially in urban anonymous societies. Control of free riding, within or between ethnies, is likewise haphazard. One cause of poor ethnic recognition appears to be the great reliance on cultural markers. Likewise, free rider control relies heavily on cultural strategies as aids to instinct, although cultural strategies are vulnerable in an era of rapid cultural change.

### *(6b) Genetic distance between ingroup and competing groups*

**Hamilton argued that within-group altruism is more adaptive in defence of the group when it is threatened by genetically distant compared to genetically close competitors**[51] (Optimal Strategy). If an individual's kin and resources are secure, there is no fitness stake in defending the ethny against replacement by a genetically identical ethny (which would be part of the same population). On the other hand, competition from a genetically distant group can potentially lead to a large loss of genetic interests. It would be adaptive to invest heavily in such competition, even if the individual's family and personal resources were not at stake. **Thus individual contribution to group competition will be most adaptive when it allocates investment in a manner sensitive to the nestedness of genetic kinship** (Optimal Strategy).

Kinship overlap complicates group interests for affected individuals. A strong type of overlap would occur when some members of an ethny had closer kinship to a randomly-chosen member of another ethny than to a random member of their own group. Ethnic altruism (toward the in-group) would be maladaptive for such individuals. However, strong overlap is most likely to occur, and perhaps only occurs, between closely related ethnies, for whom group competition is already problematic.

*Actual behaviour.* It is a fascinating question the extent to which human solidarity and conflict vary with genetic distance. Of course there is no perfect correlation. Friendship does occur across racial lines, and enmity within families. However, the broad trend is for the reverse to happen. In a sample of ethnically English individuals, male friends were more genetically similar than were random pairs.[52] In every society yet tested, people tend to befriend and marry those who are similar to them on a range of characteristics, including heritable ones such as race.[53] Certainly the family receives the most intense altruism. So relations between individuals roughly accord with the optimal strategy.

The question is more difficult to answer for ethnic relations. The issue is not as straightforward as one might suppose, and there is no systematic research on the subject that I could find. It is hardly credible to maintain that if ethnic distance were a predictor of solidarity and conflict then none of the major wars would have occurred within Europe, but between European nations and nations in Africa and Asia. Competition can be adaptive within families despite their high relatedness, so rivalry between ethnies of the same race is hardly surprising. From the perspective of Hamiltonian theory, the adaptiveness of altruism is dependent on local, not global, circumstances. Europeans did indeed fight Asians when one side managed to encroach on the other, but otherwise the main threat to European interests came from other Europeans. If the world consisted only of Europe, which in a sense it did strategically for two centuries after the industrial revolution, then genetic distances between European ethnies would become more salient. But clearly geography is an important cause of rivalry. Neighbours may have more in common genetically, but are more likely to have conflicts of interest over territory and status. Also, religious wars within the same geographic race have been at least as bloody as colonial wars between racially distinct groups. Territorial proximity and cultural differences would have triggered the group identity processes discussed in Chapter 4 (section e). In the evolutionary past, the same processes would have helped set tribes against one another whatever their relatedness.

Kin solidarity is well correlated with genetic solidarity, but ethnic solidarity less clearly so. Until a systematic study is available, I cautiously conclude that ethnic solidarity is weakly correlated with the genetic distance of competing groups. Geography and culture are stronger determinants. This leaves great scope for maladaptive patriotism.

### *(6c) Group size*

**A fundamental criterion for a good fitness investment is that it maintains or increases the size of the kin group relative to the metapopulation. This is as true of investment in families as in ethnies** (Optimal Strategy).

Competitive breeding is a core concept in Darwinian theory. Other factors being equal, lineages that reproduce more rapidly will, over generations, replace other lineages. Since ethnies are lineages, they can engage in competitive breeding. As I argued in Chapter 3, in the absence of mass immigration of genetically distant groups, a population occupying a fixed territory is guaranteed continuity at or below that territory's carrying capacity, even when its global representation falls due to high fertility overseas. But mass migration inevitably reduces the native ethny's relative fitness within its own territory, risking its continuity as a distinctive gene pool. This risk to continuity is substantially greater when the native population has below-replacement fertility.

**It follows that one important investment individuals can make in their group is to maintain or increase its numbers, whether through personal reproduction or by facilitating the reproduction of co-ethnics** (Optimal Strategy).

The causes of fertility are imperfectly understood, though cross-cultural studies indicate that low fertility is predicted by high female labour force participation.[54] Thus one pronatalist policy would be to empower the mother role to give it some of the economic and social benefits of other work, though social dysfunctions of the urban family also need to be considered. I discuss this further in Chapter 8, pp. 272–278. Related policies that could be considered are: privileging the heterosexual family as a union favoured by the state; encouraging paternal investment in offspring; restricting life-style abortions; redistributing wealth from single citizens to parents; counteracting anti-natalist ideas; and protecting religions and other traditions that support such policies.

Pronatalist policies cannot be allowed run-away success. The world's ecosystems are already under strain from unprecedented human numbers, and we may have already overshot global long-term carrying capacity.[55] In the modern world where territorial expansion is impossible or risky, a more prudent imperative than the biblical injunction is to 'go forth and perpetuate'. Eibl-Eibesfeldt makes this point when he compares the strategies of maximizing and conserving fitness. 'If we consider strategies that reduce risk . . . [sociobiological] models should be designed quite differently.'[56] **Demographic stability, including decline to an ecologically sustainable level, can also help a society defend its interests** (Optimal Strategy). As political scientist Peter Corning puts it: 'For a very small population with abundant resources, overall population growth is obviously adaptive. But for large human populations, especially those that are pressing the limits of their resources, population stability over time is arguably a more adaptive strategy in strict Darwinian terms. . . . reproduction at the "replacement" level would be viewed as optimal, and anything either above or below that rate would be less adaptive.'[57]

Low population density and therefore better prospects for long-term ecological sustainability are most readily achieved behind controlled borders. By relieving population pressure on emigrant societies, open borders will tend to delay the day that high birth rate cultures become disciplined by crowding. The cost is that all attractive societies with unpoliced borders will come to share the crowding experience no matter how disciplined their reproductive behaviour. Controlled borders prevent local population explosions from becoming global by encouraging the development of responsible reproduction practices. Moreover, restricted immigration allows populations to decline without risking ethnic replacement. A world of open borders would necessitate an elaborate international population policy enforced by legal and, in the final resort, military means. However, regions of open borders can be viable. Eibl-Eibesfeldt points out that an adaptive open border option is viable without coercion in regions of ethnic homogeneity and similar fertility patterns, so long as the region as a whole maintains restrictive immigration from the outside world.[58]

*Actual behaviour regarding group size.* There is not a strong research literature on investment in ethnic, as distinct from personal, reproduction. Competitive breeding has occurred in historical times,[59] though whether it has been an important factor in human evolution is the subject of the debate over genetic group selection. This has occurred at least since Roman times, when the Emperor Augustus approved legislation in 18 BC that promoted larger families among his own patrician class. Augustus offered honour and prizes to fathers, but to bachelors he was stern, delivering his rebuke with the gravity appropriate for matters of ethnic survival:

> . . . mine has been an astonishing experience: for though I am always doing everything to promote an increase of population among you and am now about to rebuke you, I grieve that there are a great many of you. . . . We do not spare murderers, you know. . . . Yet, if one were to name over all the worst crimes, the others are as naught in comparison with this one you are now committing . . . for you are committing murder in not begetting in the first place those who ought to be your descendants; you are committing sacrilege in putting an end to the names and honours of your ancestors; and you are guilty of impiety in that you are abolishing your families, . . . overthrowing their rites and their temples. Moreover, you are destroying the State by disobeying its laws, and you are betraying your country by rendering her barren and childless; nay more, you are laying her even with the dust by making her destitute of future inhabitants.[60]

Pronatalist policies exist in some contemporary societies, most offering inducements such as child subsidies and taxation benefits. At least in Western societies, these measures have been failing, evidence that the majority of citizens are not striving to maintain numbers. A recent coercive pronatalist policy was enforced by the notorious Ceausescu Stalinist regime in Romania that banned abortions and the contraceptive pill.

Most European-derived ethnies are shrinking, as is the Japanese population. The pan-European share of the world's population will soon fall to 10 percent from the 25 percent that it had reached after two centuries of rapid growth to 1900. This is a welcome trend for ecological sustainability, and by itself presents no serious threat to continuity. But below-replacement birth rates are unsustainable over many generations in the face of mass immigration from non-European ethnies. Such a combination can only be sustained until the shrinking group disappears or becomes so marginal that its fertility levels no longer much affect overall trends. From an evolutionary perspective, the culture and behaviour contributing to permissive immigration policy are maladaptive.

Immigration policy has been deployed adaptively in the past. As noted earlier, most nations have not allowed much immigration at all, except for specific categories of workers. This is still the situation. Only a handful of countries around the world allow large scale immigration. Australia used immigration policy after the Second World War to greatly enlarge its population after the scare of Japanese invasion. Australia's population in 1945 was overwhelmingly British in origin. The post-WWII immigration boom began in 1949 with subsidized immigration from Northwest Europe. By the 1960s the search for immigrants had extended to Eastern and Southern Europe. The 'White Australia Policy' was maintained until the early 1970s. The data on genetic distances quoted in Chapter 3 indicate that Australia's immigration policy managed to increase the population dramatically without much compromising the ethnic genetic interests of the Australian people, until Middle Eastern and Asian immigration was introduced in the 1970s. A similar story can be told about the United States, which restricted entry by non-Europeans until the mid 1960s. At the beginning of the twenty first century, these two countries are maintaining high immigration intakes generations after their most habitable territories were settled, a policy that risks ecological ruin. Furthermore, the large intakes of non-Europeans are substantially reducing the long-term relative fitness of the founding populations.

The European Union has developed a mix of restrictive and liberal immigration policies. As noted in Chapter 3, Europe is a region of relative ethnic homogeneity. Also, fertility levels are comparable between European ethnies. So far the Union's policy of open internal borders and guarded external borders has not resulted in swamping, although the borders are porous and immigrants, legal and illegal, are pressing to enter from Africa and Asia. The pressure to accept replacement migration will grow as the European population ages and industry seeks factory workers and the welfare lobby seeks to fund benefits to the elderly. In this regard both capitalists and socialists risk committing the error of short term thinking so eloquently described in evolutionary perspective by Eibl-Eibesfeldt.[61] The problem is larger than the absurdity of rushing in factory workers just in time for automation to render them redundant. Replacement migration is irresponsible because it can only put off the day of reckoning by one or two generations, while the cost to national genetic interest is permanent. Sooner or later all societies must come to terms with a static population, or else world population will continue to grow indefinitely. As the rapidly growing populations in Asia and Africa modernize their economies, like Western societies they reduce family sizes in what is called the demographic transition (see earlier discussion on p. 82). The change is linked to lower child mortality rates and especially to increased parental investment in each child, together contributing to a growth in the educated middle class.[62] Available evidence indicates that many developing countries will face their own demographic crunch in a few decades. The United Nations Population Division estimates that birth rates in less-developed nations will fall to those of developed nations by 2050. In the US it took 69 years for citizens over 65 years of age to rise from 7 percent of the population to 14 percent. France took 114 years. China is projected to make the transition in 25 years, Indonesia in 22 years.[63] These developing countries will be faced with a large welfare burden of their own. Where will these newly stabilized populations find replacement migrants? Economist Gary Becker, the 1992 Nobel laureate, believes that welfare systems can be designed that sustain ageing populations, for example by following the system proposed by Milton Friedman in 1962. An economic 'solution' that relies on perpetual migration is another case of short-term thinking. The responsible policy for societies faced with ageing populations is to find sustainable solutions.

Family reproductive strategies are intrinsically more secure than ethnic population policy, though much more is at stake in the latter. While individual reproduction might fall below replacement, the children that are raised are overwhelmingly the genetic offspring of their parents, and relative fitness is not lost when small family size is part of a society-wide pattern. In the modern world, ethnies numbering in the millions are intrinsically less secure. They are capable of imprudent fertility and immigration policies that not only squander large amounts of ethnic genetic interests within one or two generations but allow irreversible replacement migration by genetically distant populations.

### *(6d) Salience of intergroup versus inter-individual competition and the timing of ethnic mobilization*

No fixed pattern of altruism is optimally adaptive because circumstances change. **When intergroup competition is not threatening or advantageous it is more adaptive for individuals to engage in individual competition and invest in kin. Conversely, when the ethny as a whole is under threat, group mobilization becomes adaptive in the form of reduced individual competition and greater effort for the tribe** (Optimal Strategy). Providing for a family, and especially raising children, requires intense and steady investment over decades. But giving to the nation, especially self-sacrifice, is only adaptive when directed to preventing or removing a threat to the ethny. To be adaptive, ethnic mobilization must coincide with threat.

**The individual is prudent to make greater sacrifices to prevent losses to group fitness than to exploit opportunities for ethnic expansion** (Optimal Strategy). This follows from the supposition that continuity is more likely when backed by a powerful base than is recovery from a diminished base. Lost numbers or territory or status amount to reduced group strength, and therefore increase the risk of further encroachments by competitors.

A counterargument is that in a competitive world only expansionist ethnies have survived, by actively pressing against their borders and being ever ready for opportunities to grab more territory. The argument might continue that this is currently happening to many Western countries, as immigrants from Asia, Africa and Mexico press against their borders. But this phenomenon is a matter of individual and family strategies. It requires no investment in the ethny as a whole for a family to seek better conditions in a wealthy country. Ethnic expansionism that depended on an ethnic strategy would be a corporate enterprise, usually the conquest and forced settlement of another people's territory. Tribal conquest of neighbouring territory and great folk wanderings were common until, say, 1900, when all the habitable continents had been settled by agricultural and industrial cultures. But even in the heyday of European expansion, during the four centuries from 1500, few agricultural populations were replaced by European colonists, though many were ruled for a time from Madrid or London or Paris. Rather, the waves of European settlers filled empty spaces in the Americas and Australasia and parts of Southern Africa after sweeping aside the sparse hunter-gatherer populations that had survived decimation from European diseases. Those days are over. Since forced ethnic replacement is usually a high-risk strategy it will rarely be a good investment for the ethnic altruist. Another advantage of the defensive strategy is that it stabilizes the international environment, increasing everyone's security and facilitating mutual enrichment through trade.

The defensive strategy requires that individuals maintain vigilance for threats to the ethny. Ethnic competition within multi-ethnic societies is endemic, and nations sometimes go to war with one another. The risk of one ethny adopting a group strategy and gaining an advantage over a less organized ethny is analogous to the prisoner's dilemma. Prisoners can be under pressure to be the first to inform on their accomplices to avoid the severe punishment meted out to uncooperative defendants. Similarly, in ethnic politics, groups that abstain from asserting interests can lose status and influence to more assertive competitors. Also, in multi-player games, cooperation between some of the players can produce a strategy that is irresistible by individuals acting alone. Only a counter group strategy can prevent defeat. Individuals who do not show ethnic solidarity are, in principle, putting at risk those copies of their distinctive genes bound up in their ethny.

*Actual behaviour regarding the salience of intergroup versus inter-individual competition.* The most persistent ethnic groups have cultural traditions conveying 'oppositional symbols', such as tales of past victories and, especially, defeats.[64] Nothing is more likely to weld a group together than external threats. The prospect of a great conquest does not inspire as much devotion to the tribe as does fear of some peripheral encroachment by foreigners. For millennia leaders have held up the bloody shirt of a slain ingroup member, real or rhetorical, to distract attention from internal conflicts.[65] Such rhetoric mobilizes group cohesion and willingness to sacrifice for the tribe. This is the time when young men volunteer for military service and citizens donate time and material resources to the cause.

The universality of the tendency, in times of group peril, to put aside individual squabbles and unite in common defence, indicates its past adaptiveness. During times of group danger, it must have often been adaptive for individuals to risk personal fitness in defence of tribal members and territory. Otherwise, the propensity for such behaviour would have been weeded out and we would all be radical individualists immune to patriotic rhetoric. Participation in warfare is probably less adaptive for individuals in societies where the risk to personal fitness from mobilization is not shared between elites and the masses.

The psychological underpinnings of group solidarity during intergroup contests have been experimentally explored. The classic study was done by Sherif, already discussed in Section 4g.[66] Sherif randomly assigned boys to groups that were pitted against one another in various competitions. The boys rapidly identified with their new groups, expressing positive attitudes towards them and negative attitudes towards the outgroups. In another study, groups that had been competing were made to coalesce into cooperative alliances when challenged with what Sherif called a 'superordinate' goal. It seems that humans come well prepared psychologically to switch between individual and group strategies in ways sensitive to the demands of the situation. However, and this is a crucial point, adaptive timing of mobilization is dependent on information, and adaptive choice of ingroup depends on a confluence of group identity and genetic interests.

In Sherif's study the groups were not familial or ethnic. In primitive societies with low exogamy, group identification would have been almost perfectly aligned with the band or tribe within which people were embedded from birth. Because multi-ethnic societies have been rare over the course of evolution there has been little selection pressure for a more fool-proof innate mechanism able to discriminate between descent groups other than by knowledge of close kinship, shared language, phenotypic similarity, religion, or territory.[67] But in an age of mass migration the match between culture and genes is far from perfect. People of diverse backgrounds then share the same territory, speak the same language, and dress similarly, releasing cooperative behaviour towards members of the multicultural society. In the modern world this cultural and situational fooling of the ethnic-selection mechanism is often adaptive because it avoids triggering social identity processes and thus blunts mutually destructive conflict, while promoting reciprocity. Unrelated individuals and ethnies can be combined into societies possessing the valuable public goods of internal peace and extended markets, though not possessed to the degree found in homogeneous societies.[68] As shown by the present day West, these public goods can be maintained even as the founding population is being replaced by immigrant ethnies. The atavistic predisposition to invest in individuals who are phenotypically but not necessarily genetically similar can be maladaptive.

Tribal societies are vigilant in defending their borders and are able to mobilize for intergroup conflict. In modern states defence against armed invasion is the responsibility of military institutions. But regarding internal ethnic competition and immigration, civilian mobilization is highly variable. Civilizations differ in the average intensity of ethnic consciousness, Western societies being at the individualistic end of the individual-collective spectrum.[69] While ethnic identity survives, it is not salient in everyday life. Few individuals run risks to defend their ethny. Extreme individualism would be adaptive if ethnic competition were not in prospect. But in the real world of mass migrations between continents, individualism constitutes a form of unilateral disarmament that invites exploitation, intentional or (more usually) unintentional.

Gathering reliable information is a basic problem for individuals living in modern societies. The physical scale of these societies and their anonymity reduce the efficacy of personal observation as a means of assessing threats to the ethny. Citizens must rely on government and on mass media organizations for such information. But, as I argue in Chapter 7, ruling elites do not always mobilize the people adaptively. By the 1960s Western elites no longer mobilized Western ethnic majorities at all.

Humans are psychologically adapted to switch investment from individual to group solidary strategies when the group as a whole is threatened. With so many people living in ethnically mixed societies, this adaptation no longer reliably serves to optimally allocate altruism according to the salience of intergroup versus inter-individual competition.

*(6e) Maximizing the ratio of fitness benefits to costs*

We saw in Chapter 5 that ethnic nepotism can be adaptive according to the criterion of Hamilton's Rule for adaptive altruism. At the heart of that rule is the ratio of fitness payoff to cost. An altruistic act is most likely to be adaptive when it conserves more distinctive genes than it risks. More needs to be said about how this adaptive ratio might be achieved.

Chapter 3 described the very large proportions of ethnic genetic interests compared to those found within families. One might suppose that we should make every effort to minimize investment in our families, because the potential payoff is so small. But the opposite is true. Minimizing the cost of ethnic altruism is more often a salient issue than minimizing family altruism, because the former is much more vulnerable to the risks of free riders, to incomplete information about kinship coefficients, and to other uncertainties. When all the other variables bearing on the adaptiveness of ethnic altruism are favourable, it is adaptive to make large sacrifices for one's ethny. But when that is not the case, as it so often is not, it is inherently more difficult to invest adaptively in one's ethny than in one's children and other close relatives.

The risk of conferring altruism on co-ethnics declines precipitously when an investment is of low cost to the altruist's individual fitness yet produces significant benefits for his inclusive fitness. It follows that most ethnic nepotistic acts should be of this type, such as **political activity and inexpensive but repetitive acts of favouritism shown to ingroup members. Favouring many individuals at low per-unit cost also helps overcome the free rider problem** (Optimal Strategy). If group markers are statistically reliable, helping a large number of individuals eliminates the risk of allocating one large investment to a free rider.

Any situation that reduces the cost to the giver creates an opportunity for adaptive ethnic nepotism. **When a person's family is already provided for such that further investment yields greatly diminished returns, there is a rising payoff to investing surplus resources in the ethny. Resources at the discretion of an individual but derived from the public realm are well invested ethnically** (Optimal Strategy). An example is a business person or government official who has jobs or contracts or information at his discretion that his family is legally barred from receiving. Allocation of these favours to co-ethnics costs the actor little or nothing but can be of considerable benefit to the recipients.

A special case to discuss is war. Since the cost of warfare is always high for the individuals who lose their lives, can it ever be a prudent strategy? Hamilton did not consider all wars to be maladaptive, and believed that limited national wars using conventional weapons were not pathological.[70] **Hamilton's Rule for**

**adaptive altruism cautions against an aggressive war policy in this technological age, unless there is near certainty of significant gains at low risk. Modern economies are necessarily based on open and highly interdependent societies, vulnerable to conventional and terrorist attacks. More than in the past, aggressive policies are likely to be counterproductive by engendering devastating retaliation. Much greater risks can adaptively be taken in defence of territory, autonomy, or survival** (Optimal Strategy).

*Actual behaviour for maximizing the ratio of fitness benefits to costs.* This is one set of optimal strategies that does seem to be fairly consistently reflected in actual behaviour. People do seem to take the cost of altruism into consideration when allocating investment between family and ethny. High costs are borne on behalf of the family. But ethnic nepotism usually takes the form of repetitive petty acts of discrimination, ubiquitous in everyday life. A questionnaire study of Canadian university students in the 1990s found that they would discriminate in favour of their own ethnic group if the cost were low.[71] Ethnic discrimination is the bane of multi-ethnic societies as reflected in anti-discrimination laws, quotas, community relations committees, boycotts, sensitivity training, and other signs of disaffection. Petty ethnic nepotism includes voting along ethnic lines, self-segregation of neighbourhoods and schools, and ethnic discrimination in the economy.[72]

The depressing effect of multi-ethnicity on redistributive welfare, reviewed earlier (p. 80), is consistent with individuals seeking through the ballot box to reduce their investment in other ethnies. A study of Moscow beggars finds that they receive more generous donations from fellow ethnics.[73] In the Kalahari Bushmen acts of assistance are favoured that cost the giver little but benefit the receiver a great deal.[74]

Costly acts of nepotism are usually reserved to protect or advantage close kin, but there are numerous examples of great sacrifice on behalf of tribe and nation, including heroic military deeds reviewed in the next section. Although people are often moved to make charitable donations to strangers in distant lands, there is some evidence that large gifts go disproportionately to the ethny or nation.[75]

Another implication of this discussion is that petty discrimination is controllable subject to efficient monitoring and the administration of small punishments. Anti-discrimination measures can be effective because of the modest motivation for most acts of discrimination, corresponding to the small quantities of genetic interest at stake. The aggressive side of multiculturalism might help explain the under-investment in majority ethnic defence evident in contemporary Western societies.

Warfare has been practised for much of human history, at least in the form of endemic low-intensity inter-tribal conflict. Defending one's tribal territory has surely been adaptive, even for warriors who were killed or wounded. But advances in military technology have made warfare less secure as an adaptive strategy for soldiers and their families. By Napoleonic times some national wars were certainly maladaptive for those who died in them. These were not wars conducted for the national interest, but for the glorification of a leader or to advance a religious claim. That cannot be said of Russia's huge losses incurred defending the country in the Second World War—20 million deaths suffered from the Axis invasion. This huge sacrifice was certainly adaptive given Hitler's intention of displacing and enslaving the Slavs. On the German side Hitler's aggressive policy must be judged reckless considering that it resulted in the death or wounding of 45 percent of adult German males,[76] the loss of extensive territories in East Prussia, Poland, and Czechoslovakia, the loss of millions of German civilians in aerial bombing and post war expulsions, and the killing of many more millions belonging to other ethnies, especially in Eastern Europe. It is not as if the carnage could not be foreseen. The first large scale industrial war of 1914–1918 was also a huge blood letting without obvious benefits on either side. The individuals and ideologies responsible for these wars failed to minimize the ratio of fitness costs to benefits.

The riskiness of warfare for soldiers and their families is compounded by the fact that elite free riding can do as much damage to the ethny as the most implacable enemy. Two generations after millions of their sons made the ultimate sacrifice, many Western ethnies have lost more genetic interest due to mass immigration than was in prospect had they lost the War. It would seem inconsistent to assess warfare as a viable defensive strategy without controlling elite free riders in times of both war and peace. I discuss elite free riding at greater length in the next chapter.

Despite notable exceptions, there does appear to be a general trend towards apportioning altruism to maximize the ratio of fitness gains to losses. Nevertheless, the trend is unreliable, especially in environments radically altered from that in which humans evolved.

### *(6f) The availability of means by which the individual can contribute to group competition: collective goods including territory*

One condition for efficient ethnic altruism was not discussed by Hamilton. Investment in one's ethny as a whole is greatly facilitated by the existence of collective goods, such as welfare, communications and defence structures that benefit everyone (see discussion in section 6a above). All but the wealthiest individuals cannot usually benefit all of a large group by distributing a private resource, because the average benefit is too low to be significant. But volunteering to work for a national institution or voting for a welfare policy or subordinating oneself to military discipline when the nation is attacked, these all can multiply one's investment. Without such collective goods it is much less feasible to invest in one's ethny as a whole, even when all the other conditions discussed above are met. For example, an amount of money donated to separate groups of uncoordinated labourers will not be as effective in building a road network as the same amount of money donated to a local council able to accumulate expertise and systematically direct resources over prolonged periods. Similarly, in defending against invasion, individual volunteers are most effective when welded together into the collective good of an army.

Investing in a whole ethny can only be achieved by contributing to collective goods. The decisive element of these goods is their jointness-of-provisioning quality (discussed above in section 6a). That is, a large number of people must be able to utilize the investment without destroying its value to each of them. Political scientist David Goetze argues that ethnic groups have persisted because of ecologies containing resources that could be jointly provisioned, what he names 'nonsubtractable resources'.[77] Such resources include large game animals well beyond the capacity of an individual or family to consume, and mutual defence against predators. His example is that of the sentry who risks his life to warn the group; every member benefits, making the warning a collective good. Heroism in battle also qualifies as a non-subtractable good because all within the home group benefit. Goetze concludes that the advantages of joint provision to a genetically related population (co-ethnics) made the ethnic group a potentially evolutionary stable unit from tribal to modern times. As the scale and technological conditions of society changed, group members were able to devise novel collective goods able to meet new contingencies.

Organization is one powerful means for providing collective goods, and some of the foregoing strategies are more effective or possible when performed by coordinated groups. **Thus one strategy for investing in the ethny is voluntarily to participate in some type of ethnically-directed organization, or not to resist authoritative conscription to such bodies** (Optimal Strategy). Examples include groups of warriors, informal networks, community service bodies, and national armies and bureaucracies.

The availability of collective goods eases the problem outlined in section 6e above, of maximizing the ratio of fitness benefits to costs, since individuals can aid the group with small contributions. Collective goods can also facilitate the weighting of altruism in a manner sensitive to threats, when individuals are free to determine the size of their contribution. **When collective goods are available,**

**individual investment should increase with the individual's capacity to alleviate group risk** (Optimal Strategy).

The critical need for collective goods in defending ethnic interests opens up new fronts of inter-group contest. Indirect strategies can be highly effective in defence and offence when aimed at building or undermining collective goods. Inclusive fitness can be defined as the *ability* to propagate one's genes into the next generation by helping kin reproduce. In a stochastic world the loss of this ability can be defined as a lowering of the probability that individual action will result in genetic continuity. It is a forward-looking definition, according to which the loss of cohesion of family or ethny or the loss of motivation to invest in these groups amounts to a loss of genetic interests, since the likely result is reduced fitness. Similarly, fitness is threatened by the destruction or redirection of institutions that defend familial or ethnic interests. Cutting ties between such institutions and an ethny is another indirect threat.

Lumsden and Wilson modified Hamilton's original formula to make it forward looking by accounting for ego's effect on the reproductive success of kin, and the latters' effect on each other.[78] Lumsden and Wilson express Hamilton's original formula for inclusive fitness as

$$w = 1 + \partial w + e$$

where 1 is the basal reference level, $\partial w$ is the individual's reproductive fitness, $e$ is the kin effect, and $w$ the total inclusive fitness. Lumsden and Wilson suggest a more general formula that allows for nonlinear effects between ego and relatives [$f(e)$] and among relatives [$\partial w(e)$]:

$$w = 1 + f(e) + \partial w(e)$$

(Note that $f(e)$ is used here to mean a function of the kin effect, not kinship.)

The loss of group solidarity or of institutions that maintain and direct that solidarity will, in a competitive world, reduce $f(e)$ and $\partial w(e)$ by weakening the ability of an individual or a kin group or ethny to conserve its genetic interests. That is why an assault on any ethny's ability to construct collective goods amounts to an assault on its fitness. 'Culture wars' can be deadly serious. Weakening an ethny's cultural identity or religious or secular communal organizations or will to organize, tends to weaken the ethny and make it more prone to subordination or displacement. Such tactics as engendering chronic shame or guilt for ethnic identity or wresting control of the state apparatus from an ethny or whipping up class disaffection or promoting extreme individualism are all assaults on genetic interests, because they tend to sever the lines of kin altruism shown in the above equation.

The great advantages of collective goods are conditional on the control of free riders. As discussed in section 6a, effective control of free riders is costly. There are broader social controls that are efficient in multi-ethnic societies, such as taxation laws. But specifically ethnic free riding, where one group increases in relative numbers or resources, arguably is much more difficult to control without undermining other values. Ethnies in modern multi-ethnic societies faced with ethnic free riders have a limited number of escape routes, which all sacrifice the rights associated with open societies. They might replace open liberal society with the apparatus of modern totalitarianism or adopt a segmentary or caste system, but none of these belong to the Western tradition.

Ethnic monopoly of a territory largely obviates the costs of controlling ethnic free riders, making the expression of public altruism adaptive under a greater range of circumstances. Control of ethnic free riders is accomplished by defending borders rather than by invidious and inefficient state intervention in internal social relations. (However, elite free riders remain a problem; see Chapter 7.) Correspondence between ethny and territory reduces the cost of identifying one's ethny, because territory is one of the primordial ethnic group markers.[79] Members of ethnically homogeneous societies stand to benefit from any altruistic behaviours that are released by territorial proximity, while members of multi-ethnic societies stand to squander such altruism on free riders. Finally, as argued in Chapter 3, ethnic monopoly of a territory provides it with continuity even when its fitness relative to the global population is falling. But loss of relative fitness within a territory risks loss of identity and thus ability to strategize as a group, tending to lock in and accelerate the initial loss. Thus territory is more than an economic resource; it is a fundamental ethnic collective good. **Attempts to defend sole use of a territory for the ethny is likely to be an adaptive form of ethnic altruism** (Optimal Strategy).

*Actual behaviour regarding collective goods and territory.* Numerous examples come to mind of heroic behaviour exhibited at a place and time when one or a few individuals could confer a large benefit on the group, especially in defence. Heracio on the bridge blocking the Etruscan army's way to Rome and the 300 Spartans delaying the Persian at the Thermopylae pass, are both examples of heroes offering their people a collective good at the cost of their lives. These choke points presented opportunities for adaptive heroism, where the sacrifice of one or a few individuals could make the difference between victory and defeat for the ethny or a large part thereof. Had the situation been different, such that a small force could not obstruct an entire army, it is questionable whether these warriors would have been as willing to risk their lives. For what? A momentary delay that did not change the outcome? In that case their own lives would have been of some greater value. But these special circumstances, deliberately chosen by the altruists to maximize the impact of their sacrifice, allowed a single warrior to make a difference for the whole ethny, and thus self-sacrifice was a good investment. Note also that such heroic acts are often in defence of ethnic territory.

In times of national emergency many individuals do volunteer for national service, subordinating themselves in large organizations. A greater number offer little or no resistance to being conscripted to such bodies. Individual competition is temporarily suspended or reduced during emergencies, though there are free riders in all societies who respond only to coercive social controls. There is some evidence that contributions to the tribe and ethny increase with the individual's wealth. Within tribal societies high status is achieved by cycling wealth back to low status individuals.[80] Students from wealthy families in Moscow report giving more to street beggars.[81] Also, contributions to public goods in multi-ethnic societies are less generous than to collective goods in ethnically homogeneous societies.[82]

The examples given above show that the means are available, or can be contrived, to serve collective interests.

There is also ample evidence of 'culture wars' used to conduct ethnic competition. Ethnic activists behave as if they know that assaults on the organizing ability or morale of a people can disarm them. A typical method is the derogation of the opposing group's identity symbols, beliefs, and leaders.

There is much evidence of the importance to national identity of a demarcated territory and of the willingness of many individuals to invest in the defence of their tribal or national lands. The territorial component of the band and tribal strategy was so fundamental in *Homo sapiens*' evolutionary past that it might have become an innate psychological need, or one readily triggered as part of the ethnocentrism syndrome. Indeed, humans probably inherited their territoriality from their prehuman hominid and primate ancestors, since this trait is a feature of much animal conflict. Primates generally maintain exclusive use of a territory by: site attachment and avoidance of neighbouring groups' ranges; site-dependent aggression and definition of boundaries; and active defence of the territory's resources by marking or eviction.[83] It has been long known from cross-cultural comparisons that all hunter-gatherer bands and tribal peoples defend territories against other bands and tribes.[84] Even nomadic peoples establish bounded camp sites.

Whether or not territoriality is instinctive, it is a universal strategy used to 'affect, influence, and control' human societies.[85] It is also universally manipulable. Territorial bonds and the sense of ownership vary in intensity and according to culture, for example both being strengthened by tribal rituals that cause individuals to identify and bond with the group's territory.[86] Whether the association of peoplehood with a land is genetically or culturally transmitted, it appears to be universal in both the tribal and national worlds.[87] Persistent ethnic identities all include a territory as part of their identities, whether presently or once occupied by the group.[88] The evolutionary background of human tribal territoriality was discussed by anthropologist Sir Arthur Keith.[89] In tribal societies a territory is necessary for subsistence and for continued solidarity. Removed from its land, a tribe's social organization tends to break down, reducing its ability to maintain independence and continued existence. Keith argued for an intimate connection between a people as a descent group and their claim to a territory as essential constituent elements of nationhood:

> [A people constitute] a nation because they are conscious of being "members one of another" and of being different from the peoples of other lands. They are, and always have been, an inbreeding people. They have a particular affection for their native land. . . . If their country or its people are in jeopardy . . . they rally to its defence; they would give their lives freely to preserve the integrity of the land and the liberty of its people. . . . They are sharers in a common interest and in a common destiny; they hope and believe that their stock will never die out. They inhabit a sharply delimited territory and claim to own it.[90]

Keith noted the important property of a nation that it could accept immigrants: 'They have the power of assimilating strangers into their community and of making those assimilated sharers in all their hopes and fears, traditions, customs, and modes of speech.'[91] But he meant the small scale immigration that has always occurred between tribes and nations, since in the same paragraph he wrote: 'The genes or germinal units which circulate within the frontiers of their land differ in their potentialities from those which circulate in all other countries. [They] form, in a physical sense, a homogeneous community.'

## *Conclusion*

### *Why the poor match between optimal ethnic nepotism and actual behaviour?*

To summarize the above comparisons, the match between optimal and actual altruistic behaviour is best where fitness concentration is highest and more reliable, in the nuclear family. Actual investment in ethnic genetic interests is more erratic. In modern societies ethnic nepotism is usually confined to petty acts of face-to-face discrimination and avoidance, though it sometimes overshoots the optimum, for example rallying around the flag of a multi-ethnic state. An adaptive mass response to the immigration that has been inundating several Western societies since the 1970s is yet to appear. It would seem that outside the family the human repertoire of innate behaviours is an unreliable guide to fitness needs in modern settings.

It is to be expected that large-group strategies are more fragile than small-group ones. Evolutionary biologist George C. Williams observes that adaptations for defending extended groups are rare, due to the risk of free riders. Altruism for the group is confined to such species as social insects, and is rare among vertebrates. The only vertebrate species whose social organization resembles that of the honeybee is the naked mole rat, whose inbred colonies range in size up to about 80 individuals. In each colony a worker caste of moles serves the reproduction of the queen and her consorts.[92] No natural law guarantees an adaptive response to life-threatening events. Williams notes that whole populations and species do in fact go extinct, sometimes for want of behavioural changes that would be obvious to a biologist.

> What does a population do when threatened with extirpation, for instance, the reduction of the sockeye salmon stock of the Yukon River to 1 percent of its normal size? Nothing special happens at all. The individual salmon keep on with their normal activities, each trying to reproduce more than its neighbors, with no regard to effects on the stock as a whole. Individual salmon respond to individual threats in adaptive ways, but salmon populations take no concerted action to avoid being wiped out. Their populations show no functional organization like that of a bee colony.[93]

Despite the wholesale loss of ethnic genetic interests by Western societies, Corning argues that humans remain well adapted. '[M]uch of our economic and social life . . . [is] either directly or indirectly related to the meeting of our basic survival needs.'[94] Corning argues persuasively that this is true for many proximate interests. But his extensive review of the literature on biological measures of societal adaptation turns up no analysts who adopt genetic continuity as a criterion. Corning appropriately adds the interest of inclusive fitness, noting that it is usually maladaptive for individuals not to reproduce.[95] He does not extend this logic to argue for the adaptiveness of ethnic nepotism, understandable given the absence of this concept in the literature on basic needs. Consequently Corning's analysis does not include an assessment of the degree to which modern humans protect ethnic interests.

Current human behaviour is surely not a reliable guide to what constitutes an adaptive fitness portfolio. Corning notes but does not deploy Tooby and Cosmides's view that 'present conditions and selection pressures are irrelevant to the present design of organisms and do not explain how and why organisms behave adaptively, when they do'.[96] This restates Eibl-Eibesfeldt's argument that evolved behaviour patterns need not be adaptive in the man-made environment of mass anonymous societies.[97] Or, as psychologist Robert Hinde has put it: '[A]lthough basic human propensities evolved through natural selection, humans may behave in ways not conducive to their inclusive fitness in part because human behavior is largely influenced by the values and norms of their society, and by the rights and duties associated with their positions as incumbents of particular roles within the society in which they live.'[98]

It would be wrong to suppose that humans explicitly recognize their genetic interests in the family. Genes do not figure as proximate interests. Humans evolved in ignorance of genes and instead are guided in directing altruism by phenotypic kinship markers of varying reliability. The interesting question is, why do proximate interests still reliably guide us towards our familial but not our ethnic genetic interests? Let me offer an answer that combines innate and environmental causes.

We have been reproducing in families since before the emergence of humans or even of hominids. After all, maternal care is universal to mammals and is one of that class's defining characteristics. Because of the ancient origins of the family, innate proximate mechanisms have had time to evolve for identifying, bonding with and investing in offspring and siblings. Families represent such a high and reliable concentration of their members' distinctive genes that innate psychological mechanisms have evolved to monitor and protect that ultimate interest. Hamilton noted that the age-old notion of shared 'blood' is adequately precise to assess the coefficient of relatedness for relatives. 'Man has a great interest in his blood kin. Popular terminology reflects that interest. Human knowledge of human pedigrees is sometimes amazingly extensive, . . . especially . . . considering the dependence on oral tradition . . . in primitive societies. There is no doubt that an appeal to kinship in general does tend to moderate selfishness and encourage generosity in human social interactions; . . . .'[99]

Although in modern societies the family is reduced from its traditional extended size, and families are often located in novel urban and demographic settings, the nuclear family retains many elements of its ancient character: parents (or at least a mother) who bear, nurture, and raise children until puberty. Much information for kin-recognition is provided by the environment in the form of proximity and early bonding, though these usually entail the release of innate parental motivation. The importance of environmental cues is indicated by the fact that adopted children do not receive dramatically less care, though there is a consistent pattern of reduced step-parental investment.[100] So it is not genetic relatedness *per se* that has kept altruism effectively channelled towards close kin despite rapid environmental change, but proximate mechanisms. This has not occurred in the case of the tribe, where the match between optimal and actual strategy is weakest.

The tribes of the twenty-first century are still concentrations of genetic fitness, as tribes have always been. Cooperation and group strategizing between co-ethnics is still common, especially within minority groups.[101] But ethnies are often distributed in phylogenetically novel constellations of space and culture. It is the social world outside the family where modern societies differ most profoundly from the Environment of Evolutionary Adaptedness, in the realms of work, mass transport, anonymous suburbs, high residential mobility, and ethnic diversity.

The phylogenetic novelty of modern societies, especially multi-ethnic enculturation to common language, dress, and living patterns, tends to reduce ethnic identification. It can also exaggerate it when, for example, religious differences keep groups from socializing, working together, or intermarrying. Another major factor slowing assimilation is overt racial difference. Human societies have been monoracial for almost all of their evolutionary history. Our ability to detect physical resemblance was not evolved to detect genetic distances between races but between families and clans. Combined with the preference for the company of individuals with similar appearance and behaviour, this detective ability leads to the remarkable tendency of friends and spouses to be similar across many physical characteristics.[102] In comparison to intra-ethnic differences, inter-racial differences are large in colouring, physiognomy, hair form and body odour. These differences may constitute super-normal releasers of familial ethnocentrism. Alexander makes a similar suggestion buts thinks the proximate cause could be that morphological differences confound the subtle cues of paralinguistic communication learned within the family.[103] In any case the result is heightened levels of racial identification and discrimination that persist despite the Western cultural norm of individualism and pervasive attempts to indoctrinate the population to tolerate diversity and, in the multicultural mode, to celebrate it.

Hamilton also argued that novelty of environment is the reason humans have not evolved the elaborate instinctive patterns for defending fitness, but instead possess 'amorphous and variable inclinations'. '[B]ut, of course, considering what a newcomer man is to his present ecological situation . . . this is what we expect.'[104] The novelty of industrial society has tended to decouple social patterns from ethnic interests. Under the heading 'Interest Groups', the *Encyclopaedia Britannica* notes that: 'In primitive or developing societies, the most prominent type of interest group is the natural (i.e., primordial or communal) one—that is, one based on kinship, lineage, neighbourhood, or religious confession.'[105] In Western industrialized societies primordial interest groups are less prominent than secondary types devoted to advancing some shared concern about resources or the environment. The *Encyclopaedia* continues in a vein applicable to ethnic genetic interests. Some interest groups are 'latent': an interest can exist even though it is unrecognised by those who share it. There may be a common interest among certain members of the public even though these individuals may not have combined into a formal or informal organization. The individuals may be unaware that they have a common interest or, if aware, see no reason to defend or promote it. Or, even if the members consciously wish to defend or promote an interest, the law or powerful elites might restrict their ability to associate for such a purpose.

None of the foregoing detracts from Alexander's view that genetic interest can be inferred from the 'direction of [an individual's] striving',[106] so long as this claim is limited to species in their natural habitats. But beyond the family, in Western societies genetic interests are no longer reliably identified by striving or even by identification. A species in genetic equilibrium with its environment does not consciously recognize its genetic interests, but does so implicitly in its adaptive behaviour. The same is true of humans before the advent of modern genetics, though humans came closer to conscious realization of genetic interests with their kinship metaphors. However, by transforming their demographic, political, cultural and residential environments, modern humans have come to resemble more than ever before the individualists that Williams observed in sockeye salmon, including their inability to act in concert to defend the population as a whole. While organisms displaced from their natural environments still have genetic interests, they are unlikely to strive to protect interests they no longer recognize.

Human adaptation has long been a mix of instinct and learning, or rather *instinctive learning* since humans have genetically programmed learning dispositions.[107] Even aspects of mothering have to be learned, though this is the most intense and ancient investment in genetic interests and the physiological components are innate. Similarly, tribal and national solidarities are learned from cultural traditions that bestow group identity and pass on mobilizing techniques.

Culturally derived techniques are also critical for conducting exchange relationships, a vital strategy for surviving and reproducing throughout human existence.[108] If favouring kin and ethny is warranted because it serves inclusive fitness, then the same can be said for reciprocity. And if reciprocity of various kinds is an important strategy for defending genetic interests, it follows that threats to the trust and extensive friendly networks that facilitate exchange will usually be threats to the actor's genetic interests. Modern market economies are managed by sophisticated social technologies devised by domain-general intelligence that facilitate exchange, including centrally created and regulated currencies, contract law, police, conflict resolution mechanisms including litigation and criminal courts, and markets of various kinds.[109] This complex web of cultural devices for monitoring and controlling exchange behaviour is required for the same reason that novel methods are needed to regulate ethnic nepotism: because domain-specific capacities and tribal culture are insufficient to cope with the anonymity and sheer scale of transactions conducted within mass societies. General intelligence must be applied to the problem of devising collective goods able to mobilize and channel the altruism of whole peoples in adaptive ways.

Since exchange is conducted between individuals with different genetic interests, there is often tension between the urges to cooperate and to dominate. The same tension can be expected between groups. An extended market will boost average wealth, arguing for cooperation and the submergence of tribal identity.[110] But group competition is possible within a shared market, as is clear from studies of ethnic economies and ethnic middle man groups,[111] making it prudent to retain group identity and mobilizing capacity. Optimization of economic growth with democracy, social peace and equal opportunity occurs more often within relatively ethnically homogeneous nations.[112] In the next chapter I argue that the greatest international good, avoidance of mutually destructive war, can be achieved while retaining the most powerful group strategy for preserving ethnic genetic interests, the nation state.

### *Individualism, nationalism, and humanism versus a mixed portfolio*

So far in this chapter I have not discussed the genetic interest contained in the whole species, in accordance with mainstream neo-Darwinian theory that rejects explanations in terms of species survival. This was the lasting lesson of the critiques of V. C. Wynne-Edwards's theory of group selection by J. Maynard Smith and G. C. Williams.[113] An altruist who transfers resources or security from his own survival and reproduction to benefit random members of his species favours free riders by subsidizing the reproduction of genetic competitors.

One need not part company with neo-Darwinism to suggest that investing in humanity at large can be adaptive when species survival is at stake. In such a circumstance survival of smaller groupings is conditional on the survival of the whole species. It is less plausible to contend that survival of the species might be dependent on the sacrifice of one's family or ethnic interests. Preservation of the biosphere will probably involve international sharing and cooperation to block the proliferation of weapons of mass destruction and stabilize population at sustainable levels. But it will not conceivably require the suicide of families or nations.

The argument for investing in humanity begins with the question that has guided much of this book. In which group is it most adaptive to invest? Should we show loyalty to the family, the individualist position? Or should we invest mainly in some intermediate category, such as urged by extreme nationalists? Or should we invest much of our life resources in mankind as a whole, the extreme humanist position?

Emphasizing any one part of a fitness portfolio must detract from others. This point is illustrated in Figure 6.1, in which I summarize the above discussion concerning fitness portfolios. Figure 6.1 is an attempt to express ideologies as fitness portfolios. An individual's lifetime investment is set at one. The curves do not show actual expenditure, but are meant to illustrate how ideologies differ. The contrasts, rather than the actual values, are the point of the exercise. Also, I have tried to portray ideological ideals, not how they work out in reality. If data were available, they might reveal that in all societies individuals expend at least 95 percent of their life effort on maintaining themselves, the rest being divided up between family, ethny and the wider population.

The lesson that emerges from the analysis so far is that the larger the interest group, the more conditional is the adaptiveness of loyalty to it. The risks of free riders and disproportionate allocation of altruism make the family the least risky beneficiary of investment and humanity-as-a-whole the most risky. Yet faced with a threat of sufficient magnitude, the species is the largest store of our gen-

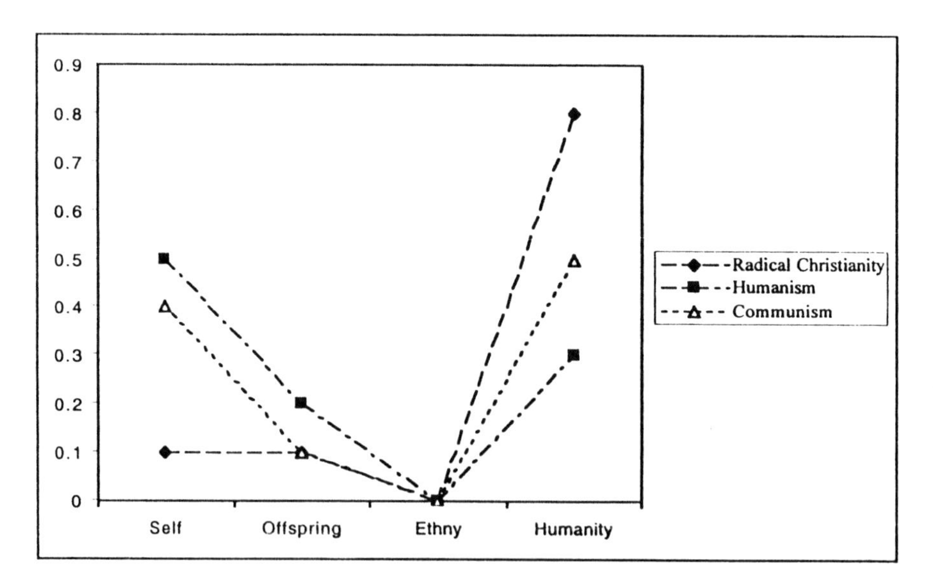

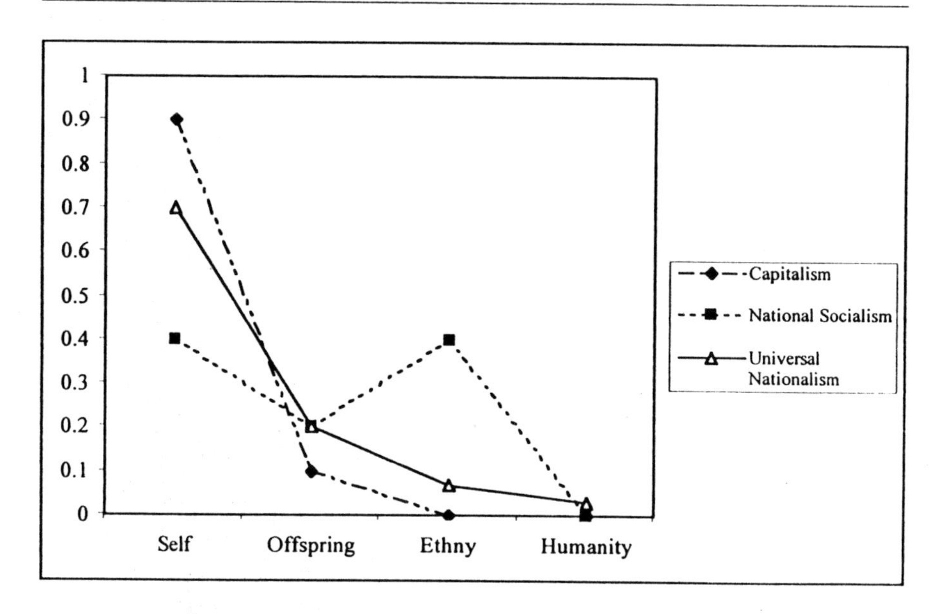


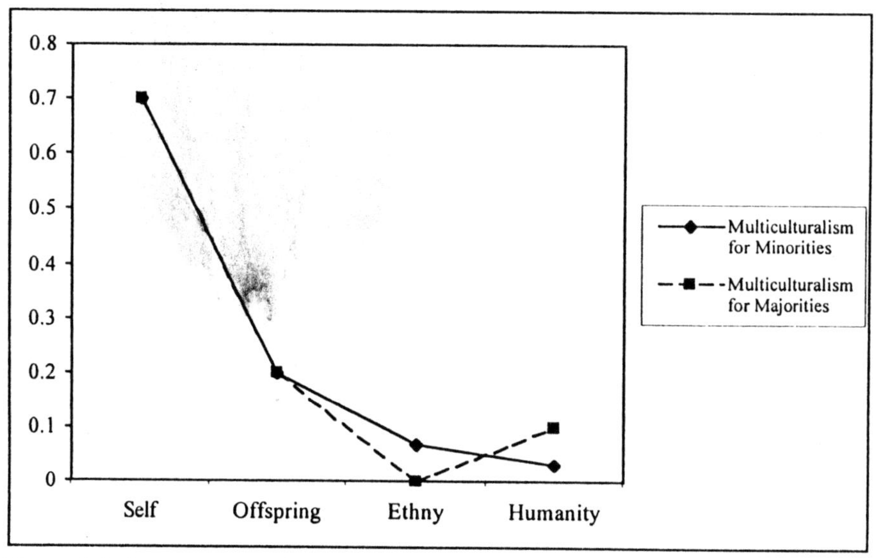


*Figure 6.1. Impressionistic distribution of life effort between self, offspring, ethny, and humanity according to different ideologies. The values are chosen not to be realistic but to show differences between ideologies.*

etic interests. Which, then, is the best strategy? Which camp has it right: extreme individualists, nationalists, or humanists? The preceding discussion in this chapter should have indicated what I think is the correct answer: all are suboptimal when exclusively pursued. Optimality depends on a balanced portfolio, sensitive to circumstances, with altruism being distributed across all levels of kinship, albeit weighted heavily towards groups carrying highest concentration of one's distinctive genes.

A well-balanced portfolio will usually include investments at all concentrations of genetic interests, a position I call 'universal nationalism' since it includes robust investment in the ethny (see Figure 6.1b and the next chapter). It includes raising children, helping relatives raise theirs, defending and enriching the community in which those children will most likely find mates, and contributing to the culture and security of one's ethny. Last in importance, though not unimportant in the modern world, is responsible global citizenship. Promoting mutual enrichment and stability through trade, cultural exchange, and the weaving of a code of civilized international conduct, all should help to prevent or minimize globally destructive trends, such as chauvinistic aggression and environmental degradation.

Choice of priorities depends on circumstances, and it is adaptive to switch priorities as circumstances change. Events will sometimes call for a skewing of investment towards one level. For example, when one's ethnic interests are threatened by territorial aggression or by subordination within a multi-ethnic society it makes sense to shift resources away from the familial and especially global spheres to activities of ethnic defence. In extremis it can be prudent to promote defensive war and to risk one's life fighting in it. One might encourage extra diplomatic efforts such as forging international treaties or building regional or global institutions. The adaptive course might be to discourage a pointless war, even if this entails personal risk.

The ability adaptively to alter the portfolio mix as circumstances change is a strategic advantage. To illustrate, ethnic group strategies can be prohibitively costly in a number of circumstances. For decades it was imprudent to be an ethnic activist in the Soviet Union, because the Bolshevik regime that assumed power in Russia in late 1917 suppressed expressions of ethnic solidarity, believing that these detracted from class loyalty. Institutional threats were made abundantly clear through the mass media and schooling system. Would-be activists held back from dangerous or wasteful behaviour. In these circumstances inability to perceive such threats or to restrain acts of ethnic loyalty would have brought down the irresistible power of the state, reducing the individual's chances of mating and raising a family. Individuals unable to switch altruism from ethny to family, unable to suppress their patriotism, would have been less successful than individuals able to adopt a flexible strategy. It would not have been adaptive to forget one's national identity altogether. If this had happened, regional ethnies would not have taken the opportunity to reassert national autonomy when the Soviet Union finally broke up in the 1990s.

An optimal portfolio will usually invest resources and risk in a manner sensitive to genetic distance, if not in proportion to it. All unimpaired individuals take care of their basic physical and social needs, avoiding danger and seeking nourishment and society. Since nurturing one's children is a day-by-day experience for a large fraction of humanity, it is the most common form of genetic investment after personal maintenance. Threats to intermediate sized groups, such as the ethny or religious community, are the next most common, while threats to broad regions and to humanity as a whole are relatively rare. Genuine concern about these latter threats and investment in alleviating them thus arises least often. Optimal portfolios will typically retain these rough proportions, based on the factors discussed so far in this volume, especially free riders and the exponential decline of kinship with genetic distance.

It is possible to be *too* flexible in switching investment between interests. If all but the most radical contingencies require decreasing investment from self to family to ethny to species, a wise rule of thumb is always to retain, in decreasing order or priority, some personal dignity, clannishness, patriotism, and humanity. The advantages of the first three standards are clear, but upholding minimum standards of compassion towards our fellow man is also important. We can safely extend to all members of our species some prerequisites of the humanist sensibility. This sensibility is a secular version of the universal element found in all the great religions. This is usually a minimal concession to humanity at large, but it can become a considerable act of altruism in extremis, when resources are stretched thin and any kind of consideration can forgo an advantage. Both secular and religious humanism are often suspended in these circumstances, when more immediate interests of individual and group survival are given priority. But whether in war or peace, efforts should be made to uphold standards of decency as an investment in the global social environment. Spreading a code of civility protects the weak, including the temporarily disadvantaged, making the world a safer place for everybody.

An effective method for doing so is to urge one's nation to become a signatory to, and respecter of, international institutions such as the Geneva conventions on the amelioration of the treatment of wounded, prisoners, and civilians in times of war, the World Health Organization, certain United Nations programs such as the International Fund for Agricultural Development, and the Universal Declaration of Human Rights, approved by the General Assembly of the United Nations in 1948. These are public goods that benefit all of humanity, although the last would be improved by clauses that defined and protected genetic interests. If it is deemed appropriate to declare as basic human rights such proximate interests as choice of work, availability of rest and leisure, education, and an adequate standard of living, surely our diplomatic representatives could throw in, as an acknowledgement of a widespread aspiration, the right to live among one's own people in a territorially demarcated self-governing community.

Not all universal declarations deserve support. For example the United Nations' *Declaration on Race and Racial Prejudice* adopted in 1978 by the General Conference of UNESCO is a confused and misleading document which, if taken seriously, would deny the peoples of the world the right to defend their vital interests. One can discern, through the heavy Soviet-like rhetoric, an element of noble intent, namely to affirm the dignity of all peoples and to condemn insults to that dignity. But by condemning 'all forms' of ethnic and racial discrimination the Declaration would prevent most nations from preserving their identities in an age of mass migration.[114] The United Nations itself has for many decades disseminated such pronouncements, perpetuating the ideological climate of the immediate post-Second World War era when the Soviet Union and its sympathizers were taken seriously as moral agents, and perhaps as an appeasement to countries emerging from the shadow of colonialism. An example is the 1975 resolution by the General Assembly that equated Zionism with racism, rather than condemning specific acts by Zionists, as if it is immoral in principle for a people to seek to protect their ethnic interests. The resolution was repealed in 1991 after the collapse of the Soviet Union, but the Soviet era UN doctrine on ethnicity and race remains in place. Every nation has an interest in retaining autonomy, an interest that is endangered by the pretence of global homogeneity. It is in every nation's interests to have a forum for discussing and deciding matters of common interest to the global community. But it would only benefit a small elite for the United Nations or any organization to gain a degree of global hegemony.

It is not always possible to have a balanced fitness portfolio. A skewed portfolio will sometimes be inevitable, though this is no reason to cease investing. For example, as discussed in Chapter 8, life circumstances will leave some individuals childless. The adaptive course is to do more for other relatives, the closer the better. The time freed from child rearing can also be turned to ethnic defence and advancement, as well as promoting common human interests. Alternatively, one might be prevented from investing in the nation state. Events such as mass immigration or agitation by competing ethnic groups might have severed the link between state and nation or reshaped the state into an instrument that redistributes resources from one's own ethny to competing groups. It would hardly be adaptive to risk one's life, or that of a son, to defend a state apparatus that presided over the replacement or subordination of one's people. When one has no nation state in which to invest—arguably the situation in multicultural societies—a prudent fitness portfolio will emphasize family and ethny and deemphasize the state in favour of agitation for political reform.

*Pathologies of Right and Left*

Mainstream ideologies are implicit fitness portfolios, since all consist of propositions and values that bear on the apportioning of limited resources available to individuals during their lives (see Figure 6.1). All ideologies therefore have implications for the adaptiveness of individuals who embrace them. For example, the hyper-individualism so common in the West today focuses effort on individual satisfaction to the exclusion of family life; many individuals fail to reproduce despite having health and resources (see 'Capitalist' curve in Figure 6.1b). Individualism is beneficial in promoting economic growth but atomizes families and ethnies and renders the latter vulnerable to replacement and subordination. The acceptance of replacement migration by hyper-individualist societies makes it not unreasonable to describe that ideology as pathological. In this section I compare some of the pathologies of two other ideologies, nationalism and humanism, which in modern affairs are usually identified with the political Right and Left respectively.

Nationalism is closer to reality than contemporary humanism in recognizing ethnic kinship, although this is usually an expression of unreflective sentiment. Still, nationalist instincts are a truer guide to genetic interests than is humanism (I am considering only those true believers who do not espouse a doctrine as a vehicle for advancing some antithetical interest). Certainly nationalists allude to an underlying genetic reality when they refer to the nation as a family or to co-ethnics as siblings. The kinship metaphor lends nationalism an emotional authenticity and depth that is harder to find among humanists. The latters' recurring calls for 'universal brotherhood' suggest that it is difficult to hold onto feelings of indiscriminate benevolence without invoking kinship. The humanists' most authentic emotion is the expression of sympathy for easily-visualized demographic categories in distress—the starving child; the distraught mother; the dead or disfigured soldier. These images also move the nationalist, but with extra pain when the afflicted are fellow nationals. The Left's dismissal of natural categories in general,[115] including the genetic reality of ethnicity, is factually wrong in the majority of instances where tribes and nations are descent groups and, more to the point, where ethnic boundaries correspond to relatively steep genetic transitions. But error does not lie only on the Left. Ethnic nationalists are often unbalanced in the intensity and communality of their claims to national kinship and thus interestedness. The 'fictive kinship' postulated by some analysts is not fictional by virtue of claiming kinship where none exists, but by exaggerating blood ties that undoubtedly are real.[116]

Sometimes nationalist assertions of a blood interest in a defensive or aggressive foreign policy are simply false, because a nation they oppose is closely related to their home nation or because segments of the home society do not share a genetic interest in a particular competitive stance. France, the home of modern cultural nationalism, is genetically diverse, its majority language a relatively recent development of Romanized Frankish domination from the early Middle Ages.[117] Despite the domination of French from its seat of power in Paris, there are still differences of dialect and language within France that correlate markedly with genetic differences. 'There is thus some obvious correlation between the genetic and the linguistic mosaics of France.'[118] Within Western Europe, regional identities might correlate more reliably with genetic interests than does France's overarching national identity. The latter achieves greatest genetic distinctiveness in comparison with non-European races.

Nationalists often ignore or down-play expressions of intra-ethnic class competition, arguably not in the ultimate interests of the poorer members of the group (this is discussed in the next chapter). Also individual genetic variation and intermarriage ensure that some ingroup members will have genetic interests in other groups, presenting conflicts of interests between members of the ingroup not addressed by nationalist doctrine of the authoritarian variety. Another strategic weakness of nationalism from the point of view of genetic adaptiveness is that the great mobilizing power of genetic metaphors can be put to mutually destructive purposes. National Socialist Germany is an example (Figure 6.1b).

An economic analogy is the speculative bubble, which can occur anywhere in the fitness portfolio, though risk rises steeply as fitness concentration declines. With ethnic and global altruism more conditions must be met to insure against free riders. The best known fitness investment bubbles are the massively destructive wars of the twentieth century, and the similarly destructive class conflicts of the same century. The Leninist-Stalinist version of extreme humanism represented a large investment of wealth (foregone) and blood (spilt) for no prospective genetic payoff at all. Fear of this nightmarish experiment among the middle classes of Europe contributed to the fascism that initiated the Second World War. Though Italy and Germany do represent large ethnic genetic interests, the general destructiveness of the war that fascist aggression initiated, afflicting both its victims and its home societies, was a speculative bubble of huge proportions. Fascism is an over-investment in national interests at the cost of individual and foreign group interests. These societies might have been spared much tragedy if those who invested in the communist and fascist experiments had not been unhinged by extreme historical circumstances and had kept to more balanced genetic portfolios.

The humanist social pathology illustrated in Figure 6.1a has expressed itself less frequently than nationalist social pathology, perhaps because humanist ideology is not as intuitively appealing. In the past it has been attractive mainly to cosmopolitans and intellectuals. Now, the power of modern media to convey images from trouble spots around the globe into every lounge room is making a once elite idea more common.

It is understandable that extreme humanists do not wish to take genetic interests seriously. To maintain their Olympian detachment from kin and ethnic ties, any humanist who accepted that humans are evolved animals with interests to match would be driven to deny particularities of ethnicity and race. Otherwise some kind of local affiliation might recommend itself. Instead the humanist opts for maintaining the absurdity that all humans are equally related within and between ethnies. Indeed, this position is sometimes advocated by those wishing to maintain a humanist stance, though understandably not much analysis is offered in support. An exception is R. Lewontin's argument that 'only' 10–15 percent of genetic diversity occurs between populations, an estimate confirmed by subsequent assays.[119] However, this statistic implies ethnic kinship equivalent to that between grandparent and grandchild or between half siblings (see this volume, Chapter 3).

The political and economic tragedy that has always accompanied attempts to institute extreme humanism across whole societies has been caused in part by the lack of human sympathy on the part of humanist elites for the people they come to rule. In *The Social Contract* Rousseau condemned the cosmopolitan as someone who 'pretends to love the whole world in order to have the right to love no one'.[120] Predictably, humanism is attractive in an era of individualism. Rousseau's words point to a great advantage of nationalism over humanism, that the true-believing nationalist feels a powerful bond with those he wishes to rule, while the humanist leader, despite noble feelings towards the concept of humanity, has no intense loyalty to any particular subset of the world's population, and so is relatively free to treat them instrumentally. Commenting on a social worker's expressed wish to open America's doors to all the world's poor, the early sociologist Edward Ross expressed his emotional particularism with regard to his nation's reproductive interest:

> Her sympathy with the visible alien at the gate was so keen that she had no feeling for the invisible children of *our* poor, who will find the chances gone, nor for those at the gate of the To-be, who might have been born, but will not be.
>
> I am not of those who consider humanity and forget the nation, who pity the living but not the unborn. To me, those who are to come after us stretch forth beseeching hands

as well as the masses on the other side of the globe. Nor do I regard America as something to be spent quickly and cheerfully for the benefit of pent-up millions in the backward lands. What if we become crowded without their ceasing to be so?[121]

Free riders are a basic evolutionary problem for humanism. Individuals who retain some ethnic cohesion and adopt collective strategies are more likely to preserve and extend their ethnic interests to the disadvantage of individuals who eschew all altruism or hand it out indiscriminately. Hence, group strategies are the enemy of humanism (and individualism). Extreme humanism, where substantial sacrifices are made for humanity at large, would only be viable if informed and moderated by knowledge of individual and ethnic genetic interests. Moderation must include political, legal, and economic mechanisms that protect those interests against free riders. The 'socialism in one state' formulated by Stalin met some of these criteria, but in fact rested on the solidarity afforded by Russian nationalism. A cautious state could maintain a low level of universal altruism, delivering some resources to the world's needy after the nation was provided for. However, strongest protection against free riders and thus highest sustainable global altruism would be possible under a world government. This government would monopolize the use of force to settle disputes and would prevent lower-level group strategies through deterrence, punishment, and indoctrination. There could be no balance of national power, but a centralized global state, since any national entity free to strategize would constitute a potential free rider on universal altruism. There would thus seem to be a choice between extreme humanism and freedom.

### *The inadequacy of human nature stripped of cultural defences*

The optimal strategies formulated in this chapter are a first try and will probably need to be revised. More certain, I think, is that if ethnic genetic interests exist they can be defended under a range of conditions, from the primitive bands and tribes in which modern humans have lived for most of their evolutionary history to the modern state. I have also shown that in modern conditions there is often a large gap between behaviour that would defend ethnic genetic interests and actual group behaviour.

Cultures, including religions and ideologies, differ in the degree to which they approximate optimal strategies for advancing ethnic genetic interests. Several strategic dimensions are involved. Nevertheless, it should be possible to distinguish different grades of adaptiveness, defined as the extent to which a culture promotes the inclusive fitness of its members. This is a large subject. I have already suggested that all traditional religions must have served the genetic interests of those who practise them. For most of their existence they must have constituted one type or another of what MacDonald calls 'evolutionary group strategies'.[122] National religions such as Japanese state Shinto mobilized large fractions of ethnies. On one interpretation Medieval Catholicism served evolutionary group goals.[123] A cluster of covenant religions derived from the Old Testament tradition promotes ethnic partitioning, a means of sustaining group solidarity.[124]

I am not suggesting that the optimal strategies formulated in this chapter are exhaustive or in any other sense the last word on the matter. But I am suggesting that it is fruitful to analyse cultures according to their evolutionary impact, since this might help explain group differences in continuity and expansion. These are the phenomena that inspired MacDonald's and Phillip's analyses, respectively, of Judaism and Old Testament traditions. MacDonald's analysis has already been discussed (p. 103). Phillips asks, how could a small sixteenth century Tudor kingdom hardly distinguishable from its European neighbours in economics or science, have grown within a few centuries to hold virtual global hegemony, in addition achieving major population expansion of the English and kindred peoples into Australia, New Zealand, South Africa, and North America? The last member of this family of nations has for much of the last century been the world's largest economy and at the beginning of the twenty-first century stands unchallenged as the world's only military super power.

The project of explaining group differences in competitiveness is not new. It motivated the German sociologist Max Weber's famous theory of the 'Protestant work ethic', in which he concluded that capitalism flourishes most in individualistic, rational societies and is therefore inhibited by traditional ethno-religious solidarity.[125] This is a stunning rebuke to tribal strategies of ethnic advancement, since it suggests that precisely those societies that care least about ethnic continuity are most likely to harness modern technology and management and thus, to be ethnically strengthened. This is surely an overstatement, since there are secular forms of ethnic solidarity. As this book has been at pains to demonstrate, ethnic solidarity can be rational. Also, there is no reason to suppose that group identity and cohesion prevent the development of a market economy. The European societies in which modern capitalism was invented and the industrial revolution flourished were nation states, often aroused by geopolitical rivalry. The United States' economy continued to grow in the late nineteenth century when Congress restricted East Asian immigration, through the 1924 closure of immigration from Eastern and Southern Europe, up until 1965 when the doors were reopened.

Nevertheless, Weber's insight is valuable because it raises the possibility that there is no permanent winning formula for defending ethnic genetic interests, outside the simple maxim of maintaining a mixed portfolio and adjusting to circumstances. In this case means are disposable, leaving only the ends of survival and reproduction as permanent features of the strategic landscape. The means slowest to change will be a set of principles, including the need for efficient procedures for recognizing fellow ethnics and indicating the relative strength of competition coming from within and outside the ethny, control of free riders, and maintenance of efficient collective goods such as organizations and territory. These must then be selectively recombined to achieve adaptive outcomes whenever socio-economic and technological circumstances change, as they have done in ever accelerating strides over recent millennia. One strategy that bears this mix of principles while allowing for flexibility is the nation state.

*Notes*

1 Hamilton (1964; 1975).
2 Hamilton (1971, pp. 79, 89).
3 Masters (in press).
4 Alexander (1987); Darwin (1871, p. 500); Westermarck (1971/1912); Wilson (1975, pp. 562–4).
5 E. O. Wilson (1975, p. 3).
6 Hamilton (1971, p. 80).
7 T. H. Huxley (1894).
8 Olson (1965/1971).
9 Fehr and Gachter (2002); and see Ostrom (1990).
10 Hardin (1968; 1999).
11 Easterly (2000).
12 Dawkins (1989, p. 8).
13 Hamilton (1969/1996, p. 196).
14 McNeill (1979).
15 Gurr and Harff (1994).
16 Tullberg and Tullberg (1997).
17 Alesina and Wacziarg (1998); Easterly and Levine (1997).
18 Daly and Wilson (1999).
19 Case et al. (2000; 2001).
20 Jankowiak and Diderich (2000).
21 Buss et al. (1999); Daly et al. (1982); Wiederman and Allgeier (1993).
22 Lewin (2001).
23 *Ibid.* This is a selected sample. The rate of cuckoldry among the general population is much less than 28 percent.
24 Hamilton (1996, p. 19).
25 Hirschfeld (1996).

26 Westermarck (1971/1912, I, Chap. 23).
27 E. O. Wilson (1971, p. 4); and see Dawkins (1978, p. 193).
28 Keeley (1996); Konner and Shostak (1986); Wrangham and Peterson (1996).
29 Vanhanen (1991).
30 Boehm (1993); Eibl-Eibesfeldt (1982).
31 Erasmus (1977).
32 MacDonald (1994).
33 Banfield (1967).
34 van den Berghe (1981); Wiessner (1984).
35 Eibl-Eibesfeldt (1998/2002); Wiessner (1998/2002).
36 Boyd and Richerson (1987).
37 Wiessner (1998/2002).
38 Alba (1985).
39 McGowan (2001).
40 E.g. Lowery et al. (2001); Rudman et al. (2001).
41 Salter (in press-b).
42 Sanderson and Vanhanen (in press).
43 Alesina and Wacziarg (1998).
44 Easterly and Levine (1997); Schubert and Tweed (in press).
45 Butovskaya et al. (2000).
46 Alesina et al. (1999); but see contrary evidence in Masters (in press).
47 Gilens (1999).
48 Hamilton (1975, p. 134).
49 Esses et al. (2001, p. 390).
50 Adorno et al. (1950).
51 Hamilton (1975).
52 Rushton (1989a).
53 See Rushton (1989b) for review.
54 Lopreato and Yu (1988); Sanderson and Dubrow (2000).
55 For a recent perspective on overpopulation, see E. O. Wilson (2002).
56 Eibl-Eibesfeldt (1989, p. 95).
57 Corning (2000, p. 77).
58 Eibl-Eibesfeldt (in press).
59 Parson (1998).
60 Reported by Dio Cassius, *Romaika*, 1925/c.225, as quoted by Betzig (1992, pp. 351–2).
61 Eibl-Eibesfeldt (1998).
62 Kaplan and Lancaster (1999).
63 Engardio (2002).
64 Spicer (1971).
65 Caton (1983/1994).
66 Sherif (1966).
67 Sachdev and Bourhis (1990); Shaw and Wong (1989).
68 Alesina et al. (1999); Alesina and Spolaore (1997).
69 Triandis (1990).
70 Hamilton (1975/1996, p. 344).
71 Silverman and Case (1998/2002).
72 Light and Karageorgis (1994); Landa (1981); Salter (in press-c).

73 Butovskaya et al. (2000).
74 Wiessner (2002b).
75 Rimor and Tobin (1990, p. 148).
76 Glantz and House (1995).
77 Goetze (1999).
78 Lumsden and Wilson (1981, p. 208).
79 Shaw and Wong (1989).
80 Wiessner and Schiefenövel (1996).
81 Butovskaya et al., 2002 manuscript.
82 Alesina et al. (1999); Alesina and Ferrara (2000); Hero and Tolbert (1996).
83 van der Dennen (1995, p. 159).
84 Service (1962); van der Dennen (1995, pp. 427–8; 564–5).
85 Sack (1986, p. 2).
86 Eibl-Eibesfeldt (1989, pp. 321–34).
87 Connor (1985); Spicer (1971); Vasquez (1993).
88 Spicer (1971, p. 798).
89 Keith (1968/1947, Chapter 4).
90 *Ibid.*, pp. 316–17.
91 Keith (1968/1947, p. 317).
92 Sherman et al. (1991).
93 G. C. Williams (1997, pp. 51–2); and see his classic 1966 critique of group selection.
94 Corning (2000, p. 41).
95 *Ibid.*, p. 70.
96 Tooby and Cosmides (1990, p. 375).
97 Eibl-Eibesfeldt (1970).
98 Hinde (1989, p. 58).
99 Hamilton (1971, p. 73).
100 Daly and Wilson (1999).
101 Kotkin (1992); Light and Karageorgis (1994).
102 Rushton (1989a); Thiessen and Gregg (1980).
103 Alexander (1979, pp. 126–7).
104 Hamilton (1971, p. 79).
105 *Encyclopaedia Britannica* (online, 2000).
106 Alexander's (1995/1985, pp. 182–3).
107 Eibl-Eibesfeldt (1989).
108 Mauss (1968); Sahlins (1965).
109 Salter (1995).
110 W. Masters and McMillan (in press).
111 Light and Karageorgis (1994); Landa (1981).
112 Alesina and Spolaoro (1997); Easterly and Levine (1997).
113 Maynard Smith (1964); G. C. Williams (1966).
114 Errors of fact and reasoning in the UN *Declaration on Race and Racial Prejudice* are too numerous to review here; some examples will have to suffice. Article 1 Section 2 states that racism includes 'the fallacious notion that discriminatory relations between groups are morally and scientifically justifiable', while Article 9 Section 2 calls for 'special measures' to ensure equality 'wherever necessary', 'while ensuring that they are not such as to appear racially discriminatory'. The im-

plication is that racial discrimination is indeed justified when it tends to produce equality, contradicting the earlier claim that no discriminatory relations are morally justified. The *Declaration* implies that discrimination can be morally good as well as bad, depending on the outcome. But, according to the *Declaration,* adaptive living is one outcome that does not justify discriminatory means, even when everyone's interests are served. The underlying assumption seems to be that biological interests carry no weight compared to lifestyle interests (Article 1, Sections 2 and 3). Lifestyle includes the 'right to be different', but this must not 'serve as a pretext for racial prejudice' (Article 1, Section 2). Neither is it moral to base value judgements on racial differentiation (Article 2, Section 1). Thus the *Declaration* implies the improbable notion that it is possible to build or preserve a distinctive way of group life without defining group boundaries and favouring insiders over outsiders. Article 2, Section 3 rejects '[a]ny distinction, exclusion, restriction or preference based on race, colour, ethnic or national origin or religious intolerance . . . which compromises the sovereign equality of States and the right of peoples to self-determination . . .'. Yet ethnic self-determination *entails* ethnic discrimination in an age of mass migration; it can also entail the break up of states according to democratic procedures compatible with the Westphalian tradition.

How to explain such slovenliness and ignorance from the world's premier body? Apart from the political compromises with dictatorships and totalitarian regimes that have always bedeviled the UN, the *Declaration*'s intellectual roots need to be examined. The *Declaration* claims scientific authority based on the assertions of 'experts convened by UNESCO' (Preamble). The language and ideas of these assertions are rooted in politicized social science, especially Ashley Montagu (1997/1942), who drafted the first UNESCO statement on race in 1950. Little wonder that genetic interests are given no weight, despite being universal. When one reviews background discussions), pertaining to race and ethnicity, one finds a lack of intellectual and ideological diversity (e.g. the minutes of the UN Commission on Human Rights, Geneva, 30 March 1999, 55th session, 8th meeting, document no. E/CN.4/1999/SR.8, 6 April 1999). Those condemned by the UN are given no voice, and empirical assertions made by founding ideologues such as Montagu are not challenged. The *Declaration* and other UNESCO initiatives on ethnicity demonstrate that a UN agency can become heavily influenced by an ideology inimical to the interests of most of the world's peoples.

115 See Singer (1998).
116 E.g. Masters (in press).
117 Cavalli-Sforza et al. (1994, pp. 280–85).
118 Ibid., p. 284.
119 Lewontin (1972).
120 Quoted by Darnton (2002, p. 30).
121 Ross (1914, preface).
122 MacDonald (1994).
123 MacDonald (1995).
124 Akenson (1992); MacDonald (1994; 1998, pp. 109–10 + Chapter 4); Phillips (1999).
125 Weber (1958).

# 7. Universal Nationalism versus Multiculturalism in an Era of Globalisation: Fulfilling the nation state's tribal promise

*Summary*
Since territory is a fundamental ethnic good, the nation state is an ethnic group strategy when it deploys the distinctive power of the state to maintain an ethnic group's monopoly of a territory. Nation states can mobilize their peoples to provide unprecedented economic and defensive collective goods. The social technologies deployed to achieve this high level of mobilization work by mimicking the traditional tribal group strategy. The nation state is thus the implicit promise of an ethnic group strategy. The traditional nation state is failing to fulfil that promise in an era of globalisation due to the pressures of mass migration and the inadequacy of ethnically neutral constitutions. Can the nation state be reformed to better serve peoples' ethnic genetic interests? Ethnic exclusiveness alone is problematic because it tends to degenerate into political chauvinism that works against others' genetic interests, risks the general good through aggressive war, and can become a vehicle for elite free riders. Ethnic solidarity's defensive function should be retained, but its aggressive side effects moderated. A moderating doctrine is that of universal nationalism in the tradition of Bismarck and Woodrow Wilson. The doctrine applies the Golden Rule internationally, respecting a general right to ethnic self rule. Implementation would include replacing warfare with international law, the limiting of free-riding national and global elites, and territorial confinement of unsustainable population growth.

## *Introduction and chapter thesis*

In Chapter 6 I concluded that ethnic monopoly of a territory is a 'fundamental ethnic collective good' because it facilitates efficient mass investment in ethnic interests and insulates the resident ethny from regional and global population changes. This fundamental good is unevenly spread around the world. The global population presently lives in about 200 territorial states, all of which possess at least nominal sovereignty over their territories. However, only a small number of the world's ethnic groups have something approximating sole use of a state territory (e.g. Iceland; Japan; the Koreas). Many more have a majority ethny or closely related grouping of ethnies (e.g. Australia, China, Indonesia, Thailand, USA, and many European countries). Many states have no majority ethny, and some of these are home to dozens of ethnies and partial ethnies (e.g. many African states). Some states, mainly a handful of Western democracies, have for decades been increasing their ethnic diversity through immigration. Others have been fleeing diversity, especially ethnic subordination, by carving out nation states from decaying empires, such as the Soviet Union that disintegrated in 1991. The number of sovereign states has grown from about 60 in 1900 to 194 in 2001.[1] At a global level, the relationship between ethnic and state boundaries is in flux. There is demonstrably much scope to redraw international borders and constitutions.[2] Knowledge of genetic interests might allow some of these changes to be directed towards constructing states that are more adaptive for their peoples.

Under which sort of ethnic regime is it most adaptive to live? The greatest concentrations of genetic interests are in the family and the ethny. In the next chapter I show that endogamy preserves familial genetic interests more than exogamy. On that basis alone it would be more adaptive to raise a family within an ethnically homogeneous community, where each generation is most likely to find mates from its own ethny. However, ethnies carry much larger genetic interests than families. It might make sense to ignore or even sacrifice family interests if doing so preserved ethnic genetic interests. Accordingly, this chapter deals exclusively with strategy at the ethnic group level, where a key strategic issue is choice of governmental system.

My aim is to find a general system or doctrine that is both stable and adaptive for most people's ethnic interests. In the terminology of evolutionary theory, this means finding a system that is evolutionary stable for citizens, or at least more in this direction than the alternatives. The argument of this chapter can be summarized as follows. As we saw in Chapter 6, some ethnic regimes favour ethnic majority, some minorities. Pro-minority regimes are inherently unstable because they change ethnic proportions; and while majorities survive, minority-centric regimes cannot serve the interests of the greatest number. I therefore begin by arguing for the territorial nation state as a vehicle for defending ethnic genetic interests. Among traditional nation states, majority ethnic interests are best preserved in the German type, where citizenship is ethnically defined. However, all existing nation states are proving vulnerable to highly mobilized and rapidly reproducing ethnic minorities and to their frequent precursor, mass immigration. The latter is often fostered by free riding elites. These trends are especially dangerous in a world running out of living space combined with uneven wealth and population growth. The two main legitimating ideologies for these trends are multiculturalism and globalism, which in the long run threaten everyone's ethnic interests.

I argue that only territorial ethnic group strategies in the form of ethnic states are able to meet these multiple challenges, a doctrine I label universal nationalism. The advantages of this system become clear when compared to the other combinations of ethnic group strategy (have or have not) and territory (have or have not), set out in Table 7.1. Combination (2) is ethnies lacking a group strategy living in territorially defined states that they established. Lacking an ethnic strategy, these ethnies are at the mercy of international migration flows. When the society is attractive due to wealth or stability, the ethny rapidly declines in relative fitness as the rest of the world floods in. Combination (3) is ethnies that do possess an ethnic strategy of some sort but lack control of a territory. These are typically mobilized and endogamous minorities. The history of dispossession and violence experienced by ethnies in this condition makes this combination unstable. Giving up the ethnic strategy, as in combination (4), is no solution because this squanders minority influence and ultimately leads to assimilation. Assimilation is most damaging to minorities genetically distant from the majority, because this ends their ability to strategize on behalf of their distinctive genetic interests. It can also harm the interests of minorities that are closely related to the majority when the latter has no ethnic group strategy as in combination (2), resulting in the minority sharing the majority's maladaptive fate.

| | Ethnic strategy | Non-ethnic strategy |
|---|---|---|
| Territorial | (1) Traditional nation states and ethnic states. | (2) Majority ethnies in multicultural states. |
| Non-territorial | (3) Mobilized minorities in multicultural states, and traditionally endogamous diaspora peoples: Armenians, overseas Chinese, Gypsies, Jews, Parsis. | (4) Immigrants who assimilate. |

*Table 7.1. Ethnic dispositions associated with combinations of ethnic group strategies and territoriality.*

Ethnic states have problems of their own in the perennial problems of internal and external conflict. These need to be resolved or ameliorated if any state is to constitute an evolutionary stable strategy for its citizens. In the second half of the chapter I discuss strategies for refining the ethnic state to reduce these risks, including ethnic constitutions and culture. I conclude by cautioning against the unrealistic expectation of finding a perfectly adaptive ethnic strategy.

*Free riding by minority ethnies: Multiculturalism*

Multiculturalism and other versions of ethnic pluralism currently in vogue in most Western societies are types of ethnic regimes that majorities should certainly avoid. They meet none of the survival conditions I suggested in Chapter 6 (p. 135). Parity of numbers with other ethnies within the state is not being maintained; no multicultural regime has policies in place for that purpose. All ethnies show some degree of endogamy, but rates of intermarriage are rising with the proportion of minority ethnies. There is no mechanism protecting the majority's share of resources or status. For example in America's best universities, staff and students of White Christian descent, the country's founding ethny, are represented well below their share of the national population. By the late 1990s students with this ethnic background at Harvard University numbered only about 25 percent of all students, well below their approximate 70 percent representation in the US population.[3] A similar trend is apparent in Australia.[4] Another survival condition that plural, high immigration societies fail to meet is that the majority of immigrants to these societies come from genetically distant ethnies. Finally, no multicultural society sets aside inalienable territory as a fall back position should the majority find itself being swamped, though the United States, Canada, Australia, and New Zealand provide their aboriginal populations with inalienable territories. There is a general trend for fellow ethnics to congregate in neighbourhoods where they feel more comfortable, a well known case in Western societies being the 'white flight' from suburbs and cities and even whole states experiencing high levels of non-white immigration. But because there are no territories reserved for whites, this behaviour does not slow the decline of relative fitness within the state.[5] Fleeing to another zone of a unitary ship of state that is permitting replacement migration is like changing deckchairs on the *Titanic*.

Majority-ethnic citizens in multi-ethnic societies, especially multi-racial ones, are faced with an invidious choice. They can do what comes naturally[6] and direct altruism preferentially to ethnic kin. This is adaptive since it promotes relative fitness but, especially when practised by the majority, engenders social conflict and can make the economy less productive. Alternatively, majority citizens can adopt the discipline of non-discriminatory behaviour, which tends to improve the economy and raise the carrying capacity of society as a whole. However, this strategy sacrifices relative fitness when minorities are not similarly constrained, as they are not in multicultural regimes.

Ethnically plural societies would seem to threaten one ethnic interest or another. Until about 1965 Western multi-ethnic societies gave the ethnic majority precedence, disadvantaging minorities. Majority free riding on minority labour was common, for example in the institution of slavery and the importation of low-cost labour. However, since the 1960s and the victory of the civil rights movement a new *modus operandi* developed. Majority ethnies restrained their own discrimination towards minorities more than the reverse, a formula known as multiculturalism. Minorities are actually encouraged to celebrate their identity and to work together to defend economic and political rights, while the same behaviour in majorities is discouraged. This unilateral withdrawal from ethnic competition arguably benefited the economy as a whole and certainly benefited minorities. But this formula is risky for the majority since it can lead to a general breakdown in ethnic group strategy and loss of control over immigration policy, resulting in demographic replacement. When the minorities are genetically distant, multiculturalism also tends to turn mobilized minorities into free riders on majority altruism as they organize to demand preferential treatment.

Minority free riding occurs in a number of ways. When there is ethnic stratification, characteristic of most multicultural societies, free riding can occur at the bottom of the class structure in the form of welfare and other benefits conferred by collective goods largely provided by the majority ethny. Redistribution via the welfare state and other public goods then causes majority-ethnic taxpayers to pay for their own loss of relative fitness by financing reproduction of families belonging to other ethnies. Ethnic majorities can also find themselves economically or culturally dominated by highly competitive ethnies, when minority free riding is liable to take top-down forms such as steering cultural, immigration, and foreign policies towards minority goals without regard for majority interests.

Multi-ethnic societies thus tend to be maladaptive for majorities under multicultural regimes and invidious for minorities under traditional regimes. Reproductively fair plural systems that aim to stabilize ethnic proportion by severely limiting immigration, such as the United States' quota system (1924–1965) or Australia's traditional restrictive immigration policy, were rejected by some minorities as discriminatory, a true claim. In fact such systems were imposed by majority ethnies against the minorities' interest of increasing their relative fitness. The critics did not care to mention that mass immigration of their co-ethnics undermined the fitness of the majority population. Maladaptiveness for one party or another would seem to be inherent to multi-ethnic societies that do not provide the guarantees set out in Chapter 6, even when the system is demographically fair.

The inherent maladaptiveness of multi-ethnic unitary states changes the meaning of a 'working' system. The endemic conflicts of multicultural societies are a source of mobilization to all participating ethnies, including the group being displaced. It is therefore a stroke of good fortune for the native born when those who are managing their replacement cannot engineer a smooth transition. For a people losing its country, the only thing more disastrous than multiculturalism that does not 'work' would be multiculturalism that did work.

### *Universal nationalism*

The family is a product of Nature. The most natural state is, therefore, a state composed of a single people with a single national character . . . for a people is a natural growth like a family, only spread more widely (J. G. Herder 1785).[7]

Perhaps universal nationalism would be optimally adaptive for the majority of ethnies. I mean a biologically informed version of the doctrine advocated in the nineteenth century by Otto von Bismarck and in the early twentieth century by Woodrow Wilson. The idea is the Golden Rule applied internationally. Universal nationalism would not only attempt to bring political and genetic borders into alignment, but within that frame establish adaptive ethnic states in which the government apparatus unambiguously served the ethnic interests of the majority. Ethnic states are closer to the traditional German than French model of the nation. The latter is a cultural conception of the nation, while the German model adopts ethnicity as shared descent as the criterion of citizenship.[8] The German model offers a constitutional barrier to replacement migration. Immigration can still occur, but citizenship rights are reserved for those with at least one native parent. Thus full membership of the polity is conditional on intermarriage, the most reliable form of assimilation, keeping political power predominantly with the ethnic majority. The system is a powerful brake on rapid ethnic change. In 1980, for example, 3.4 percent of foreigners resident in France acquired French citizenship, compared to only 0.3 percent of foreigners resident in Germany.[9]

In the second half of the twentieth century the barriers to replacement migration were removed in wealthy Western states that had adopted the French model. The founding populations in those states are now in the process of being partially replaced by genetically distant immigrant ethnies, causing large and permanent losses of ethnically distinct genes. The French system fails to protect the population's genetic interests because instead of ethnicity it adopts a set of abstract concepts as the defining symbols of the nation, such as a constitution or a set of ideals. The result is a 'concept nation' or 'creedal nation', support for which takes the form of 'constitutional patriotism'. This doctrine, espoused by anti-nationalists such as J. Habermas and M. Walzer, is bereft of biological content.[10] Indeed, such theorists positively reject any data on group genetic differences. By overlooking genetic interests, constitutional patriotism at best leaves ethnic welfare to happenstance. In fact wherever it has been adopted, this doctrine prioritizes minority ethnic interests ahead of those of the majority. It is in practice a formula for reconciling, or blinding, ethnic majorities to their own decline while serving the sectional interests of minorities and free riding elites. Yet these theorists do not explicitly attempt to justify valuing minority or elite interests over those of majority ethnies.

Every state currently managing the replacement of its founding ethny (e.g. Australia, Britain, Canada, France, the Netherlands, the USA) has adopted constitutional patriotism of one form or another. In France the defining symbols are French culture itself, which provides some inertia against ethnic replacement due to that culture's intimate links with the historical French nation. But the children of immigrants from any part of the world can learn to speak French and adopt local habits and customs. Constitutional patriotism is usually linked to the more destructive doctrine of multiculturalism, which encourages minority identity and mobilization while punishing those of majorities. A concept nation is incapable of principled defence against ethnic replacement. The doctrine is as pathological as a conception of the family that did not allow parents to show preference for their children. According to this formula a country would lose nothing if the founding ethnic group were peacefully replaced in part or altogether, so long as some set of values was retained (democracy, equality, non-discrimination, minority rights, the local language, etc.). The combination of constitutional patriotism and the multiculturalism it facilitates is, as one would expect, profoundly subversive of native ethnic interests.

The revival of ethnic nationalism would run counter to current liberal democratic opinion by prioritizing majority instead of minority rights. However, the original liberal perspective represented by a founding figure of modern liberalism, J. S. Mill, sought to maximize the good of the greater number. From this democratic perspective putting majority interests first is warranted, though with protection of everyone's individual rights. Universal nationalism is fair, in the sense of being universalisable. Most individuals are descended from a single ethny or related group of ethnies, so that a broadly applied nationalist doctrine would benefit the majority of the world's population. Putting minorities first is not universalisable, since it undermines the interests of majorities. As argued in the previous section, pluralist strategies are unstable and often indistinguishable from a policy of privileging minorities.

J. S. Mill was perhaps the first universal nationalist. He argued that people should be able to express their desire to form autonomous national communities by freeing themselves from ethnically mixed societies and forming nation states:

> Where the sentiment of nationality exists in any force there is a *prima facie* case for uniting all the members of the nationality under the same government, and a government to themselves apart. This is merely saying that the question of government ought to be decided by the governed. One hardly knows what any division of the human race should be free to do, if not to determine with which of the various collective bodies of human beings they choose to associate themselves.[11]

The rights of ethnic majorities is not a topic of much interest to contemporary liberal and minority intellectuals, who are concerned instead to oppose the tyranny of the majority. But if ethnic pluralism within the one unitary state is maladaptive for majorities, as argued above, and if the aim is to reduce maladaptiveness for the greater number, then the following call for universal tribalism by American political philosopher Michael Walzer necessarily translates into a call for universal nationalism:

> Tribalism names the commitment of individuals and groups to their own history, culture, and identity, and this commitment (though not any particular version of it) is a permanent feature of human social life. The parochialism that it breeds is similarly permanent. It cannot be overcome; it has to be accommodated, and therefore the crucial universal principle is that it must always be accommodated; not only my parochialism but yours as well, and his and hers in their turn.[12]

Minority interests would stand to suffer at the group level in a unitary ethnic state, while remaining intact at the individual level. The right to bear and care for a family would remain, and even the chance to benefit from minority status should programs be instituted to counter discrimination. However, in an ethnic state the group interests of the majority would take precedence over minority group interests in policy areas affecting relative numbers, such as immigration. It would be prudent to withdraw state support from organizations that mobilized or coordinated minority ethnic activism, while offering state sponsorship to majority activists. Thus one critical inequality would be in the provision of collective ethnic goods. The aim would be to reverse multiculturalism by structurally advancing majority citizens' ability to invest in their ethnic genetic interests, while structurally retarding minorities' ability to do so. Of course this is an invidious situation for minorities and something to avoid if possible. The problem is the unitary multi-ethnic state, which as argued above encourages ethnic competition. It is in all ethnies' interests to promote ethnic states where they are in the majority, but discourage them where they are in the minority and distantly related to the majority ethny.

Minorities have an advantage in ethnic competition in being more mobilized than majorities. Mobilization is the willingness to make sacrifices for a cause, for example by donating money, time and work. Even a small group with limited resources can exercise disproportionate influence when its members are highly mobilized and its opponents, though superior in numbers and resources, are indifferent. A plausible explanation is that a group's influence is the multiple of its resources and its mobilization.[13] A precondition for group mobilization is group identity, and the minority experience alone makes ethnic identity more salient than for the majority. This follows from asymmetries in the experiences of minorities and majorities, in which the majority forms a much larger part of the minority environment than vice versa.[14] For example, consider a minority constituting one percent of the population that is distributed throughout society. Then a minority individual would encounter a member of the majority in 99 out of every 100 interactions, while a majority individual would encounter only one minority individual in every 100 interactions. Social identity theory[15] predicts that even in the absence of discriminatory behaviour, this asymmetry will have a powerful reinforcing effect on minority identity, so long as the group identities are detectable, for example through different dress, language, accent, or physical appearance. Heightened awareness of group identities alone leads to some mobilization, for example in the positive evaluation of the ingroup and negative evaluation of outgroups. In fact ethnic discrimination is a pervasive element of all multi-ethnic societies. Even when that discrimination takes slight forms, such as hesitancy in expressing interpersonal warmth, the result can be markedly unpleasant for the minority. That is one reason that minorities often congregate in occupational and residential areas. But for majorities the problem can barely exist; a one percent minority that looks slightly different might be noticed but unless its behaviour is overtly objectionable it is unlikely to be perceived as a threat to daily comfort.

Since minorities are usually more mobilized than majorities, and keep this edge while they retain minority status, minority elites exercise disproportionate influence on state policy compared to their majority counterparts. Psychological experiments concur with Blalock's structural model by indicating that committed minorities exercise disproportionate influence over majority opinion, and often more so than in the reverse direction.[16] Minority influence is exercised unobtrusively, even in a milieu of overt discrimination against the minority. Latent minority influence is subconsciously incorporated into the majority's worldview. Therefore, without rapid assimilation even a low rate of immigration can produce disproportionate minority influence on ethnic policy. This is another reason for majorities who have succeeded in building an ethnic state to use the power of the state apparatus to: (1) keep minorities small by restricting immigration; (2) make assimilation a precondition for political, economic, or cultural engagement in society; (3) boost majority mobilization, for example through the education system and mass media; and (4) deny state support for minority efforts to mobilize their followers. The basic principle is to ensure the majority an ethnic group strategy in the form of the state, while denying this to minorities. This has been the approach of liberal nationalism in its emancipation of oppressed minorities, a well known example being the French revolutionary regime's attitude towards the Jews. Two months after revolutionaries stormed the Bastille in 1789, the Comte de Clermont-Tonnerre summed up the Enlightenment position thus: 'The Jews should be denied everything as a nation but granted everything as individuals.'[17]

One way around the swamp of the unitary multi-ethnic state is assimilation, in which gene pools become mixed through intermarriage to form a new homogeneous population. Intermarriage solves the problem of inter-ethnic competition, at least within the state. This can take as little as a few generations. It is in the majority's interest to maintain control of the state at all stages of the assimilatory process. Eventually the descendants of the immigrants become part of the majority and the process ends, with a new majority in control. The aim should be a smooth transition, one that avoids an abrupt loss of power by the original ethny. This should occur automatically in a democracy where no group maintains a group strategy outside of government. Should the majority ethny possess ethnic organizations or a tradition that prevents assimilation of newcomers, the society will become polarized and perhaps stratified by ethnicity. This is an old phenomenon, evident in a social elite in the United States and Mexico, for example, growing ethnically distant from the business elite.[18] Stratification can also occur if one or more minorities, but not the majority, maintains a group strategy outside the state apparatus, and the state becomes disinterested in ethnic affairs or partisan in favour of minorities. Whichever group manages to maintain a group strategy the longest will be greatly empowered in inter-ethnic competition.

Assimilation ends a minority's existence as a distinct, strategizing group, leaving its genetic interests to the vagaries of various selection pressures, including majority preference for majority characteristics. Racial minorities can take many generations to achieve panmixia, so for them this is not a short-term strategy.

Another approach is ethnic federalism. In an ethnic federation the central government would be limited to providing mutual defence and foreign policy, but ethnically homogeneous constituent states would retain control over immigration. This solution allows all groups to establish collective goods that facilitate defence of ethnic genetic interests. An economic disadvantage of ethnic federalism compared to a unitary state is that it constrains the free flow of labour according to market forces. Mass immigration of the last few decades has increased the number and the zero-sum character of conflicting interests. Thus federalism is not a simple or uniformly beneficial solution to the invidiousness of the unitary multi-ethnic state. But there is hope in humankind's ingenuity. For example, Eibl-Eibesfeldt has discussed ethnic federalism in the context of reproductive interests, while recognizing that no all-purpose solution exists.

> In multi-ethnic states, federalistic structures allow different ethnic groups self-government within certain bounds and in cooperation with the other groups sharing a superordinate interest on the basis of reciprocity. This can work as long as such a social contract implies that differential reproduction at the cost of the other is avoided . . . .
>
> Since the state is historically a very recent development, it is no wonder that man is still in the experimental stage regarding governmental forms. No one could provide a ready-made governmental recipe, but there are a number of guidelines available. Unless we remain receptive to new ideas and adaptations, we will face serious problems.[19]

An informed ethnic nationalism would be compatible with regional cooperation and international trade, for example in the form of the European Union. Even substantial population flows need not imperil ethnic genetic interests when the destination country is ethnically similar to the source country. Freedom of movement between closely related ethnies could reduce the maladaptiveness of ethnic altruism produced by kinship overlap. It would not challenge the doctrine of the efficiency of free markets, but would consider any economic cost of preventing large migrations between genetically distant populations to be money well spent. While homogeneity usually enhances economic performance and several other social goods in the long run (see pp. 196–199), some populations have been willing to trade material wealth for the ideal of living in a national community. Australians felt this way at the end of the nineteenth century, as expressed in 1901 by the new federation's first prime minister, Alfred Deakin. The prime minister was aware that many Americans had found African slavery to be profitable, but was also aware of the tragic consequences, including the War Between the States (1861–65). He was determined to put nation building before short-term economic profit:

> However limited we may be for a time by self imposed restrictions upon settlement—however much we may sacrifice in the way of immediate monetary gain—however much we may retard the development of the remote and tropical portions of our territory—those sacrifices for the future of Australia are little, and are, indeed, nothing when compared with the compensating freedom from the trials, sufferings and losses that nearly wrecked the Great Republic of the West [the USA].[20]

Ethnic states need not be more aggressive than present states, though staunch in matters of defence. In many instances they would provide the underlying conditions needed to demobilize the militarism and chauvinism expressed by some multicultural 'concept nations' such as the United States. Ethnic states have the advantage of allowing relatively individualistic behaviour to be adaptive, by muting ethnic interests as a factor in intra-state politics and economics. Perhaps this is why the industrial revolution was the product of relatively homogeneous nation states in the European tradition.

### *Liberal nationalism and collective liberty*

In principle the ethnic state, once established, is conducive to liberalism in its non-utopian political and economic forms. Adaptiveness is the functional underpinning of the 'state of nature', on which the founding liberal thinkers predicated the concept of natural rights. The freedom of speech, movement, and contract found in band societies was considered by philosophers such as John Milton and Marchamont Nedham in the seventeenth century to be gifts of God, but they can also be seen as natural in the anthropological sense, as conceived by Jean-Jacques Rousseau in the eighteenth century. The egalitarian ethos pervading band and pre-chieftain tribal societies[21] is adaptive because it prevents petty tyrants from increasing their rank and hence relative fitness at the expense of other group members. Democracy, socialism, and nationalism are modern ideological expressions of the equality, sharing, and solidarity universal to band societies. It is an idea worth exploring that in order for a modern ethnic state to fulfil the tribal promise of protecting the people's genetic interests, it must necessarily bear central features of the primordial polity, the egalitarian band.

J. S. Mill maintained that national communities are conducive to representative democracy. '[I]t is in general a necessary condition of free institutions, that the boundaries of governments should coincide in the main with those of nationalities.'[22] Mill observed that bringing political and ethnic borders into alignment can be complicated, for example by 'geographical hindrances'. Nevertheless, he approved the general principle as a liberal one. Even approximating ethnic homogeneity assist liberty. 'Free institutions are next to impossible in a country made up of different nationalities. Among a people without fellow-feeling, especially if they read and speak different languages, the united public opinion necessary to the working of representative government cannot exist.'[23] Mill thought that the final resort against despotism was the military, and that soldiers were most likely to protect their own people. 'Soldiers to whose feelings half or three-fourths of the subjects of the same government are foreigners, will have no more scruple in mowing them down, and no more desire to ask the reason why, than they would have in doing the same thing against declared enemies. An army composed of various nationalities has no other patriotism than devotion to the flag. Such armies have been the executioners of liberty through the whole duration of modern history.'[24] Philosopher David Miller, agrees with Mill.[25] The advantage of the nation state does not justify all demands for secession, he concedes. 'What it says is that national self-determination is a good thing, and that states and their constitutions should be arranged so that each nation is as far as possible able to secure its common future.' It is appropriate to search for a second-best solution rather than abandoning altogether the great advantages offered by the nation state.

The nation state advances liberty, a central liberal value. Modern liberalism has focused on securing citizens' freedom from the state Leviathan, but philosopher Quentin Skinner argues that a more important freedom is the collective liberty of society from external rule.[26] This is the classical view taken up by Machiavelli and seventeenth century English political thinkers. When a state is subordinated to an alien power, the people that it rules are effectively enslaved. Thus the liberal precept of self government is predicated on the autonomous state.

Political philosopher Margaret Canovan surveys the place of the nationhood concept in modern political theory.[27] She concludes that liberal theory, for example dealing with social justice, takes for granted the existence of a national community. The communitarianism on which redistributive and other altruistic policies are based would be strengthened in an ethnic state. Voluntary public altruism is needed if redistributory institutions are to operate with the minimum of coercion. Historian William H. McNeill argues that liberal theorists have not thought enough about the consequences of multiculturalism, ethnic stratification, and high levels of immigration. 'Polyethnic lamination—clustering different groups in particular occupations and arranging them in a more or less formal hierarchy of dignity and wealth—is again asserting itself in the Soviet Union as much as in France, Germany, Great Britain, and the United States. . . . Such social arrangements do not accord well with liberal theory.'[28]

The nation state also has economic advantages. Strong institutions reduce uncertainty of exchange in any society, thus lowering the cost of enforcing contracts, or 'transaction costs' in economic jargon.[29] Building collective goods including public institutions is easier in ethnically homogeneous societies, because these show higher levels of public altruism and cooperation,[30] less civil war,[31] greater democracy[32] (itself the most powerful counter to civil war[33]), less corruption,[34] higher productivity,[35] and accelerated social and economic capital formation as well as economic growth.[36] Furthermore, ethnically homogeneous societies suffer less economic damage from external shocks.[37]

Relatively homogeneous states might also make capitalism more evolutionarily stable. When competition is between individuals, and not between ethnies, individuals are behaving adaptively when they maximize (individual) utility as modelled by econometricians. But in multi-ethnic states individual economic rationality can be a losing strategy in the face of group competition. Collusive economic strategies can establish ethnic monopolies[38] that disadvantage individualistic actors, and make demands on government that lead to suboptimal economic decisions[39] including quotas, set asides, and subsidies to ethnic groups. Together these pressures segment the economy along ethnic lines that do not conform to optimal market processes.

Economist William Easterly finds that high levels of trust and cooperation are achievable in multicultural societies that have 'high quality' institutions. These are the institutions of the modern Western state: rule of law, bureaucratic quality, and freedom from government expropriation.[40] However, he also finds that ethnic diversity is negatively correlated with quality of institutions. In other words, once high quality institutions are in place, a multi-ethnic society can, in principle, run as efficiently as a homogeneous society; the problem is developing a sufficiently high level of cooperation to reform a country's institutions. Ethnically homogeneous societies are thus advantaged in having relatively high organic public trust, and in thus being able to build institutions that maintain that trust in the face of changing conditions. A remaining question is how resistant high quality institutions are to rising ethnic diversity.

Sadly Mill's view that national states are preconditions for free institutions is being confirmed by a trend in 'liberal' politics to suppress opponents of multiculturalism. Often at the urging of minority lobbyists, governments began in the latter part of the twentieth century to bring legal pressures selectively to bear on majority ethnic activists, for example legislation barring ethnically exclusive associations and living areas, and banning speech critical of other ethnies even when it does not qualify as incitement. The process is limited in the United States by the constitutional guarantee of free speech, but is relatively unrestrained in Europe. Liberalism of this variety deserves quotation marks, because it violates basic tenets of original liberal thought concerning tolerance and the importance of open discussion to the democratic process.[41] Many modern 'liberals' tolerate or actually support multicultural regimes that rely on large scale indoctrination practised through the schools and a cooperative mass media.[42] Dissident voices appear only sporadically, usually in local forums, and lack the integration at the national level that would make them a serious alternate regime. Integration of political education by schools, media, film and music studios, universities, and major foundations is an under-researched phenomenon, but it is plausible to suppose that this combination overwhelms isolated advocates. The process does not violate constitutional protection of free speech because it is not administered by the state; neither is anti-trust legislation used to break up ideological and ethnic monopolies.

The political risks of multicultural institutions should not be underestimated. A reliable system of indoctrination requires nearly total 'milieu control' in which the indoctrinatee has few or no alternate sources of information and values.[43] Indoctrination of the public thus necessitates the collaboration of cultural and political elites. Once such elite collaboration is in place and backed by the coercive power of the state, multicultural societies might be made to operate as smoothly as homogeneous societies lacking this apparatus. The danger—I suspect the certainty—is that this apparatus amounts to elite tyranny, an exploitative system of powerful social controls. This tyranny can afford to retain democratic processes because the social controls at its disposal allow little chance that an election will overthrow the political elite when it is allied with the cultural elite.

We thus arrive at another benefit of the nation state, which is democracy, or a closer approximation. Even if consensus in homogeneous societies were no better than in a multi-ethnic society, the former would enjoy the great advantage of coexisting with relative freedom—including the ability of individuals to strategize for family and ethny—while multi-ethnic consensus is difficult to sustain without totalitarian-like social controls on the pursuit of vital interests. The connection between nationalism and democracy is probably stronger still, based on L. Greenfeld's analysis of the emergence of the English nation:

> Democracy was born with the sense of nationality. The two are inherently linked and neither can be fully understood apart from this connection. Nationalism was the form in which democracy appeared in the world, contained in the idea of the nation as a butterfly in a cocoon.[44]

### *Free riding by co-ethnics: Tyranny and class conflict*

An evolutionary problem remains with ethnic states. A homogeneous nation state precludes ethnic free riding on collective goods, but homogeneity is no guarantee against free riding by co-ethnics. This is because individual differences of genetic interests still exist between individuals in ethnically homogenous populations, though they are less pronounced than in ethnically mixed societies. Intra-ethnic free riding should be less maladaptive than inter-ethnic free riding, because co-ethnics share more of their distinctive genes than do individuals from different ethnies. Free riding between kin is less harmful still, for the same reason. However, differences of genetic interest exist between kin and between members of the same ethny. Co-ethnic free riding still endangers the inclusive fitness of the majority of the ethny. At the minimum, genes and culture that produce group altruism will come under selection pressure, undermining group solidarity. Thus altruism directed towards the ethny is evolutionarily unstable when free riding is not controlled.

Liberal and Marxist critiques of state power can be read as cautionary tales about co-ethnic free riders. Liberal theorists are suspicious of overbearing state power, since it can switch from paternalism to exploitation. Many liberals adopt a 'night watchman' model of the state endowed with limited powers. The *laissez-faire* 'Manchester school' of liberalism goes further by advocating a minimalist state with powers sufficient only to protect fair market processes, which then maximize production of wealth by efficiently allocating resources according to supply and demand. The implausible assumption here is that Darwinian economic processes among millions of strangers are always self-regulating, provide safety nets for unfortunates, and protect individual actors from aggressive group strategies. A more sophisticated liberal view extends its suspicions to the market. It recognizes that the state can be a force for protecting the individual against invidious market and political processes. The further step advocated in this chapter is to extend protection to individuals against group strategies, especially ethnic ones, by giving the majority ethny or some constellation of ethnies a monopoly over the state apparatus.

The Marxist view of the state is largely naïve in a sense parallel to *laissez-faire* liberalism, since it asserts that mass anonymous societies will be self-regulating following the withering away of the state. More realistic is the Marxist view that the state apparatus can be in coalition with the capitalist class against the lower classes. The greater influence of elites on state policy is a source of risk to citizens who provide the tax revenue and public service that underpins state institutions. The risk is that these contributions are expropriated by selfish elites instead of being redistributed and thus becoming collective goods.[45] The economic role of the state is therefore a critical issue in evaluating a national group strategy, and is discussed at the end of this chapter.

The evolutionary biology of free riding in modern societies is poorly researched. It seems plausible that the two main types of non-ethnic free riders parallel those of minority ethnic free riders: life-style welfare recipients and elites. A democratic state will lose legitimacy if it fails to provide welfare by transferring resources from rich to poor. But inequality cannot be eliminated altogether if the great economic engine of competition is to be kept running. A modern economy will lose its competitive edge if it does not reward large economic contributions with substantial advantages in prestige and resources.

Free riding on welfare will not have long-term deleterious effects if recipients are randomly distributed across society. When it goes to the genuinely needy, welfare is a valuable collective good, acting as an insurance policy by spreading risk. As a safety net it facilitates risky undertakings, including the elaborate division of labour on which high civilizations depend. The risk comes from lineages that draw on welfare to increase their reproduction over generations. Likely effects include a weakening of society's work ethic, and an increase in any heritable characteristics contributing to welfare dependency. Welfare dependency as a lifestyle drains resources away from productive activities. Welfare free riding is likely to reduce productive citizens' relative fitness as well as everyone's ethnic genetic interests. However, volitional welfare dependency could be self-limiting. For example, the relatively generous welfare states of the 1970s in English speaking societies had by the 1990s been superseded by regimes granting lower benefits and tighter monitoring. There is considerable evidence that some of this contraction in welfare has been due to the reluctance of ethnic majorities to subsidize minorities over-represented on welfare rolls.[46] The possibility of co-ethnic free riding on welfare needs to be addressed by any social program that seeks to establish an evolutionarily stable social order.

Elites can free ride by increasing their own relative fitness. This has been the general pattern since societies began to become stratified.[47] While depressing lower class fitness, elite free riding would have increased the frequency of heritable traits contributing to elite status. In societies that allowed some class mobility based on merit, this would have tended to strengthen the ethny's work ethic and strategic intelligence. Thus in meritocratic societies elite free riding has the potential to preserve ethnic genetic interests while depressing the familial genetic interests of the lower classes. However, since the mid nineteenth century elites in developed economies have had lower fertility than other classes.

Elite free riding is possible despite small elite families. An elite can treat the state as a vehicle for personal fulfilment, as monarchs did when they engaged in wars of honour and religion. A leader with a small family who uses the nation as an instrument for self aggrandisement or moral crusades is as surely a free rider as the despot with a harem. Elites can also free ride as proxies for minority free riders, when they serve other groups' ethnic interests to the disadvantage of the home group. Majority elites disposed to free ride would arguably be vulnerable to the latent influence exercised by mobilized minority elites. In a world made small by global transport there is great scope for elites to transform their societies, for example by opening the floodgates to mass immigration.

In his 1941 book *The Managerial Revolution* James Burnham argued that a new type of elite free rider, constituting the 'managerial class', had emerged in industrial societies by the mid twentieth century. This class manages state and corporate bureaucracies but, in contrast to the entrepreneurs and great capitalist families of the eighteenth and nineteenth century, does not own large amounts of capital. Managerial power derives from a near monopoly of expertise as technicians, financiers, lawyers and bureaucrats. That expertise can arise from the personality and good luck of gifted individuals, but is most reliably produced by the university system, which is a necessary component and bastion of managerial power. This elite group has privileged access to the levers of cultural and political power and tends to develop self-serving ideologies and institutions that put its interests ahead of the national interest.[48] Institutions staffed by expert managers include the major corporations, the university system, public foundations, senior levels of the union movement, the mass media, banks and other financial bodies, and the decision-making strata in the entertainment industry.

By the 1980s the consensus among economists was that the managerial class would act in the interests of the corporations and other organizations they managed because remuneration was linked to corporate profits. However, a new image of this class as prone to free riding emerged from the series of scandals that afflicted the American corporate sector in the first years of the twenty-first century. It was in the United States that the managerial class had achieved most autonomy and influence over mass culture and government. In case after case it was collusion between elite managers, lawyers, and analysts both within and outside the corporations that enabled them to bypass controls and parasitize shareholders on a massive scale. The most spectacular case was the collapse of Enron Corporation, a tragedy for employees and small shareholders that led to revelations of criminal conspiracy by the products of America's most prestigious universities. Accounting scandals were revealed at other companies, and many supposedly independent analysts were found to have fraudulently advanced corporate interests at the expense of the public who relied on their advice. These criminal cases involved a minority of corporate executives and other experts. Legal exploitation on a much larger scale had been escalating since the 1980s in the form of remuneration of senior executives. Between 1988 and 1998 the average executive pay rose from 93 times that of production workers (already high compared to other industrial economies) to 419 times, usually in the form of share options.[49] With average income across the US in the vicinity of $50,000, average pay for chief executive officers of major US corporations rose from $2 million in 1990 to over $10 million in 1998.[50] In 2000, despite a slowing economy and large scale retrenchment of rank and file employees and middle managers, average CEO pay rose to a 'stupendous' $13.1 million. The highest paid CEOs and some other executives were paid hundreds of millions, mainly in stock options. In 2000 the 20 highest paid CEOs received an average 'compensation' of $112.6 million.[51] These trends reflect growing inequality between families. In 1980 the top five percent of families earned 14.6 percent of all incomes received by US citizens. By 1999 this share had grown to 20.3 percent.[52] Accumulated wealth was even more unevenly distributed. In 1979 the wealthiest one percent of US families held 22 percent of the country's assets. By 1989 this had risen to 39 percent.[53]

Exorbitant executive pay was excused by the doctrine that managers' interests should be aligned with those of the companies they serve, a doctrine that signally failed in the case of several corporations. If the managerial class free-rides on a corporate system that is protected by an elaborate system of internal controls, what hope can the public have that this class will not put its interest before theirs? To fully understand this phenomenon, one should perhaps add ethnic alienation to Burnham's analysis. It is worth noting that widespread elite corruption occurred in a post-national state where growing ethnic diversity and a semi-official ideology of multiculturalism had alienated elites from the majority culture. Since ethnic diversity tends to degrade public altruism,[54] it is plausible that multiculturalism will undermine paternalistic feelings towards the masses, loosening moral constraints on predatory motivation. An authoritative nation state with legitimacy based on interests that transcend profits might be better able to keep experts 'on tap' rather than 'on top'.

In evolutionary perspective class competition is complicated by reproductive interests at the familial level, in contrast to ethnic competition which involves group level interests. The challenge is to strike a *bio*social contract between the classes that balances individual reproductive opportunities with the need to conserve jointly held ethnic genetic interests. If a solution exists to this problem it will probably involve treating ethnic genetic interest as a collective good,[55] one that is managed by the state as an evolutionary group strategy.[56] Collective goods are already managed by states, such as group defence, education, and communication infrastructure. When an ethny's genetic interest is the collective good, one term of the social contract is to maintain that interest down the generations. For an interest to outlive us, it must be heritable, retained by a lineage. The ethnic state is a contract entered into by a people for their posterity. It simultaneously serves the interests of generations past, present, and future, protecting the full nature of society as conceived by Edmund Burke: 'Society is indeed a contract . . . it becomes a partnership not only between those who are living, but between those who are living, those who are dead, and those who are to be born.'[57]

Contemporary states already exercise some control over free riding on these collective goods through such means as law-based policing of public behaviour, universal taxation, and compulsory national service. One challenge is to maintain the transparency of both the strategy and its management by the state as a means of protecting against elite free riding of the form that manipulates state power. A related challenge is to keep the double-edged sword of ethnic nepotism selectively blunt. States can secure their borders without destabilizing the international order, but over-mobilization of national culture and institutions into aggressive militarism has proven to be destabilizing and thus a threat to everyone's interests. The challenge is to prevent the double-edged sword of ethnocentrism from cutting both ways. Partial remedies include such strategies as balance-of-power diplomacy, participation in international institutions, reducing national military forces through regional security arrangements, and fostering an elite culture that celebrates defensive patriotism and repudiates aggression. The challenge is to balance this agenda with the building of institutions that husband national solidarity and its multiple benefits.[58] A genuine universal nationalism would celebrate other people's independence and dignity as well as that of the home ethny. That would be sustainable humanism.

Since all nations have an interest in preserving ethnic interests, designing a benign nation state would seem the appropriate domain for a constitutional idealism in which abstracted values complement ethnic nepotism as a basis for legitimating social arrangements. A principled resolution of class and national conflicts could become a feature of a universal nationalism that respected all ethnic genetic interests. This would resemble what Canovan calls 'liberal nationalism'.[59] Working out the details of such a strategy would be a vital contribution to developing a sustainable political ecology to complement the efforts for an ecologically sustainable global economy.

### *Free riding by globalist elites*

*Globalisation* is a fact. It is the progressive integration at the global level of international markets and politics, with growing awareness of that process among those involved. Globalisation has been aptly characterized as the compression of space and time, and as the formation of a global village. In addition to rising flows of capital and goods, tourism and membership in international organizations increasingly engender a global overview of events.[60] What has not been pointed out is that globalisation intensifies genetic competition between populations, precisely because it shrinks the world; everyone gains potential access to everyone else's territory, making replacement migration feasible. In 1980 1.5 percent of the world population were migrants, a figure that had risen to 2.2 percent by 1995.[61]

*Globalism* is a doctrine or ideology that urges some form of globalisation on the world as good or inevitable or both. Globalism has neo-liberal and neo-Marxist versions (hereafter called Right and Left Globalism), in addition to other versions such as environmentalism and cross-border advocacy networks that do not necessarily fit this dichotomy. Both argue that the nation state is losing—and ought to lose—effective sovereignty as its policy options become ever more constrained by pressures to harmonize economic and social programs. Right globalists seek to subordinate non-economic values, including national interests, to the values of the international market. The final goal is an era of unprecedented and perpetual efficiency, profitability, and economic growth. Left globalists seek to subordinate non-universal values, including national interests, to international humanist values. In the Marxist variant, globalisation is seen as the logical culmination of capitalism, which will usher in an age of international socialism where exploitation vanishes and wealthy regions (nations no longer) will be taxed to bring poor regions up to the world standard. Right and Left globalism both encourage the 'denationalizing of national territory'.[62]

If globalisation did not provide material benefits, Right Globalism would have little impetus among the wealthy economies presently advancing it (chronically poor societies are actually unreliable supporters of the redistributionist ideals of Left Globalism). Globalism is a movement of colourless economic and intellectual elites. By the nature of their creed, such would-be leaders are attractive mainly to that small constituency for whom globalisation feels like an unambiguous triumph, and whose sense of security is preserved within gated communities, shielded from the international proletariat. Another constituency, at least in principle, is any minority that would benefit from the weakening of majority ethnic power caused by the nation state being subsumed within regional and global governments. However, the evidence for a significant minority impetus to globalism is poor compared to capitalist and leftist impetus. If the integration of markets did not offer real benefits, globalism would be a rich man's club entertained by intellectuals. But global trading routes have been developing for centuries because they do in fact help generate wealth. International accords, part of the armamentarium of globalism that includes free trade zones and regional defence pacts, not only facilitate trade but reduce the spectre of war that so marred the twentieth century.

Understanding globalisation's merits helps explain its growth. Lowering trade barriers, freeing up the movement of goods, capital, and labour, do indeed boost overall wealth by removing impediments to efficient market allocation of resources. From the perspective of the world economy as a whole, it is profitable (to the owners of capital) to move a labour-intensive factory from a high-income country to a low-income one. And if one's top priority is to reduce conflict, then establishing international governance is indeed a good idea. A large number of business people and ordinary citizens have a stake in these aspects of globalisation.

However, globalisation has costs, due partly to the inequality exacerbated by global capitalism, and partly to collateral damage inflicted on other interests. There are values other than increasing overall material wealth and avoiding conflict. Pope John Paul II introduced some related issues in a speech to the Pontifical Academy of Sciences in late April 2001, when he called for the adoption of an ethical standard able to safeguard diverse cultures from the homogenizing impact of globalisation.[63] 'Globalism must not be a new version of colonialism', he declared. The Pope expressed concern that globalisation left people helpless by destroying lifestyles and cultures. 'Globalization often risks destroying these carefully built-up structures, by exacting the adoption of new styles of working, living and organizing communities . . . The market imposes its way of thinking and acting and stamps its scale of values upon behavior. The Church will continue to work to ensure the winner in this process will be humanity as a whole, not just a wealthy elite that controls science, technology, communication and the planet's resources to the detriment of the vast majority of its people.'

The Pope's speech was consistent with concern about ethnic interests. Most local cultures belong to an ethnic group, or a cluster of related ethnies. Protecting cultural identity is tantamount to protecting ethnic identity, a critical first step in the process leading to defence of ethnic autonomy, continuity, and resources. International diversity of cultures roughly corresponds to another treasure of mankind, our ethnic genetic diversity.[64] The Pope's message concerned the cultural side of ethnicity while ignoring the genetic side. Combining the two would have brought the Pope's caution close to that of Eibl-Eibesfeldt who argues for the benefits of 'international multiculturalism'.[65] Eibl-Eibesfeldt warned against the 'grey uniformity' and social discord risked by intra-state multiculturalism, which pushes diverse ethnic groups together. The Pope could also have mentioned mass immigration, a major threat to the integrity of local cultures. The mass immigration and 'replacement immigration' urged or tolerated by advocates of globalisation is the surest way to obliterate local cultures as well as swamp their genetic interests. The home countries of the world's multinationals are also victims of this irreversible form of homogenisation, though the elites often live in gated neighbourhoods whose pleasantness is assured by immigration control in the form of private security guards.

The political risk to local cultures was not touched upon in the Pope's speech. If local cultures are to retain their integrity they cannot rely on the benevolence of external actors, including Left globalists. They need to be equipped with an effective group strategy for making their way in a turbulent world, namely the nation state. Possession of a state sovereign over the territory in which a people lives allows a people to control immigration, regulate business activity, protect vulnerable elements of society, and take other measures to defend local interests.

Retention of national independence is all the more important in light of the apparently inevitable inequalities that globalisation produces between nations. The wealth gap between rich and poor countries has grown since modern globalisation began at the start of the colonial era. There is an international hierarchy of wealth, with the industrial, capital-rich economies forming an exclusive core and the dependent, poor, underdeveloped economies forming the periphery. Changes to a country's status in the international hierarchy do occur, the most spectacular being the United States' rise from a peripheral colonial outpost in 1700 to membership of the industrial core by 1900. But such changes are infrequent and predictable. For example, Anglo Americans—ethnically, scientifically and technically an outgrowth of Britain—participated in the industrial revolution soon after their Old World cousins invented it. Usually economic dependency is self-reinforcing, as the core nations accumulate capital at the expense of the periphery.

The problem is minimal for the elites of underdeveloped societies, who as individuals can participate in the global capitalist economy. But the nation as a whole suffers from structural disadvantages. For peripheral nations the best chance is an enlightened group strategy led by those national leaders who have not been recruited by the global elite. Wallerstein proposes three national strategies for improving a country's position in the international economy.[66] Two of these strategies—'seizing the chance' and 'self-reliance'—involve a collective effort in which state elites lead their people out of the peripheral gloom and elbow their way closer to the warm glow of the international capitalist hearth, or build their own. The third strategy is 'development by invitation', in which the core economies find it in their interests to pass some productive tasks to outlying economies. For a nation at the periphery of the international system to squander solidarity by allowing ethnic diversity to increase, winding down its institutions of national mobilization, dismantling industrial policy, and releasing its elites from accountability, would be to forgo these national strategies and much of the chance to bring the nation into parity with the wealthy economies. International cooperation does not preclude national solidarity, but dissolving a nation would effectively abandon its people to the subservient role of waiting on others' interests.[67]

*Regionalism* is a stepping stone to globalism. Like globalisation, the aggregation of national economies into regional trading blocks such as the European Union can make good economic and diplomatic sense. All can benefit when a group of nations lowers its internal trade barriers and establishes common political and legal institutions able peacefully to resolve disputes. However, it can be dangerous to extend the process to full political integration, since this tends to blur national identities and national mobilization. Regional elites are liable to identify less with their home populations than are national elites. If so they will be attitudinally halfway to forming a global elite, drawn closer to class interests and away from ethnic ones. Also, a nation state's formal integration into a regional federation would necessarily result in the dismantling of the institutions that make it an autonomous society. As Romano Prodi, European Commission President, admitted in 2001: 'Do we realize that our nation states, taken individually, would find it far more difficult to assert their existence and their identity on the world stage?.'[68] At present the danger is clearest in Europe, where elements of the corporate elite are allied with the anti-national Left in pushing to homogenize the European market under a regional super state legitimized by constitutional patriotism.

When peace is assured, immigration controls can be safely dismantled, if the ethny's closest neighbours, or more precisely those ethnies able to enter and stay as immigrants, are closely related. Strong kinship overlap between ethnies creates a minority interest in lowered borders, and weakens the risk to the majority. This is true for all continents, as is demonstrated by the curves in Figure 7.1 showing the levelling of genetic diversity with increasing geographical distance within the six inhabited continents. The curves apply only to precolonial populations. This means that regions occupied by autochthonous populations are better able to maintain long-term security and genetic continuity after surrendering some sovereignty to a regional federation despite low national mobilization, than are colonial societies established far from the mother country. Despite some complications, large regions in Africa, Europe, the Middle East, South Asia, North East Asia, and South East Asia are better situated to relax their internal border and immigration controls than are the states of the New World in North America and Australasia, as well as Russians in Siberia and Chinese in Singapore.

Since ethnicity is a nested phenomenon, regional states and federations will also have collective ethnic interests. The advent of mass long-range transportation increases the risks of opening regional borders. Genetically distant migrants might gain entry to the whole region through one free-riding partner country. This risk is reduced if membership in the regional bloc is made conditional on controlling immigration according to regional policy. The institutional challenge for those seeking adaptive regional integration is to find a sustainable national political culture that lowers ethnic walls within the region but keeps them high at regional borders. One approach might be to demarcate regional blocs along the boundaries of distinct civilizations, which have an ethnic dimension.

This solution would bear some resemblance to the international constellation described by political scientist Samuel P. Huntington.[69] Although Huntington does not discuss genetic interests, he includes shared ancestry, tribe, ethny, and nation as identities that people value. In Huntington's view, after the Cold War the peoples of the world will realign along the lines of ancient civilizations. He identifies nine civilizations: (1) Western (Western Europe, the United States,

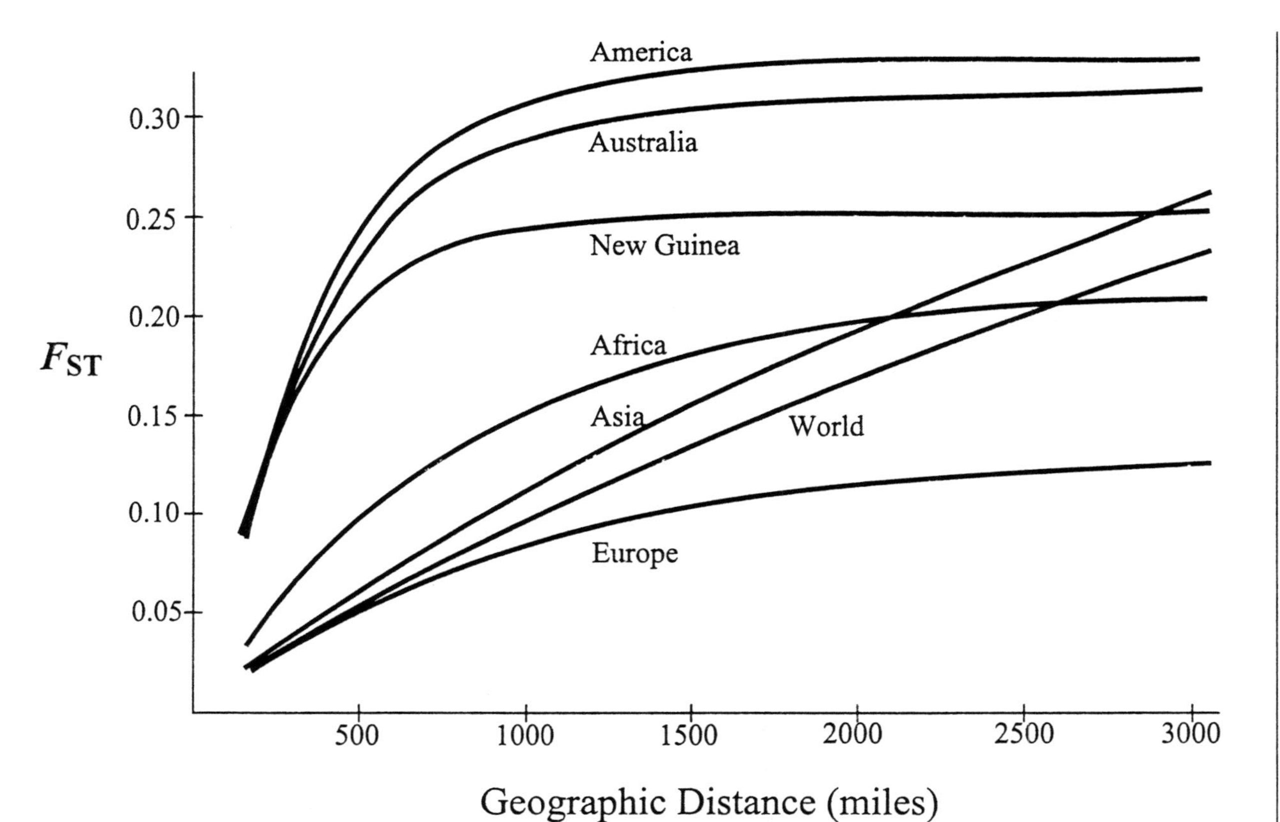


*Figure 7.1. The relationship between genetic distance and geographic distance within continents. Note that the curves are based on pre-colonial populations. Europe is the most homogeneous of all continents, reducing the fitness loss due to intra-regional immigration (from Cavalli-Sforza et al. 1994, p. 122).*

Canada, Australia and New Zealand); (2) Orthodox (Eastern Europe, though this might be a subset of the Western civilization); (3) Latin American; (4) Islamic; (5) African; (6) Sinic (China and related cultures); (7) Japanese (but this might be a subset of the Sinic); (8) Hindu (India); and (9) Buddhist (Tibet, Mongolia, plus Burma, Thailand, Laos, Cambodia, though this might be too nebulous). Huntington believes that geopolitical alignment will no longer be determined by ideology but by a return to something like nationalism, though at a higher level of aggregation. 'In the post-Cold War world, the most important distinctions among peoples are not ideological, political, or economic. They are cultural. Peoples and nations are attempting to answer the most basic question humans face: Who are we?'[70]

The answer offered by Huntington is consistent with recognition of ethnic genetic interests, though he also emphasizes cultural traits: 'People define themselves in terms of ancestry, religion, language, history, values, customs, and institutions. They identify with cultural groups: tribes, ethnic groups, religious communities, nations, and, at the broadest level, civilizations. People use politics not just to advance their interests but also do define their identity. We know who we are only when we know who we are not and often only when we know whom we are against.'[71] His analysis is conducted at the level of ethnic group markers and group sentiment rather than the genetic interests these might serve. Elsewhere Huntington advocates something similar to universal nationalism, by calling on Western leaders to cease intervening in the affairs of other civilizations; instead they should 'preserve and renew the unique qualities of western civilisation'. As for the United States, '[I]ts interests will be advanced if it . . . adopts an Atlanticist policy of close co-operation with Europe, one that will protect the interests and values of the precious and unique civilisation they share'.[72]

Protection of regional confederations and civilizations from mass immigration is a necessary condition for mutual ethnic continuity, but it is unlikely to be sufficient to achieve that end. The fact is that within every region, except Japan, there are already large racial minorities. The English, for example, are on track to become a minority in England by the end of the century.[73] An ethnically homogeneous country that lowers its immigration barriers to a regional neighbour has effectively decided to share its neighbour's ethnic destiny. In a world of burgeoning transcontinental migrations, regional population policy lies disproportionately in the hands of defectors, be they ethnic groups, individual states, or the centralized regional government. Dismantling border controls creates an added risk. Defector states are, in principle, freed of the costs of their immigration policies when migrants choose to move to greener pastures within the region. It would appear imprudent for any country to abdicate control over immigration policy except in the course of building an ethnic nation.

The above view of regionalism resembles parts of the conservative British critique of European political union that was maintained through the 1990s.[74] The most important difference is the theoretical underpinnings of the two analyses—ethnic kinship is not part of the Tory position, at least not in its formal guise. Nevertheless, the resemblance is striking at the functional levels of culture and political structure. The conservative argument against the building of a centralized European state is predicated on the same belief that underpins leftist and corporate support for that state: both believe that a United States of Europe would submerge constituent nations in the same way that the American states became progressively subordinated to Washington after 1788 despite constitutional safeguards to the contrary. This would prevent the defence of vital national interests, including local political, economic, and cultural autonomy. Since national autonomy preserves some scope for an ethnic group strategy, and since Britain is no longer an expansionist state, the conservative position converges to an extent with a policy of universal nationalism.

*Globalisation opens avenues for elite free riding.* When the elites that manage a multinational corporation assert corporate independence from the nation state, for example by moving manufacturing offshore or incorporating in another state, they are requisitioning the huge investment made by the nation in establishing and promoting the company. Large corporations often grow out of cooperation between state and business elites, in which the state represents corporate interests in managing labour relations, educating the labour force, building and maintaining the nation's infrastructure, subsidizing research and development, securing overseas markets, and guaranteeing domestic markets in the face of international competition.

Economic, cultural and genetic interests are interconnected. For example, employers stand to benefit from increased immigration of cheap labour and the removal of manufacturing to low-wage countries, both of which undermine the employment prospects of the native born. The availability of a vast pool of labour in poor countries, rendered cheap and docile by their governments' failure to bargain collectively, tends to reduce all wages to subsistence level. This was the insight of the eighteenth century English economist David Ricardo, expressed in his 'Iron Law of Labour'. When wages rise above that of the poorest country, manufacturers are tempted to move operations, or import more cheap labour. However, the law relies for its validity on poor societies being disunited. Globalism thus undermines some local economic and ethnic interests, making it adaptive for a large number of people to resist it. The appropriate target of the people's ire is the political elite, which is responsible for making concessions to globalism. These concessions amount to an abdication of the state's responsibility as the bearer of the nation's group strategy. Allowing mass immigration from genetically distant populations is the most egregious betrayal, because it confuses identity and dilutes ethnic kinship, thus reducing the people's ability to act as a solidary group. This reverses the state's role as defender of the people's ultimate interest to an enemy of that interest. The people lose their historical investment in the nation state, which is effectively hijacked for private purposes. Globalism can thus strip ethnies of their most powerful instrument for pursuing ethnic interests.

In a democracy voters can, in principle, replace treacherous and incompetent officeholders with new representatives. In reality the political mobilization and coordination needed to sweep a party from office requires leadership from alternate elites in the media, academe, professions, and business. Yet globalism tends to affect these elites. The centripetal force in business is towards transnational corporations, a few hundred of which dominate the global economy. The same is true of media and mass entertainment of films and music that influence audiences' understanding of world and domestic affairs. A successful business career often brings individuals into the orbit of a cosmopolitan corporate culture that can view national cultures as parochial. Prestige is associated with projects and capital of global scale. Similarly, academic performance is often judged by peer groups with an international outlook, when advancement becomes contingent on affirming globalist values.

*The puzzle of Left support for globalisation*[75]

Right Globalism is a straightforward matter of corporate selfishness, the drive by business leaders to find ever-larger markets and secure a stable trading environment. But it is remarkable that globalism is also advocated by elements of the intellectual Left. Both parties are visible, for example, in pressing for the dissolution of the European nation, and the erection of a concept-type superstate. Both support replacement migration, at least into Western societies.[76] From this twin Right-Left perspective the new regional state is not intended to be adaptive for ethnic families, but profitable for corporations and politically correct for neo-Puritans. Since the aim is not to do positive harm, things might turn out well enough, for a time. From the genetic standpoint, a political union between European countries has more chance of being adaptive than other regional experiments, because Europe is the most genetically homogeneous of all the continents (Figure 7.1). But the enterprise carries real risks.

The intellectual Left has become largely alienated from mankind as an evolved species. It is remarkable that in an age where biological science informs us of the genetic dimension of ethnicity and of the general principles underlying stable altruism, leftist intellectuals have no place in their doctrines for genetic interests. This is, after all, an interest shared by the mass of ordinary citizens, the recognition of which is liable to advance ecological sensitivity and constrain elite exploitation. For example, as discussed earlier in this chapter, the German philosopher Jürgen Habermas opposes traditional patriotism and argues instead for *Verfassungspatriotismus*, constitutional patriotism. Habermas fails to account for genetic interests in his criticism of 'quasi-natural people'. He echoes contemporary multicultural theory by arguing for a concept nation in which each state works out 'a distinctive interpretation of those constitutional principles that are equally embodied in other republican constitutions—such as popular sovereignty and human rights—in light of its own national history'.[77]

This view is accepted by Basam Tibi, a German citizen of Persian origin who studied with Max Horkheimer and Theodore Adorno.[78] Like Habermas, Tibi believes that Europe should be defined by a 'community of values', not of ethnicity or culture or religion, but one consisting of belief in the Enlightenment values of human rights, separation of religion and politics, democracy, civil society, and pluralism.[79] The only difference with multiculturalism is that these values should become the *Leitkultur* or leading culture of Europe, which immigrants are expected to embrace. In Tibi's view, Europe as a whole should adopt a cultural definition of national identity, and reject the ethnic definition. This is hardly a complete reading of Enlightenment ideas, which include the birth of modern nationalism, the democratic privileging of majority ethnicity, and the linking of minority emancipation to assimilation. The Enlightenment also celebrates empirical science including biology, which culminated in man's fuller understanding of himself as a part of nature.

Many 'greens', who present themselves as the most reliable defenders of the environment, are so denatured with regard to biological theory that 'adaptiveness' is a word absent from their manifestos. Even some scholars familiar with evolutionary biology, who understand that all life has genetic interests, do not extend this understanding to their fellow man. For example, social scientists Robert Cliquet and Kristiaan Thienpont, in an otherwise farsighted analysis of the sociobiology of ethnic conflict, conclude that 'in modern culture and a globalising world, the traditional in-group/out-group syndrome has become inadaptive', and that consequently efforts should be increased to 'counteract the innate drives towards nepotism, tribalism, patriotism, ethnocentrism, racism, xenophobia, etc.'.[80] It is right to point out the risks posed by tribal instincts in the modern world, but the benefits should also be recognized. Expressions of these group-preserving drives can be safely counteracted only if they are replaced by institutional protections that perform the equivalent adaptive functions.

Another scholar familiar with behavioural biology is the well-known ethologist Robert Hinde, who writes: 'As the interdependence of nations becomes more crucial, the pursuit of national self interest acquires a less positive and more negative value.'[81] Hinde seems to be confusing an interest with strategies for achieving it. The family analogy is again useful. Imagine that the elders of a family locked in a feud with neighbouring families realized that this was counterproductive, and that instead it would be adaptive to coalesce into a larger interest group. They make peace with their neighbours, and perhaps establish an accountable legal system to dispense justice instead of revenge. These measures might work, but one strategy surely not worth trying would be to dissolve the family, since the adaptive point of the exercise is to preserve everyone's familial interests. Means, but not vital interests, are expendable. In the international arena, if war is no longer adaptive then it is prudent for citizens to oppose militarism; it is not prudent to dissolve the nation as a strategizing unit if the nation has a genetic or other adaptive significance. It is a matter of education and politics to redirect patriotism from warfare to technology and trade.[82] But national interests remain.

The modern Left opposes ethnic and racial discrimination (by majorities) even when this would advance core Leftist values such as equality. Left social philosophers such as Michael Walzer and Brian Barry will not countenance protecting welfare rights by limiting diversity using racially discriminatory immigration policy, even though they suspect there is a causal connection.[83] Democracy and popular sovereignty are other core Left values subordinated to anti-racism. It seems beyond these thinkers' imagination to envisage their ideal polity leading, via democratic process, to defence of ethnic interests. The recent call by a leading leftist intellectual for the Left not to allow the neo-Darwinian revolution in the social sciences to pass it by, fails to discuss ethnic nepotism.[84] The anti-nationalist element in leftist ideology has trumped all others values. The phenomenon of a denatured political Left has emerged in Western societies among some urban university-educated professionals, sometimes called the 'new class'. Sociologist Katharine Betts observes that in Australia '[t]he left-leaning sections of the new class ostensibly care about economic protection, welfare and solidarity with the . . . poor but these values lack moral urgency to them beside the new crusades for racial equality and international human rights'.[85]

The priority accorded anti-racism among the cosmopolitan Left has taken on bizarre dimensions. For example, Italian semiotician and novelist Umberto Eco hopes for the swamping of Europe by African and other Third World peoples because this will demoralize racists![86] Eco admits that this moral purification will come at some cost, including endemic communal violence across the continent. But the turning of Europe into a 'coloured' continent cannot be stopped and therefore should be accepted. Eco tries to spice this view with a democratic flavour by likening racists, those who would resist the coming of multiracialism to

Europe, to the elite patrician class of Ancient Rome. Patricians were intolerant of foreigners and deserved to be forgotten by history, Eco thinks. They were superseded by the splendour of Roman proto-multiculturalism, where all subjects of the Empire became Roman citizens and mixed freely. Eco omits to mention that the patricians were the source of Roman democracy, that it was the patricians and their tribal vassals who painstakingly built the republic over centuries starting from a minor city state. The destruction of patrician power was linked to the reduction of the senate's prerogatives by god-emperors. Eco rejoices that one such emperor was African, evidence of imperial Roman virtue. Is this now a hallmark of the multicultural Left, that it finds tyranny acceptable so long as it presides over the destruction of ethnic homogeneity?

The analogy between racists and patricians is further misconceived from the democratic perspective because elites are bound to be the last to suffer personally should they permit their people to be replaced. In reality it is the unskilled workers who are the first to suffer from unrestricted immigration, and it is their 'racism' that is crushed by multicultural regimes. If Habermas, Tibi, Walzer, Barry, and Eco are any guide, the Left now sacrifices democracy, once its core value, when a people would act collectively to protect its vital interests. Whose interest does this new leftism serve?

The Left, as it has evolved over the course of the previous century, looks down on the ordinary people with their inarticulate parochialisms as if they were members of another species. Perhaps it is the result of 'pseudospeciation' described by Erikson.[87] These leftist elites behave like visitors from another planet, Olympian leaders possessed by a cosmic perspective who have no attachment to the fate of any particular ethny. From their point of view investing in ethnic continuity detracts from other values without discernible benefit, since they care nothing for the preservation of national communities. Ethnies are considered irrelevant to the welfare of people in general. It would be understandable for Martians to be so detached from particular loyalties. But it is disturbing to see humans doing so, especially humans who identify with the Left. So far has the anti-national Left drifted from its democratic rationalist roots, that it no longer accepts the legitimacy of democratic support for nationalist politicians.[88] This intolerance of popular will could be legally enforced under the European constitution if the radical 'European Union Charter of Fundamental Rights' is made law.

The Charter of Fundamental Rights, approved in Nice on 8 December 2000, would outlaw any national legislation that attempted to control immigration with respect to ethnicity and a long list of other social categories. Article 21 reads: 'Any discrimination based on any ground such as sex, race, colour, ethnic or social origin, genetic features, language, religion or belief, political or any other opinion, membership of a national minority, property, birth, disability, age or sexual orientation, shall be prohibited.'[89] This provision would, if made law, expose Europe's peoples to genetic replacement in their own countries. A biologically informed constitution that aspired to be democratic would set the population's ultimate interests in concrete. Such a constitution would outlaw ethnic discrimination between citizens in the public realm, but guarantee ethnic discrimination in choosing immigrants, and convey the presumption that such discrimination serves the vital interests of all ethnies. The European Union Charter of Fundamental Rights is not the sort of product one would expect from an informed and compassionate movement of the Left. Rather, it resembles an inexplicable gift to Right Globalism, an instrument of surrender to be imposed on peoples who have been temporarily defeated and are about to be permanently dispossessed.

### *The ethnic state as the people's champion*

Despite the rise in influence of international corporations and institutions that owe allegiance to no nation, states still possess considerable authority. A shrinking world might actually cause national authority to rise.[90] National legitimacy based on emulation of the tribal group strategy gives the nation state the collective power of mass altruism that cannot be matched by profit-making enterprises. Business people take commercial risks calculated to increase wealth, but only families and nations and their simulacra can lead people freely to accept risks to life and limb. Heroism and solidarity are not optimal business practices, a motivation deficit that disadvantages those who would translate financial into political power. The nation state has a surfeit of altruistic capital that authorizes it to improve the operation of the internal economy and create competitive advantage of citizens *vis-á-vis* the outside world. The ability of a nation state to thrive within the international market depends on the ability of the government to induce cooperative relations between labour and capital and to provide the collective goods needed by citizens to compete internationally.

There is room for nation states to help their people benefit from the efficiencies of global markets while jealously safeguarding national identity and genetic interests. The ethnic state, especially when it is protected by an explicitly ethnic constitution that has the ethnic prerogatives of the majority at its core, would be a powerful vehicle with which to fulfil this mission. It would retain national cohesion by limiting immigration to the readily assimilable.

Controlling elite free riding in a global economy is a challenge shared by all nations. As argued above in this chapter, the problem is not inequality caused by differences in productive effort. This must be allowed if creative talent and initiative are to be rewarded. The problem is the narrower one of avoiding elite parasitism on the average citizen. This occurs if a state, supported by the taxes and altruistic service of a broad base of citizens, comes to promote sectional interests in ways that amount to a net transfer of resources from the latter. One solution is redistributive taxation, which causes all citizens to benefit from international exchange. That a substantial tax burden need not discourage productive activity is shown by the states of Western Europe, many of which have comparable average incomes to that of the United States, but unlike America also support generous welfare states and foreign aid programs. These societies are ethnically homogeneous compared to the United States, or have devolved many governmental functions to ethnically defined regions as in Belgium. They also benefit from the efficiencies of ethnic homogeneity, for example in education and law enforcement, combined with greater cooperation between labour organizations and corporations.

A steep progressive income tax might minimize the risk of elite free riding, but it can also drive many of the most creative citizens and their firms to relocate overseas. From the point of view of business elites that have recourse to such methods, paying taxes is optional. Paying taxes then becomes either a matter of profitable business practice (e.g. by earning the good will of a friendly state) or of genuine philanthropy—altruism. Altruism in general is more adaptive when directed towards a collective good that selectively benefits one's ethny, such as an ethnic state, than towards public goods. However, voluntary taxes are liable to be modest in scale compared to coerced taxes and thus allow free riding by national elites and undermine the national group strategy. It is therefore prudent for nation states to cooperate to curtail tax-free havens, minor states that have seized on a method for exploiting the taxpayers of the most productive national economies. Those taxpayers provide the education and early opportunities for entrepreneurs and help nurture their nascent businesses, only to have the profits of those businesses funnelled out of the country without significant redistribution.

The most destructive form of free riding by business elites is promotion of mass immigration of low-wage workers. If this cannot be directly blocked, a fall-back strategy is to make the businesses that profit from this immigration pay all of the costs associated with them. By virtue of its large scale, mass immigration is made up of poorly educated manual workers rather than professionals and skilled workers. As such it typically exerts a net cost for native born taxpayers due to extra investment in language training, welfare, housing and basic infrastructure such as sewage and water, and education for their children.[91] When the businesses that lobby for liberal immigration laws do not pay these costs, ordinary taxpayers must, while simultaneously entering into wage competition with the newcomers. The behind-the-scenes dealings that drive the process can be sordid, as when a businessman donates to a politician's election fund in return for a vote in favour of liberalizing immigration. The businessman is, in effect, bribing the politician to give him taxpayers' money and give foreigners the jobs of the native-born. Ending this free riding would shift the economic burden of immigration to those who profit from it.

International organizations such as the United Nations are potentially dangerous, because they can be used as vehicles for elite free riding. The greatest potential danger is global government, which could enforce extreme humanism by redistributing wealth, territory, and populations. It is too optimistic to assume that popular will alone is sufficient to prevent the consolidation of political and economic power in a global government, that '[p]eoples will not switch their loyalties from the nation-states with which they identify to international organizations they cannot control . . .'.[92] The UN and other global and regional bodies are valuable as public goods for promoting peaceful relations. If they did not exist nations would be advised to invent them. For example, the establishment in 2002 of the UN International Criminal Court, a permanent war crimes tribunal, promises to deter and punish the perpetrators of atrocities. This builds on the regulatory thrust of the Charter of the United Nations, a mainstay of international law that respects the sovereignty of states, explicitly forbidding the 'threat or use of force against the territorial integrity or political independence of any state'. Ethnic groups have no obligation to states that threaten their vital interests, and constitutional change and secession can be expected to remain features of the states system. However, all states, whether new or old, homogeneous or heterogeneous, have an interest in supporting the United Nations only while it remains a bulwark of the state system formalized by the Treaty of Westphalia of 1648. This treaty ended the anarchical savagery of the Eighty Years' War between Spain and Holland, and the German phase of the Thirty Years' War. By solidifying state borders, the Westphalian System hinders the development of ethnic nations. But the same system is a necessary condition for the continuation of ethnic nations, once established.

By the same token, any international body should be opposed that would undermine the ability of nation states to strategize, within reasonable limits, to defend their peoples' interests. Ethnic interests in general would be undermined if the United Nations or any other agency were able to enforce extreme humanism, demanding not only civilized conduct but redistribution of resources. A world government would give elite power maximum leverage and profit. It would consolidate and expand the trend towards a global elite, recruited from wealthy and talented individuals from across the globe.[93] This elite is at present a category rather than a cohesive group, united by status and economic interests but also riven by conflicting interests. Communication between national elites has for centuries been valuable as an informal channel of diplomacy and cultural exchange. But now there are social and economic selection pressures for cooperation between elite individuals. The institutional environments of bodies such as the UN reward cosmopolitan, not national, loyalties. And coalitions drawn from elite ranks can attempt to circumvent the national group strategies that constrain their business and ideological projects. For example, a meeting of chief executives will canvas ways to separate their corporations from loyalty to the nation states that nurtured them, concluding that 'only nationalism can stop us'.[94] A world government could lead to the institutionalisation of these *ad hoc* coalitions into a permanent strategy.

The UN has demonstrated its fallibility and corruptibility time and again. To call the UN policy on refugees fallible would be overly diplomatic. The policy is not only absurd but also damaging to those few countries that accept refugees. The 1951 UN Convention Relating to the Status of Refugees defined a refugee in the context of post-Second World War Europe, a definition extended worldwide by the 1967 protocol.[95] The Convention is administered in such a way as to legitimise illegal immigration. It was designed to handle relatively small numbers of refugees, yet in the year 2000 the UN estimated the world refugee population to exceed 22 million. Taken literally, the definition of who is a refugee applies to hundreds of millions of people in war-torn and repressive parts of the world.[96] Yet UN officials continue to condemn nations that seek to defend their borders against the mix of genuine refugees and illegal economic migrants.[97] When numbers are large it is impracticable to give full individual treatment to every unlawful arrival, including the right to appeal court decisions at the highest level funded by the host nation's taxpayers. Receiving countries thus automatically run foul of UN inspectors if they succeed in defending their borders. In effect, only national capitulation satisfies the UN when a nation is confronted by large numbers of refugees. The Convention is especially perilous for liberal democratic societies that honour their international commitments, because such societies attract economic and political migrants from the distant corners of the earth.

While insisting on idealistic treatment of refugees by the West, the UN's international membership ensures that it is far from incorruptible. It is the level of government most removed from the peoples of the world and therefore least sensitive to their aspirations. Brutal dictatorships are represented in the General Assembly, participate in debate, condemn the 'racism' of democratic societies, help staff UN agencies, and otherwise contribute to 'world opinion'. One notable result was the UN's inaction in the face of the mass murders committed by the Kmer Rouge in the 1970s, continuing to recognize this regime as Cambodia's legitimate UN representatives until 1991, despite widespread knowledge of the atrocities.[98] Yet one voice or another will usually have an interest in praising the UN as a moral force. The same can be said of the European Union, another supra-national body the chief administrators of which are far removed from popular mandate and, perhaps, sense of community. In March 1999 the European Commission, the EU's appointed executive body, resigned *en masse* following the release of a report that found some individual members responsible for fraud, mismanagement and nepotism and the Commission as a whole to have lost political control of the use of EU funds and the appointment of staff.[99]

Nations should not engage with the United Nations because of its supposed nobility, but as a means of urging it towards the practical purpose of peacefully resolving and preventing violent conflict between independent nation states. Membership should be sought when it is more advantageous being inside than outside the organization. As one advocate of enlightened nationalism has argued, '[I]nternationalism should not be viewed, like charity, as a badge of good intentions. Nor is it, like empathy, an absolute good in itself. It is simply a method to advance the interests of people organized into national societies under particular circumstances'.[100] Participation in politics is necessary because it is a substitute for war. But it should never be forgotten that politics, before all else, is the effort to accumulate and exercise power. The most stable political solutions are just, but power and morality are in perpetual tension. Statesmanship is the rare synthesis of ethics and *Realpolitik*, but most politics is less than honourable and truthful. Thus the UN is safest—though never safe—as an association of nations representing the group strategies of their peoples in the framework of the Westphalian tradition. It should adjudicate relations between states, not individuals. That might lead to world citizenship and the establishment of a global elite that owed loyalty to no ethny or to a small subset of ethnies. Direct rule by a world government would risk allowing a global majority to override the prerogatives of nations, opening them to more overt domination by big business and dissolution through mass immigration from the larger ethnies.

It would be short sighted for a nation enjoying temporary dominance (even when lasting for centuries), to bring about a world order that overrode states' rights, since power does not last. Foreign policy should anticipate future vulnerability.

Global government is not the only danger. Signatories to international treaties risk having elite free riders turn their states' foreign affairs powers into instruments for breaking down immigration barriers. The United Nations' unrealistic definition of refugees discussed above makes this a real possibility. An ethnic constitution would expressly override foreign affairs powers in matters of immigration and national autonomy. And an enlightened ethnic state would be active in advancing the globalisation of universal human interests, for example by promoting independence of other nations and encouraging the reciprocal defence of ethnic interests. It would work to strengthen the UN as a guarantor of national independence and a vehicle for promoting adaptive humanism while subverting attempts to shape it into an instrument for globalism.

The world population will gain from globalisation of markets and from international institutions that minimize conflict, so long as free riders and maladaptive ideologies are prevented from exploiting the process. Since the majority stand to lose from the latter, there is general interest in keeping the nation state alive as a basis of authority that stands apart from the global elite and is able to mobilize local populations in defence of their vital interests. There is no guarantee that the majority interest will prevail without perpetual vigilance.

The foregoing discussion has raised some important issues that require further treatment.

### *Refining the Ethnic State*

It stands to reason that it would be prudent for a population to defend its most precious collective interest—distinctive genes carried by the ethny—with the most powerful means at its disposal. In this chapter I have asserted that the state is that instrument, a proposition I shall presently defend. In the following sections I conclude that the modern state is not only a powerful instrument but derives much of its power from mimicking the primordial ethnic group strategy. However, no state yet developed has reliably kept its promise as an adaptive ethnic group strategy. Sooner or later states have become maladaptive for their citizens in one or more of the ways discussed earlier in this chapter. Since no state fulfils the promise of an ethnic group strategy, all states so far conceived are deceptive to some extent. States have been both an asset and a threat to ethnic interests. An ethnic constitution would correct some of the weaknesses in the traditional nation state. However, even an ethnic state cannot guarantee the founding ethny perpetual control over the administrative apparatus or the territory it controls. That insufficiency implies the need for adaptive changes in political culture, and raises the issue of whether it is prudent to include state symbols and rituals in ethnic identity. I devote the remainder of this chapter to these issues.

*State power, legitimacy, and nationalism*

Since its inception the state has been, and remains, the most powerful actor within and between societies. The specialization and social stratification necessary for state formation requires large societies that only emerged following the Neolithic Revolution. Archaeologists believe that spontaneous state formation—the pristine state—occurred rarely. The innovation then spread rapidly as neighbouring societies scrambled to defend themselves against the state's unprecedentedly powerful military forces. Either they emulated state structures or were conquered and incorporated into the new system. As advantageous modifications were made to the state, these also spread for the same reason.

A state's power derives partly from its constituent institutions. First of these is the means of violence and coercion. The difference between the state and earlier forms of rule is that these means are possessed by institutions that last longer than the tenure of a single ruler. The ruler or ruling group appoints loyal officials to head these institutions, but the tenure of most personnel is independent of changes in leadership. Institutional longevity allows for the accumulation, in traditions, of expertise in violence and coercion and other administrative methods. These traditions are reproduced within the institution despite changes of personnel and increases in size. The result is superiority in training and tactics over non-state forces. The British captured and controlled a disorganized Indian subcontinent with a few regiments. The Roman Empire, starting from the Roman city state, showed a remarkable ability to scale up its republican institutions until it dominated the entire Mediterranean basin and beyond.

The second source of state power is drawn from its control of a bounded territory whose population provides the taxes needed to maintain the state apparatus. Specialized military and administrative institutions allow a centralized authority to control a territory orders of magnitude larger than tribal territories, even with primitive means of communication. The great ancient empires maintained states spanning several thousands of kilometres using foot and horse-drawn transport. Thirdly, the state monopolizes law-making within its territory. In most societies the state is considered legitimate by the population due to religious or ideological mandate, so that the state exercises sole political authority, or legitimate power, exercised through the promulgation of laws.[101] A law is a standing order to do or not do something. When backed by the authority of the state, a standing order is far more efficient than a face-to-face command due to its territorial application and longevity, derived from the corresponding characteristics of the state. Comparing states with equivalent communication and policing techniques, the state whose rulers enjoy greater legitimacy will rule more efficiently. Gains to efficiency in coordinating mass activity yield greater state power.[102]

Improvements in one or more of these three bases have increased the power of the state, as indicated by its ability to control its own population and conquer others. The rationalization and professionalisation of bureaucracies, initially in ancient China and much later in France and Prussia in the eighteenth and nineteenth centuries, is one such development. Another is the efficiency of territorial control due to improvements in transportation and communication. Finally, techniques for legitimating state power have undergone profound developments. These changes are of special interest when comparing ethnic policies, because ethnicity is a powerful source of legitimacy.

Rule by brute force is widely considered a moral failure. It is also a practical failure, because to maintain their rule despots must interfere with creative social processes. The efficiencies of the market place cannot be realized if there is a ban on meeting in public places, yet allowing the latter can lead to networks and organizations able to counter the state. Technical and scientific innovation requires a market place of ideas that can also disseminate political critiques. Despotisms are even failures in the military realm. Since the time of Alexander the Great highly motivated armies led by inspirational men have prevailed by dint of initiative and altruistic sacrifice over much larger armies commanded from the rear by despots.[103]

The power gained through legitimacy has led to refinements in institutional techniques for winning hearts and minds. The earliest technique was organized religion. Evidence that religions can be treated instrumentally by leaders comes from the horticultural economies of Highlands New Guinea. These are 'big man' societies in which leaders lack command authority and must rely on persuasion to achieve group goals. At least since the introduction of the staple foods of the pig and yam several hundred years ago, tribal big men have bought and sold rituals as means of producing fiercer warriors or encouraging productive work.[104] Unlike big men, state rulers possess the authority to command, an authority that remained dependent on religion until the modern Western state separated church and state. In many early states the ruler simultaneously exercised political and religious authority, as in ancient Egypt. The Roman Catholic Church was dominant in Medieval Europe, to the extent that kings could be disciplined through excommunication. Loss of religious mandate made it more difficult to command obedience, and efficiency of rule could fall so far that a king could also lose his head to a rival who possessed the Pope's blessing.

Nationalism is the most powerful legitimising force in the modern era. It is a modern cultural and political movement with origins in democratisation and more generally in the Enlightenment. Overtly nationalist ideas date from as recently as the late eighteenth century, in the intellectual climate preceding and accompanying the American and French revolutions. It is compatible with a wide range of modern economic systems, legitimating state power in capitalist and socialist societies alike. It appears to have universal appeal, taking hold once a society has developed mass education, a mass circulation vernacular press, job-market education, the industrial mode of production and consumption, and mass internal transportation.[105] The resulting enlargement of standardized linguistic and cultural units, with unprecedented rapidity of information flow, prepared the way for nationalist elites to develop mass constituencies. The emergence of irresistible nationalist mass movements began with the activities of cultural activists such as the brothers Grimm. The resulting homogenized national memories were disseminated via the new public education system to emerging literate elites, who then led the final political phase of democratic nationalism.[106] These emerging national elites' ethnic concerns were given practical impetus by economic and status concerns. Local elites in business, the professions and culture, found nationalism to be a powerful common strategy that could simultaneously push governments to build tariff walls and induce cultural-national brand loyalty among the consuming public.[107] The nation state developed in a system of contending armed polities locked in shifting balances of power. The nation as group strategy is implicit in the classical theory of liberty found in the ancient Roman legal tradition, as well as in the 16th century Florentine political theorist Machiavelli, the early modern English neo-roman thinkers, the German realist theory of the state, and most recently in the political philosophy of Quentin Skinner.[108] In this tradition, individual freedom is conditional on the freedom of the nation. The state is thus an actor able to represent the interests of its citizens, primarily by protecting their collective autonomy, in the anarchic international environment as well as in the orderly world of alliances and global markets. The economic and cultural developments underlying nationalism disrupted the international power balance. Mobilized nations afforded military establishments unprecedented resources in material and manpower as the new form of legitimation unleashed tsunamis of patriotic altruism. The confluence of economic and sentimental interests tended to vertically integrate the classes in the nation building enterprise.

The widening franchise in the 19th and 20th centuries was preceded and perhaps accelerated by the dissemination of ideas about man's fundamental interests. A crowning example is the Scottish Enlightenment of the 18th century, in which Lord Kames and his student Adam Smith developed a view of man as guided by self interest.[109] Another Scottish philosopher from this era was David Hume, who argued for the primacy of empirical knowledge. Although not yet joined with Malthus's, Darwin's or Mendel's discoveries, the conception of humans as self-interested creatures driven by material needs influenced economic and political thought in a direction compatible with incorporating genetic interests. Six members of the American Continental Congress and a future president, James Madison, were influenced by Humean thought during their Princeton days. That influence is apparent in the sixth of the Federalist Papers in which Madison argued that individual liberty can only be secured by balancing the power of different interests, including those of the federal and state governments, the executive, the legislature, as well as economic interests.[110] Any theory of interests is preadapted, as it were, to incorporate genetic interests. Perhaps the balancing of group genetic interests, as recommended in this chapter, is necessary for preserving individual interests, including liberty.

A final example of state legitimacy being founded on the implicit promise of an ethnic group strategy is what followed in the 19th century, as the rise of mass democracy led to the politics of economic interests. Bismarck, one of the great advocates of universal nationalism, acknowledged the fairness and practicality of interest-based politics in breaking with the National Liberals in 1870s Germany:

> Political parties and groups based on high policy and political programmes are finished. The parties will be compelled to concern themselves with economic questions and to follow a policy of interests. . . . They will melt like ice and snow. Voters with the same interests will co-operate and will prefer to be represented by people of their own instead of believing that the best orators are also the most skilful and most loyal representatives of their interests.[111]

Bismarck was referring to economic interests, but the same principle applies to any shared value, so long as that value is recognized and incites partisanship. Nationalism was one such democratic movement of shared interests, able to motivate cooperation and the demand for leaders loyal to the national interest.

Nationalism thus had, from its inception, the content of a group strategy. Its material preconditions in mass education, mass vernacular press, and regional transport system, were collective goods, which it could protect and advance by virtue of its power to produce mass solidarity. These processes would have been ineffective without behavioural universals evolved within the tribal milieu. Emerging public education systems exploited the universal propensity for indoctrinability to ethnic identity that is pronounced in childhood and adolescence.[112] And all age groups are influenced by social identity processes, for example by reacting patriotically to attacks on the group.[113]

Moreover, nationalism is in spirit an *ethnic* group strategy, whose emotional force derives from kinship and tribal markers. All nationalisms identify with a delineated territory either occupied or once occupied by the group. The nation state justifies its existence primarily with the promise of territorial defence. All the emerging nationalisms had origins in premodern nations and thus in tribal history and identity.[114] The kinship dimension of nationality is implicit and explicit. Much on the implicit side is fictive. Modern transport creates wider circles of familiarity than ever before, reinforced by the homogenizing effect of a single market and vernacular press which, combined with mass education, enculturate the population to a common dialect and shared memories. The resulting phenotypic similarity mimics tribal homogeneity. (Implicit fictive kinship can also be a factor in multicultural societies, but only to the extent that the multicultural ethos is compromised, for example, by enculturation to a common language, dress, and habits.) Nationalism's initial folkloric expression distilled family tales of ancestors and kindred peoples. The political rhetoric of national identity and mobilization is rich in kinship metaphors such as the founding fathers, the motherland, brothers-in-arms, and fraternity.[115] Moreover, the institutions of a nation state take the semblance of 'a kind of extended family inheritance, although the kinship ties in question are highly metaphorical; . . . it gives to cold institutional structures an aura of warm, intimate togetherness. . . . The existence of a nation makes it possible for the state that governs, coerces and taxes to do so in the name of the same collective "people" who have to put up with these ministrations'.[116] Canovan[117] attributes her insight to Edmund Burke's description of the English nation in comparing it with the French revolutionary state:

> In this choice of inheritance we have given to our frame of polity the image of a relation in blood; binding up the constitution of our country with our dearest domestic ties; adopting our fundamental laws in the bosom of our family affections; keeping inseparable, and cherishing with the warmth of all their combined and mutually reflected charities, our state, our hearthe, our sepulchres, and our altars.[118]

Thus the nation state as it originated in Europe in the nineteenth century and spread to the rest of the world, marshalled the instinctive elements that had originally evolved to make the individual a ready participant in family and tribal group strategies. Recall that an ethnic group strategy is a cooperative group effort among members of the same ethny to defend themselves from or otherwise compete with members of other ethnies. A tribal group strategy is the same thing, only set in the tribal milieu.

By virtue of the behavioural systems it manipulates, the nation state is the implicit promise of an ethnic group strategy, even though it imperfectly performs that function. The nation state is a psychological substitute for the primordial band and tribe. Put in evolutionary terms, nationalism succeeds in galvanizing mass altruism towards the state because it emulates key elements of the tribal group strategy. All those patriotically-motivated citizen soldiers who gave their lives in defence of the nation did so in the spirit of altruism to the tribe, not as a gift to humanity or in fulfilment of contractual obligations.

Historically the nation state developed in the direction of becoming a majority ethnic group strategy, at least until the mid twentieth century. It did so because elites could more effectively mobilize military and economic collective goods the closer the state resembled a tribal strategy. Global *laissez-faire* capitalism was perhaps more effective at maximizing the growth of the international economy as a whole; at least that is what formal econometric models predict. But nationalism was the most efficient means for harvesting the public altruism of a population, whether for the purpose of defence, empire building, economic protection, or legitimising socio-economic hierarchy. Nationalism was and remains a powerful political force partly because the social technologies most efficient in mobilizing mass anonymous societies are constrained by the evolved human behavioural repertoire to mimic kin and tribe.[119] It also mobilized elites. Sincere patriotic leaders were among those who supported graduated income taxes and sent their sons to the battlefields in company with the sons of the lower classes.

For these reasons, at the mid point of the twentieth century the typical Western nation state approximated in outline an ethnic group strategy. The state could be said to administer what approximated an ethny (or closely related set of ethnies), because European nation building culminating in the nation state system of the late nineteenth century worked by deploying kinship markers that did in fact correlate to some extent with ethnicity, markers of territory, language, culture and religion. The nation state partly satisfies all of the six criteria of an adaptive group strategy set out in the previous chapter. Its chief advantages in this regard are as follows. (1) A majority of the population it administers is drawn from one ethny or closely related ethnies, providing some confidence of ethnic relatedness and reducing the impact of ethnic free riders. (2) It exercises sovereignty over a territory. (3) It wields unprecedented power to defend borders from unwanted immigration, violent or peaceful. (4) Finally, its administrative apparatus is backed by a monopoly of legitimate coercion that allows the provision of significant collective goods partially proofed against free riders.

### *Shortcomings of the traditional nation state*

Despite these strengths, no state yet developed has for long kept its promise as an adaptive ethnic group strategy. Sooner or later states have become maladaptive for their citizens in one or more of the ways discussed earlier in this chapter. The nation state is an imperfect vehicle for ethnic interests, as demonstrated by the massively destructive world wars and genocides of the twentieth century and the current rapid decline in relative fitness of the founding ethnies within many Western states. Deviation from an ethnic group strategy is manifest in the haphazard correspondence of state and ethnic boundaries, state-sponsored hyper-mobilization leading to mutually destructive fratricidal wars, state-sponsored under-mobilization, and elite free riding. One failing, predictable from the ethology of ethnic solidarity, is that the fatal formula of the 'concept nation' can be instituted by elites in place of the ethnic nation, since the human behavioural repertoire can be tricked into extending solidarity to individuals and groups on the basis of group markers that have lost their efficacy as extended-kinship markers.

These weaknesses apply to the mildly collectivist nation state, with some redistribution and welfare provisions. They also apply to the libertarian state model. A recent advocate of the libertarian state is Hans-Hermann Hoppe, who agrees with de Tocqueville about the tendency of states, especially democratic ones, to grow in power and intrusiveness until they override individuals' natural choices of ethnic community.[120] His policy recommendation is ethnic *laissez-faire*, in which individuals are allowed unlimited private choice and town-level self rule. But this would offer no defence against large scale immigration on the national or continental level. Hoppe's libertarian rejection of state power (he wants most functions privatized) fits with his omitting to discuss collective interests. Neither does he recognize the problem of ethnic mobilization. In his world the two levels of action are the state and individual households. Needless to say, a libertarian state would offer little protection for ethnic interests.

The traditional nation state's unreliability as a strategy for the great mass of citizens derives from its symbolic nature: redistribution, equality of reproductive options, and boundary defence can have a large symbolic component and still elicit much public altruism. To elicit support for the state they do not need to be effective to the extent that citizens' altruism towards the state is generally adaptive. Participation in the traditional nation state need not preserve citizens' inclusive fitness for that state to be robust. To repeat the most recent evidence of this, many Western nation states have changed into multi-ethnic states that are replacing or tolerating the replacement of their founding populations. Such a state can hardly be considered an adaptive group strategy, at least on the part of the founding majority ethny. From the perspective of genetic interests, even a robust and wealthy state, one that makes its way in the political and economic worlds, is maladaptive if it fails to preserve its citizens' fitness relative to other ethnies, at a minimum within the state boundaries.

The traditional nation state's failure as an adaptive group strategy is tragic for the founding ethny, because it is the only group strategy they have. In Western societies the majority ethnies have lost much of their original tribal identities. In the historical process of nation building the members of many small tribes pooled their identities and territories. In effect, if not by intent, they swapped their small tribal group strategies for larger national group strategies, drawn by the implicit promise of a tribal group strategy. In the same process they yielded control of culture production and distribution and became consumers of media products manufactured and marketed by specialized elites. Ethnic culture was thus cut adrift from the sentiments of its original tribal controllers, and vested in a branch of the new state elite. But in modern societies, especially Western ones, there is no mechanism for ensuring the loyalty of cultural elites. The same process is inevitable wherever nation building entails aggregation of populations, for example in sub-Saharan Africa and Melanesia. For the emerging states to then abrogate their tribal promise leaves ethnic majorities largely defenceless, at least for a time, since they cannot revive the myriad small tribes from which they are descended or the premodern tribal cultures that nurtured them.

Since no state perfectly fulfils the promise of an ethnic group strategy, all states so far conceived are deceptive to some extent. The state has been both a huge asset and an inexorable threat to ethnic interests.

### *Ethnic constitutions*

An ethnic constitution would correct some of the weaknesses in the traditional nation state. Existing constitutions are limited to defending proximate interests. But the ultimate interest is not happiness, nor liberty, nor individual life itself, but genetic survival. A scientifically informed constitution that takes the people's interests seriously cannot omit reference to their genetic interests.

An ethnic state would have an ethnic constitution, one that explicitly provided for the protection of existing ethnies' interests, or at least for the majority ethny's interests. Such a constitution would impede any attempt or tendency to abrogate the nation state's adaptive promise. It would contain provisions that defined citizenship in ethnic terms, and establish group rights designed to protect relative fitness. In the federal version each constituent ethny would receive the same guarantees, applicable within ethnically homogeneous regional territories. The federal government would not possess the authority to override regional immigration or ethnic laws. Territories could also be retained by those who prefer multiculturalism. An ethnic constitution might prescribe other components of an adaptive ethnic strategy, such as administrative methods for preventing free riding. Regional federations with open internal borders would be constitutionally prevented from accepting membership of states whose populations were genetically distant from that of the founding members.

There are precedents for ethnic states. In its *Staatsbürgerschaftsrecht* legislation of 1913, the German federal parliament defined ethnicity-as-descent as a sufficient condition for citizenship, adopting the principle of *jus sanguinis* rather than *jus soli*. Children with one German parent were automatically eligible for citizenship. The country accepted non-German guest workers in the 1950s and 1960s, but full citizenship for non-Germans entailed a lengthy period of residence. The law was reformed in 2000 to make it easier for individuals of non-German descent to gain citizenship. Like Germany's 1913 law, Israel's 'Law of Return' makes it mandatory for the government to accept Jews wishing to immigrate from anywhere in the world. Immigration by non-Jews is discouraged. Although the Australian Federation, established in 1901, did not specify the nation's ethnicity, this provision was implicit in the first legislation passed by the new parliament. The so-called 'White Australia policy' was a central pillar of Australian nationality, that drew on the restrictive immigration legislation adopted by the individual states before federation to protect against large scale Chinese immigration during the gold rushes of the mid 1800s. The United States was to a significant degree an ethnic state until the 1960s, when the Civil Rights movement and immigration reform swept away special protection of the white majority. Arguably the convergence of these two movements—the breaking down of external and internal protections for the majority ethny—transformed these Western societies into different kinds of ethnic states, ones that privilege minorities in various ways.[121] The United States began as an implicit ethnic state, whose Protestant European identity was taken for granted. As a result, the founding fathers made few remarks about ethnicity, but John Jay famously stated in 1787 that America was 'one united people, a people descended from the same ancestors',[122] a prominent statement in one of the republic's founding philosophical documents that attracted no disagreement. Soon afterwards, in 1790, Congress passed the new republic's first naturalization law, which limited benefits to 'free white citizens'. In 1870, following the Civil War, new legislation expanded the right to citizenship to include individuals of African descent, retaining ethnic particularism. Further elements of an ethnic state were added over time in the form of immigration laws. Asian immigration was barred in the late nineteenth century and in 1921 and 1924 legislation introduced a quota system that severely limited immigration overall and allocated quotas based on national origins, effectively reserving most immigration for Western Europeans. The 1924 legislation further stipulated that 'no alien ineligible to citizenship shall be admitted to the United States', in effect defining the country as composed of citizens of European and African descent.

Malaysia is a modern ethnic state that gives special protection to the Muslim Malay majority at the expense of the Chinese and Indian minorities. A wide ranging system of affirmative measures favouring Malays was introduced in 1970 following bloody riots against Chinese businesses which practically monopolized the economy and were greatly overrepresented in higher education. Under the system a Malay, or *Bumiputra* ('son of the soil'), can benefit from quotas at the country's universities and in the corporate world. Long-serving Prime Minister Mahathir Mohamad summed up the aim of the measures as 'making my race a successful race, a race that is respected'.[123] Since the system was put in place, ethnic Malays have improved economically and educationally, though not all goals have been met. The Chinese and Indian minorities have suffered from the rigid system of preferences.

Another example of an ethnic state is Macedonia, which until 2001 had a constitution that defined the country as the national homeland of the Macedonian people. The general ban on large-scale immigration that still applies around the world has made discriminatory immigration policy superfluous and obscured the strongly felt and rigidly imposed tribal and ethnic conceptions of society.

The best known modern ethnic state was National Socialist Germany (1933–1945), discussed earlier on page 159. This state derived many strengths from its nationalist character. Its accomplishments included a revitalized social policy, full employment, rapid economic growth, an egalitarian class structure, and the salvaging of national pride after the humiliation of the Versailles Treaty. Furthermore, some economic and health benefits flowed from the Nazi ideology's biological orientation, compared to Marxist-Leninism and in some respects even liberal democracy. For example, retention of elements of the free market economy allowed the German economy to become a run-away success while communist experiments became grim affairs of quotas, heroic labour, and deprivation, and Western economies languished. Unlike the Soviet regime, the Nazis did not ban genetics; German agriculture did not decline. Nazi Germany was decades ahead of the West in recognizing and blunting the dangers of tobacco products.[124] The majority ethny in Nazi Germany did quite well until 1939, while ethnic Slavs were killed in large numbers by Lenin's and Stalin's security forces. The Hitler regime was popular, uniting Germans across social classes. This allowed it to out-compete the powerful German Communist Party, a branch of the seemingly unstoppable Bolshevik revolution that since 1917 had been killing, enslaving, and terrorizing large numbers of citizens wherever it came to power.[125] If the Nazis had not themselves killed and enslaved millions, their reputation would not be what it is today. Nazi Germany became justifiably notorious for practising aggressive war and genocide against eastern neighbours and minorities. If an ethnicised constitution necessarily resulted in catastrophes such as Nazi Germany initiated, that would be a sufficient reason to abandon the idea altogether.

This is unlikely to be any truer than the possibility that all socialism must necessarily regress into Stalinism or that the benefits of free markets can only be realized if slavery is also accepted. Ethnocentrism is a double-edged sword, but there is good reason to believe that human ingenuity can devise social technologies for keeping the aggressive edge blunt.[126] Also, in a growingly interdependent world, nationalism can only be sustained and prosper if it respects other national interests. Ethnic constitutions are compatible with universal nationalism because they are universalisable. The world population could conceivably live in several hundred ethnic states, some independent, some belonging to federations, participating in international trade, cultural exchange, and the give and take of limited immigration. Such an international system would optimize the genetic interests of most humans by assisting continuity rather than expansion.

The right to citizenship in an ethnic state would be a fitting plank in a biologically informed universal declaration of human rights.[127] Such a declaration would not be out of place in the constitution of any ethnic state. Like the freedom to raise a family, it is in everyone's interest to have his ethnic interests protected by the power of the state and to be free to invest in his ethny by contributing to collective goods that are proofed against free riders. Conflicts of interest would still occur. But it is in most states' interests to unite to contain cancerous cells that threaten neighbour states. In a crowded world there is much more to be gained by respecting others' interests and benefiting from peaceful trade, than contributing to an endless war of all against all.

This is not to deny that nationalism is associated with violent conflict. It is an ideology both of national liberation and aggrandisement, both goals producing conflict. Bringing political borders into alignment with ethnic ones is often a zero sum game. Moreover, as the most potent legitimating force in the modern world, nationalism arouses intense emotions when national integrity is threatened. Mobilizing a people to defend its vital interests is nationalism's most precious characteristic. The human cost of a war cannot be condemned without taking into account the interests thus preserved. But tribal passion can blind communities to peaceful alternatives, causing unnecessary misery.[128] Adaptive nationalism would work to bring patriotic emotions into alignment with real interests.

To concede that war can be adaptive is not to advocate its glorification as found in fascism. The huge scale and destructiveness of modern warfare are only adaptive for participants, which now includes whole populations, under special circumstances of real threat or opportunity for risk-free expansion. The former has been greatly reduced by diplomacy, international trade, and the spread of democracy[129] while the latter has all but vanished in a crowded world awash with surplus weaponry. Fascism, including the Nazi variant, did not meet the criteria of an adaptive ethnic state as defined in this chapter because it was not democratic and thus put the people at risk of free riders. This might seem counterintuitive, since this ideology laid mystical emphasis on 'blood and soil', metaphors for genetic and territorial interests, and advanced the state as the champion of these interests.

> The soil on which generations of German farmers can one day beget powerful sons justifies the investment of our sons of today and will some day acquit the responsible statesmen, of bloodguilt and sacrifice of the people, even if they are persecuted by their contempories (Adolf Hitler 1925).[130]

Except for the willingness to sacrifice millions of lives in a reckless military adventure, these values did not distinguish fascism from conservatives of the time elsewhere in Europe and America.[131] Both correctly identified the nation and its territory as vital interests. The distinguishing elements of fascism included the unscientific components of its ideology and, of special importance, its defective political institutions. National Socialist ideology had at its core a mystical conception of race that contributed to an erroneous view of ethnies as almost distinct species with disjunctive rather than statistical differences. This was compounded by an extreme ethnocentrism that evaluated the ingroup as possessing superlative values not found in other ethnies. Struggle and competition were ripped from Darwinism and roughly pasted at the head of social policy as semi-religious goals. These categorical and hierarchical conceptions are at best naïve in light of modern biological and social science, and in practice translated into brutal chauvinism. In contrast, a nationalism that was attractive to all societies would advocate dignity for all, as a necessary condition for favouring the ingroup. It would be a demystified set of propositions based on objective truths revealed by science, truths concerning group identity and group interests, equally valid for all ethnies.

Neither did fascism possess a mechanism for preventing elite free riding by co-ethnics or for moderating ethnic mobilization. The latter escalated to dangerous levels, partly due to the historical circumstances produced by the First World War, including the very real threat of communism. Escalation was also pushed to dangerous levels by fascist elites as a means of consolidating power. These two institutional failures combined to produce aggressive foreign policies that resulted in futile wars. In Germany this institutional failure allowed the ethnic cleansing and genocide of other ethnies at the word of the dictator. Fascism represented in biological terms a mass strategic blunder, a misdirected and overblown investment by citizens in their ethnies that forced other nations to unite against them. That speculative bubble was brought on not only by historically bounded rationality leading to imprudent democratic choice, but by undemocratic state propaganda. Had Germany remained democratic it is unlikely

Hitler could have risked his and other peoples' lives in a reckless geopolitical gamble to resurrect a Medieval peasant society. Rummel's historical and cross-national survey of conflict finds that democratic regimes are significantly less warlike than authoritarian ones.[132]

The military historian Martin van Creveld is not as optimistic about the possibility of forging peaceful nation states, by which he means bureaucratic rule by a corporate entity over a territorially defined population. Van Creveld tracks the development of the state from its initial role as an instrument of efficient administration for the aristocracy to its invocation of nationalism from 1789. He concludes that the nation state has become an end in itself, powered by the superhuman force of bureaucracy. But it is the state, not nationalism itself, which is to blame for the horrors of modern mass violence and oppression, van Creveld argues. He therefore agrees with William McNeill's characterization of state bureaucracies as macro-parasites.[133] Before it was championed by the state, nationalism had been 'a harmless preference for one's native country, its language, its customs, its modes of dress, and its festivals'.[134] After it was institutionalised as a mobilizing device, state nationalism became 'aggressive and bellicose'. The state deploys any ideology that increases its control, including socialism. It is a collective Frankenstein monster run berserk, though comprised of ordinary human beings. Borrowing his opening phrase from de Balzac, van Creveld condemns the state thus:

> Born in sin, the bastard offspring of declining autocracy and bureaucracy run amok, the state is a giant wielded by pygmies. Considered as individuals, bureaucrats, even the highest-positioned among them, may be mild, harmless, and somewhat self-effacing people; but collectively they have created a monster whose power far outstrips that of the mightiest empires of old.[135]

Van Creveld continues by arguing that among the evils of the state is welfare funded from general taxation, which is the basic instrument by which the state disciplines the populace, and turns it to aggressive war against other peoples. He maintains that the state is now declining in power as other corporate entities compete with it, evidenced by the waning of major war, the decline of welfare, the rise of international organizations and alternate economic forces (especially the multinational corporation) and loss of confidence in bureaucracy.

The very moderation of state power to which van Creveld refers as evidence of its failure, can also be taken as proof of its adaptability. There is no inexorable trend towards absolute power. Rather the state is a complex social technology that has been used to human advantage as well as disadvantage.[136] It is reformable. Thus the state is not inherently war-prone, as van Creveld himself observes. The differences between state societies show that states are not necessarily destructive of individual rights. For example, the Swiss federal government is not the leviathan of the British unitary state. One route of reform implicit in van Creveld's analysis is to return the state to its role as an instrument, not of kings but of peoples. The ethnic state would be closely identified with its subjects, ideally having no legal existence apart from them or their representatives. The humanitarian benefits of managing such a reform would be considerable if van Creveld's prognosis is correct. He believes that the breakdown of state power and sovereignty is leading to the reemergence of a politically disenfranchised underclass, even in the wealthy societies. This underclass is likely to be large, because it will include all those who have so far benefited from the implicit group strategy of the nation state, namely 'people and organizations who are limited to individual states and dependent on them for their defence, livelihood, education, and other services'.[137] In such a world there would be a need for compassionate group strategies able to unite ethnic kin of diverse economic class in mutual support.

Van Creveld's analysis looks weaker and more heartless when viewed from the perspective of national interests. Nationalism is in fact more than a 'preference for one's native country'. It is the extension of tribal feeling to large ethnies, and as such is capable of having adaptive consequences. In a world of no free lunches, the adaptive benefits of nationalism might be worth some sacrifices of wealth and individual safety. But van Creveld sees the nation state as nothing more than a run-away bureaucratic juggernaut no longer able to serve individual citizens. And since that was its justification, allowing the state to subside will cause no harm, even though whole populations will lose representation in a competitive world.

Ethnic constitutions that do not result in over-mobilization can also have shortcomings. One problem is that no document can guarantee the behaviour of a polity. For example, despite its ethnic definition of citizenship, Germany accepted large numbers of non-European 'guest workers', mostly from Muslim Turkey, as it ran out of workers to power its 'miracle economy' from the 1950s. Although they were admitted for short-term mutual economic benefit, many of these workers remained as a slowly-assimilating minority. In the 1990s large numbers of refugees, mostly from Europe but also from Africa and Asia, were taken in. By the end of the century Germany had settled some seven million foreigners, making up 9 percent of the country's 81 million inhabitants. This minority is reproducing much faster than native-born Germans.[138] This dilution of German homogeneity happened despite the country's ethnic definition of citizenship. A major contributing cause was changes in Germany's political culture. The post-war denazification program was an understandable and healthy reaction to the extreme National Socialist regime. However, the program became an arguably maladaptive campaign of systematic institutional shaming of the nation's identity, prosecuted through the education system and mass media. Despite being legally entitled to repatriate guest workers and their families, the political will has not been found to do so. Changes to political culture, whether induced by external pressure or by internal elites, can bypass ethnic constitutions.

Another way that ethnic constitutions can fail is through external pressure, as demonstrated by Macedonia. This is a Christian Slavic nation with a rapidly growing Muslim Albanian minority. Following the breakup of the communist Yugoslavian state in the 1990s Albanian guerrillas began an armed struggle against the Macedonian state, which denied them equal rights.

The dispute fitted a wider pattern of Albanian nationalism and demographic expansion. In the Serbian province of Kosovo the ethnic Albanian birthrate was over three times that of the ethnic Serbs in the second half of the twentieth century.[139] The result was that the Albanian majority grew from 68 percent in 1948 to 90 percent in 1994.[140] A guerrilla campaign against Serbian police and terrorist outrages against ethnic Serbs aimed to ethnically cleanse the province and make it part of a greater Albania.

In Macedonia the ethnic Albanian minority was rapidly growing and by 1994 had reached about 23 percent of the country's 2 million people.[141] Ethnic Macedonians were down to 66 percent, with the other 11 percent made up of Turks, Vlach and Serbs. Western leaders were worried that Macedonia would descend into bloody civil war as had other ethnically mixed provinces of Yugoslavia (Croatia, Bosnia and Kosovo). In October 2001 Western diplomats were pressuring the Macedonian leadership to change the country's constitution from an ethnic Macedonian republic to an ethnically neutral one. The country's constitution already stipulated tolerance, though in practice ethnic rivalry was manifest. The constitution posed a symbolic problem for European and American diplomats, since it proclaimed that 'Macedonia is established as a national state of the Macedonian people'. Yet it was known to all concerned that the relatively high ethnic Albanian birth rate would soon make that national group a majority in Macedonia. The Western position was thus tantamount to demanding that the Macedonians hand over their state to a different ethny and be replaced on their historic territory.

Alternatives existed that would have secured peace, but were not urged by the West. These included granting ethnic Albanians equal rights within an ethnic federation, with local autonomy and restrictions on inter-state migration. Another option was outright secession, in effect cutting the ethnic Albanians free to form their own state or join Albania proper, and retaining the rest of the ethnic Macedonian state for the Macedonians. Either solution would have insulated ethnic Macedonians from the fierce Albanian birth rate. Instead the fundamental ethnic problem was not addressed, and the Macedonian government effectively ceded sovereignty to Western monitors in matters of ethnicity to allow the external 'verification of the constitutional and legal acts on [ethnic conflict], and standardization of the minority demands within the framework of European standards for minority rights'.[142]

Unwise counsel on ethnic matters also occurs between allies. For example, the United States and Israel have urged the European Union, in the throes of formulating a constitution, to accept Turkey's application for membership. This is a *quid pro quo* for Turkey's strategic cooperation, including membership of NATO during the Cold War, cooperation with Israel from the late 1990s, and the provision of bases for wars against Iraq from 1990. The Turkish population belongs to the Caucasian race but is genetically distant from the majority of European ethnies.[143] If the advice from the United States and Israel was followed, a large fecund Islamic Turkic population would gain free access to Europe. The probable outcome would be the replication across the continent of slow-to-assimilate Turkish ghettos now evident in Berlin, Frankfurt and other Central European cities. Conversely Turkey would be opened to immigration from anywhere in the European Union. A European Union that included a large Islamic population would necessarily embrace Turkey's preference for Europe, as a 'secular and open entity'[144] rather than a cultural zone unified by a Christian tradition, albeit translated into secular humanism among much of the intelligentsia. Christianity and Islam have been the closest approximations to ethnic group strategies serving the European and Turkish populations, and forcing them together would likely undermine any force for cohesion and identity still conveyed by those religious traditions. Certainly it would be another obstacle to including reference to Europe's Christian heritage in the EU's constitution. Such a reference would affirm the Union's cultural and historical identity as the descendent of Christendom and thus much more than a matter strictly of economic and political convenience.

Valery Giscard d'Estaing, ex-president of France and head of the European constitutional convention, sees Turkish membership in the EU as a deadly threat to the Union. Turkey is an Islamic society with a high birth rate, 'a different culture, a different approach, a different way of life'. Its entry would set a precedent making it impossible to refuse membership to other Middle Eastern and North African states, starting with Morocco. '[I]t would be the end of the European Union', he concluded. He might have added, 'or the end of Europe', considering the likely swamping of the European peninsular by mass migration once the borders were lowered.[145]

As part of its *quid pro quo* for Turkish cooperation, the US and Israel turn a blind eye to Turkey's continued violent suppression of its Kurdish minority, belonging to an ancient nation divided up between Turkey, Iran and Iraq. Thus relatively short term strategic goals can threaten to compromise diverse long term ethnic interests.

Non-ethnic and ethnic constitutions belong to civil and ethnic societies respectively. The former is most compatible with ethnic diversity because citizenship is defined in strictly contractual terms. Citizens' social obligations are limited to behaving in a law-abiding manner. There is no obligation or bond to society as a whole, but rather an obligation to respect other citizens' autonomy and difference. Tolerance of religious and ethnic diversity is therefore a core value of civil society. Citizens of ethnic societies are also expected to behave lawfully, which in modern nation states includes limiting expressions of intolerance of religious and ethnic diversity to political debate and the electoral process. But much more is demanded of them because the legitimating assumption is that the society is an extended tribe; citizens have mutual interests beyond the golden rule. Patriotic duty is a core value of ethnic society.

Between the late nineteenth century and the 1960s the United States' changed from being a nation state, in which the ideal citizen regretted having only one life to give for his country, to a civil society that cannot legitimately demand sacrifice from its citizens.[146] Most of the founding fathers of the Republic took for granted the nation's ethnic basis as a self-evident virtue. John Jay was an exception, believing it important to describe and praise this ethnic dimension.

> Providence has been pleased to give this one connected country, to one united people, a people descended from the same ancestors, speaking the same language, professing the same religion, attached to the same principles of government, very similar in their manners and customs, and who, by their joint counsels, arms and efforts, fighting side by side throughout a long and bloody war, have nobly established their general Liberty and Independence.[147]

Less than 200 years later, Michael Walzer, a leading intellectual advocate of civil society and ethnic pluralism working at the elite Princeton Institute for Advanced Studies, could write that '[t]here is no country called America. . . . It is a name that doesn't even pretend to tell us who lives here'.[148] 'The United States is an association of citizens [not of nationalities or states]. Its "anonymity" consists in the fact that these citizens don't transfer their collective name to the association'.[149] Further, since the United States is not a nation state, and not a Christian republic, Walzer agrees with Horace Kallen's view that the primary political duty of citizens is to protect their democratic freedoms, rather than to protect their nation (ethnic group).[150] 'This commitment is consistent with feelings of gratitude, loyalty, even patriotism of a certain sort, but it doesn't make for fellowship.'[151] Since public altruism requires a sense of fellowship and community, reminiscent of the family, civil societies cannot expect much patriotic sacrifice by their citizens. Walzer admits that 'the hard truth about individualism, secularism, and toleration is that they make solidarity very difficult'.[152] Walzer is not opposed to ethnic identity or solidarity *per se*, arguing that tribal feeling is universal and deserving of respect according to the golden rule.[153] He also has expressed pride in the young people of his own ethny.[154] But he does not believe the United States ever was, or should be, an ethnic nation, or any kind of nation. 'The kind of natural or organic loyalty that we (rightly or wrongly) recognize in families doesn't seem to be a feature of our politics. . . . [T]he United States isn't a "homeland" (where a national family might dwell), not, at least, as other countries are, in casual conversation and unreflective feeling. It is a country of immigrants.'[155]

Civil society is a precondition for multiculturalism, while ethnic society is a precondition for the nation state. A universal civil society would dissolve the nation state,[156] but universal nationalism would be a global society of nation states. Ironically, the civic model would seem to be adaptive only in an ethnic state, where citizens can relax their ethnic guard and treat each other as individuals without losing fitness to ethnic free riders. But in multi-ethnic societies ethnic demobilization is maladaptive because of multiple risks to relative fitness. To be evolutionarily stable, civil societies must keep up their external guard in the form of military defence and control of immigration. Since the citizens of a civil society lose ethnic mobilization, group defensive functions must be motivated by institutions—the constitutional prescriptions and associated administrative apparatus that make up the ethnic state. The history of the surreptitious, undemocratic dismantling of ethnic institutions in the US, Australia, Canada and elsewhere indicates the need for an ethnic constitution to carry the following provisions. (1) Any change to ethnic policy requires a referendum. (2) All referendums bearing on ethnic issues necessarily trigger a constitutionally-mandated process of mass mobilization that is completed before any vote is conducted. (3) In the mobilization process, the citizenry is well informed of ethnic issues, perhaps by a bureaucracy charged with monitoring these issues and constitutionally authorized to disseminate knowledge via the mass media and education system. Agitation by would-be free-riding elites would run up against this constitutional wall, leading to cycles of ethnic mobilization and demobilization. A beneficial side effect might be to keep the ethnic constitutional machinery in working order. The alternative would be to arrange for high levels of ethnocentrism to be made a permanent fixture, though this would call for non-destructive ways to discharge the resulting patriotic energy.

Ethnic constitutions do not guarantee the continuity of the protected ethny. However, an ethnically-tied constitution would greatly empower ethnic loyalists by giving them legal recourse in opposing imprudent immigration legislation and urging enforcement of existing laws. The delicacy of ethnic interests calls for powerful defences. A nation can take centuries to form. But as several Western societies have experienced, it takes a lapse of only one or two decades in immigration control for an economically successful society to find its unity broken and heading for genetic replacement. Even if the gate is quickly shut, the resulting social and political problems can take many generations to resolve themselves through intermarriage.

*Ethnic cultures*

Protecting a balanced portfolio of genetic interests might be impossible without a profound change in political culture. De Tocqueville warned in *Democracy in America* (Vol. 2, 1840), that modern democracy progressively incapacitates citizens' ability to defend their rights. He thought that democratic equality produces individualism, an attitude of self regard and disregard for the community. The result is the inevitable centralization of power in the state. 'It does not tyrannize, it hinders, compromises, enervates, extinguishes, dazes, and finally reduces each nation to being nothing more than a flock of timid and industrious animals of which the government is the shepherd.'[157] Any political system that protected genetic interests in perpetuity could not allow citizens to let down their guard against threats to familial or ethnic interests, including the freedom to strategize on behalf of these interests. De Tocqueville thought that the aristocratic spirit best kept up an individual's guard and that a state could be so organized as to inculcate this spirit in citizens. He observed that the United States had weakened central power by establishing checks and balances in the form of the federal division of administration, local self government, a free press, and the promotion of voluntary associations. These checks and balances also act to educate citizens in freedom, de Tocqueville thought. In modern parlance, de Tocqueville advocated the importance of a democratic *political culture* for retaining liberty.

One critical requirement for sustainable ethnic states would be strong social science research bearing on national interests. The ethnic state could fund research into the evolutionary dimension in sociology, anthropology, economics, and politics. It is in every people's interest to possess the means for making accurate, balanced analyses of their own interests, as well as social trends and causal processes affecting them. Yet since the Second World War mainstream social science and the humanities have effectively censored ideas supportive of the nation state and other issues concerning nationality.[158] Canovan remarks that 'it is true that the deep taboos associated with such subjects [as nationhood] in the British and American academic worlds have led to the neglect of important questions to do with the membership and perpetuation of political communities as well as to over-simplified caricatures of nationhood'.[159] Biological approaches to social phenomena, essential for understanding ethnic interests, have also been discouraged for the better part of a century, partly as a broad Western reaction to the Nazi German regime of the Second World War but also as a trend preceding the rise of fascism. The latter trend was driven in part by minority ethnic fear of majority ethnic solidarity.[160]

Arguably the minority-liberal orthodoxy that rose to a dominant position in the social sciences by the 1950s has helped lower the defences of Western nations against the elite-promoted mass immigration that threatens to replace them. There seems to be an homogenizing, monopolistic trend within academe as within other industries. Just as some governments 'bust' commercial monopolies, it would be prudent to prevent any ideology, including nationalism, from dominating discourse on issues critical to adaptive decision-making. Thus it would be in the interests of all states dedicated to the welfare of their peoples to free the social sciences of ideological and ethnic bias, or at least maintain a balance of biases, since in the long run adaptive policy requires a thorough understanding of social processes. It is prudent for nations to maintain the intellectual diversity able to provide citizens with alternate interpretations of cultural and scientific developments that are pertinent to their national interests.

Despite its Constitution, the United States might well have remained a relatively homogeneous European-derived nation for centuries had its elites been disciplined by an electorate mobilized by a robust ethnic culture and empowered by unbiased social science. Could the following assertions made on the floor of the US Senate in 1965, none of which were true, have survived scrutiny in a society where free intellectual debate had ensured that the public had a sound grasp of nationhood and its preconditions? And would the man who spoke these words have had a political future?

> What the bill will not do: . . . First, our cities will not be flooded with a million immigrants annually. Under the proposed bill, the present level of immigration remains substantially the same. . . . Secondly, the ethnic mix of this country will not be upset. . . . Contrary to the charges in some quarters, [the bill] will not inundate America with immigrants from any one country or area, or the most populated and deprived nations of Africa and Asia. . . . In the final analysis, the ethnic pattern of immigration under the proposed measures is not expected to change as sharply as the critics seem to think (Edward Kennedy).[161]

Senator Edward Kennedy, still a United States Senator, has never retracted or apologized for this statement, which he made to allay fears that the 1965 immigration reform legislation inevitably would result in a Third World flood and the transformation of America's ethnic makeup. Neither has he been pressed to apologize or retract by a public informed about their interests.

A fundamental issue that this volume cannot hope to answer is whether the guarantee of citizens' reproductive interests is best provided by the ethnic state or by the people armed with an ethnic political culture. Different answers would imply the need for radically different polities. If the state is the best guarantee, then individuals should accept the protection of a powerful state along collectivist lines. Democratic process would be perfunctory and outweighed by the authority of state elites unimpeded by the division of powers. The problem would then be keeping the state elites motivated to protect the population's genetic interests down through the generations. If, on the other hand, the people are the best guarantee of their own interests, then the state should be geared to more participatory forms of democracy as found in the Swiss referendum system (but not the atomistic civil society model). This would retain the aristocratic spirit of citizens jealously guarding family and ethnic lineage. Such a state would take seriously the Anglo-Saxon tradition of dividing state powers. However, this approach has already failed in America, Britain, Canada and Australia, since it evolved into civil society and multiculturalism. What would prevent it from failing again? Removing the state's ethnic-defence function from populations with territorial traditions leaves them vulnerable to minorities empowered by an ethnic culture. This makes advocacy of civil society a sound strategy for mobilized minorities.

Perhaps both options can be pursued simultaneously, though this would entail compromising both, since the two arrangements of collectivism and individual decision-making are in tension. Perhaps state power can be made an amplifying link in a feedback loop, such that the constitution orders the maintenance of institutions the routine functioning of which keep majority ethnic mobilization at an adaptive level. The need for constitutionally mandated methods for mobilizing the people for their own defence was one conclusion reached by the English neo-roman political thinkers of the seventeenth century.[162] If the commonwealth's freedom is to be defended, they reasoned, citizens or their representatives must be willing to 'devote time and energy to acting for the common good'.[163] This willingness, which Machiavelli called virtue, is equivalent to the sociological concept of mobilization. Usually it does not naturally occur in modern societies, since people put their private interests before the common good. Thus, according to theorists such as Milton, there must be laws that coerce the people away from attending solely to private interests, if their public liberty is to be maintained.

A related issue is whether it is prudent to allow state symbols and rituals to be incorporated into ethnic identity. Is it possible to prevent this while relying on the state apparatus to maintain ethnic identity and mobilization? Prevention would be desirable, since states are temporary compared to ethnic interests, and the fall of a state would injure ethnic identity if the two had become fused.

If constitutional and cultural options are in fact mutually exclusive, then relying more on ethnic political culture is probably the safest option. Though no method is failsafe, the most stable and self-correcting system would vest the power necessary to protect an interest in those who embody that interest, namely the people. Thus much of the solution lies, I suspect, in shaping political culture and its material processes and constraints, rather than in some mechanized legal-constitutional order alone. That is not to say that ethnic constitutions are not necessary. However, as indicated by the failures of previous constitutions, some ethnic, it would appear that none has been sufficient to guarantee genetic interests, even over periods of many generations, or to produce the political culture that best approaches that goal.

Neither is an informed political culture sufficient to defend genetic interests forever. Culture is conditioned by factors of production and distribution. A people can lose control of the means of reproducing their culture, even when they are the majority ethny in an ethnic state. The trend towards a global economy extends to the production and consumption of cultural products. Even when cultural production and distribution remains in friendly hands, there can be a failure to adapt to changing circumstances.

It is possible that evolutionary stability of ethnic genetic interests cannot be guaranteed by any strategy. Perhaps all that can be achieved in a world of strategizing human competitors is avoidance of manifestly maladaptive policies, rather than an ever-closer approximation to some permanent and perfect solution. If such an ideal does not exist even in principle, then individuals' genetic interests will continue to be subject to the unpredictable dynamics of society and environment. That should not be taken as recommending adaptive nihilism. On the contrary, it would mean that those who care about their fitness and that of their children should invest in some level of permanent vigilance, because they can never assume that their interests are being perfectly preserved by some external system. It would become more imperative to sustain a healthy ethnic political culture. A biologically informed political culture has the great strength of generating new principles for constructing ethnic group strategies in changing circumstances. Moreover, it makes itself difficult to subvert by focusing the people's attention on their ultimate interests.

*The diffusion of adaptive state institutions*

The reform of some existing states towards ethnic states would necessarily have modest beginnings and would require an extended period of adjustment. Except in a few countries, there could be no overnight adoption of associated policies, because of the intermingling of peoples within them. Statistics on recent trends have not been reviewed recently, but Walker Connor analysed the 132 recognized states in 1971 and found the following:

1) Only 12 states (9.1%) can justifiably be described as nation-states.
2) Twenty-five (18.9%) contain a nation or potential nation accounting for 90% of the states, total population but also contain an important minority.
3) Another 25 (18.9%) contain a nation or potential nation accounting for between 75% and 89% of the population.
4) In 31 (23.5%), the largest ethnic element accounts for 50% to 74% of the population.
5) In 39 (29.5%), the largest nation or potential nation accounts for less than half of the population.[164]

Since 1971 the 132 states have increased in number to over 190, but the mix may not have changed appreciably. For example, while Slovenia is over 90 percent homogeneous, the Letts are only 53–57 percent of the population of Latvia, the Kazakhs are a minority in Kazakhstan, and East Timor is multiethnic.[165]

Establishing an ethnic state while respecting human rights will often be difficult, requiring investment and compromise to find just solutions. Despite difficulties, universal nationalism respects the wish of a people to have its own state or autonomy. Many societies might wish to remain multi-ethnic, and the democratic impulse underpinning universal nationalism respects that too. By the same token, respect is not owed oppressive multicultural regimes in which a coalition of elites imposes replacement migration on their societies against the wishes of the majority.

If one or more ethnic states was established, how might the institution spread? The most likely mechanism would be emulation of a small number of pioneering states. Emulation could be top-down, as other state elites copied ethnic constitutions, or bottom-up, as universal-nationalist political culture spread to new populations, which then voted for new political elites willing to make the necessary reforms. Existing multi-ethnic states are likely to resist devolution, as they have in the past. International law does not recognize the right to secession for ethnic groups, since many of the states forming the world community have a vested interest in keeping minorities and their territories within a unitary system. Some principles of self determination have been widely accepted as an expression of anti-colonialism, namely the ban on territorial aggression and the requirement for a plebiscite to gain a population's permission before transferring sovereignty. These principles are frequently not respected when a great power's interests are at stake or when conflicting claims prove intractable.[166] Nevertheless, since it was first formulated by French philosophers in the second half of the eighteenth century, the right to self determination of peoples has gained repeated acceptance and widespread sympathy. This norm has frequently been violated, but it has also contributed to the proliferation of nation states, often through the break-up of empires. The idea that ethnies have a right to secede from unitary multi-ethnic states is not yet a principle of international law, but it is a powerful rhetorical and mobilizing device for national liberation movements.[167]

Which features of ethnic states would be attractive model to emulate, apart from their preservation of genetic interests? The closest approximations to such states at present are nation states with relatively low ethnic diversity, and ethnic federations. A *de facto* version of the latter is the Swiss Federation, with Indonesia moving in the direction of devolving some governmental functions to the provinces. Both types of proto-ethnic state are generally more stable. Homogeneous nation states especially have lower levels of corruption and civil war, greater collective goods including social support, higher economic growth and overall standard of living, and higher levels of democracy.[168] The number of these states is growing as democracy spreads worldwide. When given the choice, people prefer to live as the ethnic majority in a nation state. The adaptive ethnic state would improve on these qualities by making them evolutionarily stable, especially by increasing the founding ethny's ability to retain possession of its territory indefinitely, and by improving state control of free riders.

It is often argued that small nation states are not viable economically or militarily. This argument was long ago rendered obsolete by developments in international diplomacy and economics. The decline of tariff barriers and the growing efficiency of international transport and telecommunications have allowed many small states and semi-autonomous territories such as Finland, Hong Kong, Luxemburg, and Singapore, to thrive, even beside lumbering giants such as Russia, China, and Indonesia, whose large populations should have generated higher economic growth, according to conventional economic theory. Furthermore, these efficient economies have retained their independence by an assortment of diplomatic arrangements, including alliances with powerful allies, neutrality, and limited incorporation into a larger state. National independence is becoming more viable for more peoples as international law, trading blocks, and regional security arrangements temper the anarchy of international society.

The adaptive ethnic state is also likely to be attractive because of its defensive capability, especially when this is combined with democracy and a healthy economy. As a type of nation state, it would receive strong support from citizens, support that, because it was evolutionarily stable, could continue over a long period. However, the *adaptive* ethnic state would not be attractive to elites wishing to build militaristic juggernauts. Because of the need for homogeneity, an ethnic state is limited in size to that of the ethny it administers. In addition, a state ideology that is adaptive for its people will be less efficient at extracting resources and sacrifices for belligerent purposes. An adaptive state could only increase its power within society by increasing state legitimacy in harmony with majority fitness. Its main aim is continuity, itself a competitive advantage in the long run but generally incompatible with the roulette of aggressive war. Because of the imperative to optimize interests distributed across the familial, ethnic, and species levels, the adaptive state cannot concentrate on maximizing military power. Investment must also go to making citizenship adaptive over the long term by deepening the state's democratic and welfare characteristics.

Greater state power within the international system, if it resulted from adaptive state institutions, would be a side effect of mobilizing and sustaining relatively high levels of investment from citizens. Cooperative social systems that produce collective goods have not often evolved due to the free rider problem, but when they do appear they are very successful.[169] This is because cooperation unlocks synergies such that joint effort can be more productive than the sum of individual production.[170] So it is possible that as a side effect of protecting genetic interests, adaptive states would replace nonadaptive states in an economic and social arms race in which elites made reforms that were adaptive for their people the better to compete economically and socially on the international scene.

### *Conclusion*

If one thing emerges clearly from this chapter, it is the distinction between investment in family and ethnic genetic interests. Adaptive family systems appear to be fairly self-organizing such that altruism shown towards children pays off in the children's better survival and reproductive prospects. Parents nurture their own offspring more than others with a high degree of reliability. There are risks of course, such as paternal desertion and maternal extra-pair conceptions. But these risks are mitigated, if never completely eliminated, by evolved behaviours such as pair bonding and jealousy. Furthermore, these behaviours show considerable insensitivity to radical environmental change, such has occurred since the

Neolithic. People bear and nurture children and attempt to monopolize mates almost everywhere. And they prefer governments that allow them to form families. But people do not nurture their ethnies nearly so effectively as they do families, despite possessing a set of specialized behaviours for tribal defence (ethnocentrism; indoctrinability; young male warrior syndrome; territoriality). The adaptiveness of these behaviours is sensitive to changes in societal and technological environments.

The reason for this poor nurturance of the ethny in modern social settings is the loss of certain situational cues that were reliably present in primordial environments. Consider identity formation. In primordial bands and tribes individuals could rarely choose group identity. They were born into a family that was embedded in a band and a tribe and would remain in that social frame all their lives. They were never presented with a choice. Yet in modern societies there is a plethora of groups and organizational entities with which individuals can identify: clan, suburb, ethny, socio-economic class, professional association, school, university, sports team, country, culture, and so on. Although we have large interests vested in our ethnic groups, we are not genetically equipped to reliably distinguish co-ethnics from other categories, especially those that have some of our tribe's markers such as common language, religion and dress. Stripped of the circumstantial information conveyed by primordial environments, including band and tribal cultures evolved over millennia, we can reliably identify and invest in our ethnic genetic interests only with the aid of novel cultural adaptations.

A major problem is free riders. Outside the bounds of small face-to-face communities, in the modern world of mass anonymous societies, patriotism is too easily exploited by free riders of various kinds. In stratified societies elites are the most endemic form of free rider. Unless controlled by a system of checks and balances, elites divert public resources to favour some group other than the ethny as a whole, whether it be family, a subset of the ethny such as a particular socio-economic class, or a different ethny altogether. Control of free riders must involve the democratic mechanism, which gives exploited majorities an edge over their minority rulers. It will also involve the social technology known as the state, the most powerful form of organization yet developed and an emulation of the primordial tribal group strategy.

Given our growing understanding of basic human interests, a strategy is needed that can satisfy the need for the ethnic monopoly of a homeland while avoiding domestic ethnic free riding and encouraging participation in the global village, one that acknowledges both the need for autonomy and the reality of interdependence. In this chapter I have argued for a revised form of the nation state as that strategy, reinforced by an ethnicised political culture, a state whose constitution explicitly acknowledges and defends the peoples' ethnic, as well as individual, genetic interests. To be adaptive for its citizens, ethnic nationalism in the crowded modern world would be compatible with other defensive nation states. The shared interests of nation states—to minimize conflict and free riders—allows for a universal nationalism that might optimize global genetic interests. Abstaining from opportunities for expansion would constitute an investment in humanity, especially ethnic diversity.

In arguing for universal nationalism in the space of one chapter I have necessarily adopted a short-hand style that might read as overly confident. So let me state that the case I have presented is anything but complete. I have failed to deal with many important issues. A complete analysis would work through the many permutations of minority-majority relations, looking for ways to protect every group's interests. Yet for all its shortcomings, I believe this chapter is warranted by the genetic analysis in the first half of the book. I am persuaded that reproductive interests are not adequately treated in contemporary political theory. If the adaptive implications of behaviour and policy are to be accounted for, as they must be in any political theory pretending to deal with human interests, it would be inconsistent and irresponsible to omit ethnic genetic interests from consideration; and that subject is inextricably linked to the theme of the nation state. I had to have a go.

A final note of caution is also warranted on the subject of strategy. As noted earlier, it is possible that there is no perfect evolutionary stable strategy for protecting citizens' ethnic genetic interests. This could be due to the unpredictability of cultural evolution and historical events, or to an open-ended arms race between ethnic group strategies and free riders of different kinds. Is this ground for adaptive nihilism? Doing nothing to defend one's territory or maintain group identity will surely lower ethnic genetic interests more rapidly than if one makes a small contribution to these collective goods. In an imperfect world we must often choose the least harmful course from a set of undesirable alternatives. For the time being the best achievable outcome might be to minimize the rate at which relative genetic interest is lost within state territory. It would be prudent to continue searching for a better strategy, one that makes altruism viable over many generations. In the meantime it would still be adaptive to steer one's community away from short-term disasters. In the breathing space so gained, perhaps an evolutionarily stable ethnic strategy could be put in place before precious genetic interest was much degraded. In evaluating an ethnic arrangement the practical criterion is not whether it perfectly safeguards the genetic interests of participating groups, but whether it is better in this regard than available alternatives.

*Notes*

1 *Statistical Abstract of the United States: 2001*, Table 1327.
2 Tullberg and Tullberg (1997).
3 Unz (1998); *Statistical Abstract of the United States: 2001*, Table 15. Other ethnic representations in 1998 at Harvard were: Blacks 8%, Hispanics 7%, Asians 20%, Jews c. 30%. A further 10% were foreign students. This leaves about 25% of student places for White Christians.
4 Sheehan (1998, pp. 199–201).
5 Switzerland is often held up as an example of a viable multi-ethnic society. It is. But it is not an example of a pluralist or multicultural society. Switzerland's constituent ethnies occupy their own, largely self-governing cantons. These ethnies are closely related, and have similar reproductive rates. Switzerland does not accept many immigrants. Also, there is minimal redistribution of wealth between cantons, and thus not between ethnies. The central government is small with limited powers. The country thus satisfies the conditions set out in Chapter 6 for protecting ethnic fitness in multi-ethnic societies. By contrast, Communist Yugoslavia did not meet these conditions, because there was uneven population growth and much redistribution of wealth between ethnies.
6 Ethnic and racial discrimination is common wherever diversity occurs. Social psychological studies indicate that prejudice and stereotyping are automatic, occurring under minimal initiating conditions (Ashburn-Nardo et al. 2001). Brain scans of individuals confronted with images of individuals from different races indicate that the same brain areas are activated in all subjects, irrespective of declared attitudes towards other races (Golby et al. 2001).
7 J. G. Herder quoted by Canovan (1996, p. 8).
8 Brubacker (1992).
9 Jacobson (1996, p. 25).
10 Habermas (1998); Walzer (1995).
11 J. S. Mill (1861/1960, p. 381).
12 Walzer (1994a, pp. 199–200).
13 Blalock (1967); Blalock and Wilken (1979).
14 Schelling (1978).
15 Tajfel and Turner (1986).
16 Mugny and Pérez (1991); Pérez and Mugny (1990).
17 Quoted by P. Johnson (1987/1998, p. 306).
18 Nutini (1997).
19 Eibl-Eibesfeldt (1989, pp. 622, 624).
20 Birrell (1995, p. 252), quoting from *Commonwealth Parliamentary Debates*, 12 September 1901, p. 4806.
21 Wiessner (2002a).
22 J. S. Mill (1861/1960, p. 384).
23 *Ibid.*, p. 382.
24 *Ibid.*, p. 383.
25 Miller (1993, p. 4).
26 Q. Skinner (1998).
27 Canovan (1996, p. 37).

28 McNeill (1984, p. 18).
29 Coase (1937); North (1990); Williamson (1990).
30 Alesina et al. (1999); Alesina and Spolaore (1997).
31 Easterly and Levine (1997); Rummel (1997); Tullberg and Tullberg (1997).
32 Alesina and Wacziarg (1998).
33 Rummel (1997).
34 Mauro (1995).
35 Knack and Keefer (1997).
36 Easterly and Levine (1997); Salter (in press-c); Alesina and Spolaore (1997); Knack and Keefer's (1997).
37 Rodrik (1999).
38 Light and Karageorgis (1994).
39 Easterly and Levine (1997).
40 Easterly (2000) adds the institution of 'freedom from government repudiation of contracts', but I take this as equivalent to rule of law.
41 For example, King (1976, pp. 17–18) called for racial vilification legislation to protect minorities ('distinct sub-sovereign groups') that would allow individuals to bring lawsuits against those who slandered their ethnies, without mentioning defamation of majorities. This is an unexceptional proposal in the case of abusive language likely to incite. But King recommended strict control even of scientific discourse, such that statements deemed factually incorrect or disrespectful of groups would be subject to legal recourse (p. 221). The aim, he states, of such measures would be to 'eliminate in an even-handed fashion all corporate hostilities directed by one group towards any other as defined by more or less "primordial" criteria'. (p. 221). This is surely a formula for totalitarian control by one group or another, since law courts are not competent to judge scientific claims, and the legal system is most readily used by the dominant ethny or elite. The effect would be to prevent weaker ethnies from asserting their interests in the public square. One wonders what has become of the Left as the bearer of the Enlightenment value of fairness through reason. Traditionally, the moderate left perspective identified with the liberalism of J. S. Mill has sought to resolve social problems rationally through the greatest consensus among interested parties. The consensus must be unforced and must not involve deception. All the parties' interests are put on the table and a solution—usually a compromise—is worked out by parties possessing equal information concerning the factors bearing on their interests. Such a resolution is likely to be more robust than one based on force or deception, because 'the truth will out' and parties will, sooner or later, come to realize what their interests are and wish to renegotiate an agreement once they realize that it is inimicable to their interests. A solution that so debilitates one party that it is prevented from seeking recourse, is considered ethically unacceptable to liberal sentiment. Revision is little problem if the initial agreement deals openly with the most important interests. Since genetic interests are the most fundamental, liberals should support social policies that take these vital interests into account. Certainly they should not resist an open discussion of genetic interests, if they are genuine liberals.
42 Lichter et al. (1994); McGowan (2001).
43 Salter (1998/2002).
44 Greenfeld (1992, p. 10).
45 van den Berghe (1981; in press).
46 Salter (in press-b).

47 Betzig (1986); Vining (1986).
48 Burnham (1964/1975); Francis (1984/1999).
49 Phillips (2002).
50 *Business Week*, 19 April 1999, p. 34.
51 *Business Week*, 16 April 2001, p. 70.
52 *Statistical Abstract of the United States: 2001*, Table 670.
53 Phillips (2002).
54 Salter (in press-b; in press-c).
55 Goetze (1998).
56 Boyd and Richerson (1985); MacDonald (1994).
57 Burke (1790/1968, p. 14).
58 Salter (2002; in press-a).
59 Canovan (1996).
60 In 1980 tourist arrivals amounted to 3.5 percent of the world population and there were 14,273 international organizations. By 1995 tourist arrivals had ballooned to 9.9 percent, and by 1998 there were 48,350 organizations (Guillén 2001, p. 239).
61 Guillén (2001, p. 239).
62 Sassen (1996, pp. 25—30).
63 'Globalisation Hurts Cultures, Pope Says', *International Herald Tribune*, 28–29 April 2001, p. 2.
64 Schiefenhövel and Salter (1994).
65 Eibl-Eibesfeldt (2001).
66 Wallerstein (1979).
67 The world-systems theory on which these remarks are based is closely related to dependency theory and Marxist modernization theory (Sanderson 1999, pp. 203–218). All are radical anti-capitalist analyses which, like mainstream social science for the last century, ignore psychological and biological variables (Sanderson 2001). There is now some evidence that average intelligence, as measured by IQ tests, significantly affects a society's ability to modernize its economy and enter the core of global capitalism (Lynn and Vanhanen 2002). However, there is also good evidence that ethnic diversity depresses economic growth by reducing national solidarity (Easterly and Levine 1997). Acknowledging these effects leaves much scope for macro-level theories, such as those concerned with world-systems and dependency, to explain some of the variance in economic development. It also indicates that collective national strategies can improve local economies.
68 Mann (2001).
69 Huntington (1996).
70 *Ibid.*, p. 20.
71 *Ibid.*
72 Huntington (1997, p. 39).
73 Browne (2000).
74 E.g. David Howell (2002).
75 It is more important to understand the Left's antagonism towards people's genetic interests than, say, the religious Right's position, for two reasons. First, the Left in one form or another dominates academic, media, and entertainment culture in the West. Secondly, the Left is meant to be the child of the Enlightenment, unlike the conservative, especially the religious, Right. Why do leftist intellectuals reject an idea that is simultaneously the product of science, reflective of humanistic self-acceptance,

and conducive to popular sovereignty? The combination of power and nihilism within modern leftism recommends it as an object of analysis.

76 Champions of the corporate sector who exemplify support for replacement migration are the *Wall Street Journal* in the United States and *The Economist* in Britain. Both publications assess immigration on purely economic criteria, criticizing or ignoring public figures who consider impacts on other interests (e.g. see *Economist* editorial, reprinted in the *International Herald Tribune*, 2–3 Nov. 2002, p. 6).

77 Quoted by J. Johnson (2000, p. 449).

78 Horkheimer and Adorno were the two leading members of the Frankfurt School whose critiques of German and Western culture in general deployed a fusion of Marxism and Freudian psychoanalysis. Thus it is not surprising that Tibi, as a product of that School, should urge on his fellow German citizens a type of patriotism that would not be disturbed by their peaceful genetic replacement.

79 Tibi (2002).

80 Cliquet and Thienpont (1999, p. 281).

81 Hinde (1989, p. 60).

82 Hechter (2000); Salter (2002).

83 Walzer (1983); Barry (1996); Salter (in press-d).

84 Singer (2000).

85 Betts (2002, p. 41).

86 Eco (2002).

87 Erikson (1966).

88 In the year 2000 social democratic and socialist governments around Europe boycotted the Austrian conservative government after it entered into coalition with the Freedom Party headed by Jörg Haider, which had received the second largest vote in the election.

89 Quoted in Mattei (2001, p. 24).

90 Guillén (2001).

91 Borjas (2001).

92 Steel (1995).

93 The phenomenon of an emerging global elite was raised by Hiram Caton (personal communication, 14 April 2002).

94 *Business Week* Executive Programs (1996–1997).

95 Collinson (1993). The 1951 UN Convention defines a refugee as a person '[who] owing to [a] well-founded fear of being persecuted for reasons of race, religion, nationality, membership of a particular social group or political opinion, is outside the country of his nationality and is unable or, owing to such fear, is unwilling to avail himself of the protection of that country' (quoted by Millbank 2001, p. 2).

96 Millbank (2001).

97 For example, despite offering illegal immigrants high standards of sanitation, accommodation and nutrition by world standards, Australia was condemned by a representative of the UN High Commissioner for Refugees in July 2002 (*The Australian* 29 July). Australia's system was ruled to be invidious because it detained all asylum seekers lacking a visa. The system was 'mandatory, indefinite, and non-reviewable', the representative claimed. No comment was offered by the UN on how Australia could change these three elements without losing control of its borders.

98 Rajagopal (2002).

99 V. Miller and Ware (1999).

100 Steel (1995).
101 For further reading see Hall (1993).
102 See Salter (1995, Chapter 12), regarding the relative efficiency of laws versus face-to-face commands.
103 Keegan (1993).
104 Wiessner and Tumu (1998).
105 Anderson (1993, p. 617); Gellner (1983).
106 Hroch (1985).
107 Nairn (1977).
108 Q. Skinner (1998).
109 Herman (2002).
110 *Ibid.*
111 Quoted in Taylor (1967/1955, pp. 161–2).
112 Eibl-Eibesfeldt (1998/2002).
113 MacDonald (1998); Salter (2001).
114 Armstrong (1982); Greenfeld (1992); Smith (1986).
115 Connor (1993); Eibl-Eibesfeldt (1970); Holper (1996); G. Johnson (1987).
116 Canovan (1996, pp. 69, 71).
117 In addition to fictive kinship, Burke and Canovan also observe fictional elite status in English identity. That is, English political culture portrayed every Englishman as the member of an exalted caste. Canovan goes on to note that the fiction of an aristocratic nation did not prevent English society from being steeply hierarchical, evidence that this form of political fiction might also have maladaptive consequences for those who harbour it.
118 Burke (1790/1968), quoted in Canovan (1996, p. 70). Burke and after him Canovan identify the semblance of kinship as the glue binding the English nation together in a non-coercive order of altruism. However, neither examines the potential for deception leading to maladaptation in this semblance, this fictive element of nationality. Indeed, the issue of adaptiveness is not raised even by political philosophers who assert the superiority of the fictive kinship type of nationality over the real thing on the basis that this fiction minimizes conflict and makes the nation more inclusive (e.g. Canovan 1996, p. 75; Miller 1995). Even van den Berghe (1981), who takes an explicitly sociobiological approach, does not discuss adaptiveness, despite his assumption that ethnic nepotistic ideologies such as nationalism are always exploitative.
119 Salter (1995, Chapter 1; 2001; 2002).
120 Hoppe (2001).
121 Graham (2002).
122 *Federalist* No. 2, 31 October 1787.
123 Shari (2002).
124 Proctor (1999).
125 Rummel (1996) estimates that the Soviet authorities killed over 60 million between 1917 and 1987, mostly its own citizens, while most of Nazi Germany's 20 million victims (1933 to 1945) were non-Germans.
126 Salter (2002).
127 L. Tiger has argued for a 'Biological Bill of Human Rights', but did not refer to ethnic interests (speech to the annual meeting of the Association for Politics and the Life Sciences, Washington D.C., 31 August–3 September, 2000).

128 Tishkov (1997).
129 Rummel (1997).
130 Adolf Hitler in 1925, quoted by Maser (1973/1976, p. 230).
131 In recognizing ethnic boundaries conservative and fascist ideologues of the 1920s and 1930s, including Churchill and Hitler, had a better grasp on reality than did the liberal and Marxist intellectuals who were already well represented in Western universities in the 1930s and 1940s. The Right's error was to conceptualize ethnic boundaries as disjunctive and think of ethnies in mystical and essentialist terms. Leftist ideology committed a larger and simpler scientific error when it denied the very existence of ethnic interests. Contemporary liberal intellectual opinion has still not shaken off the misconceptions that began entering that tradition at the beginning of the twentieth century, misconceptions that underly multicultural myths such as the non-existence of ethnicity and race as biological categories. Examples include J. Huxley et al. (1939/1935) and Montagu (1942/1997), the latter republished without apology in the face of modern genetics.
132 Rummel (1997).
133 McNeill (1979).
134 van Creveld (1999, p. 259).
135 *Ibid.*, p. 258.
136 Salter (1995).
137 van Creveld (1999, p. 419).
138 Eibl-Eibesfeldt (1991).
139 Ramet (1992, p. 141).
140 *Brockhaus Encyclopedia* (1995).
141 http://www.macedonia.org/.
142 http://faq.macedonia.org/politics/, September 2002.
143 The genetic distance between Turkey and European populations is not provided by Cavalli-Sforza et al. (1994). However, Turkey clusters with Iran in the genetic tree of West Asia (Cavalli-Sforza et al. 1994, p. 242). Taking Iran as a proxy for Turkey, the genetic distance from the closest European nation, Greece, is comparable to distances found within Europe (Greece-Iran distance is 0.0075), though larger than that found between most neighbouring ethnies. But Greece is already on Europe's periphery, putting Turkey at a substantial genetic distance from most of the European population. The population of over 60 million is six times larger than that of Greece. The high fertility and endogamy of Islamic populations would magnify and prolong the genetic impact of Turkish immigrants on Europe, while the pronounced religious and cultural differences would undermine organic solidarity.
144 Oguz Demiralp, Turkey's ambassador to the European Union, quoted in the *International Herald Tribune*, 9–10 Nov. 2002, p. 3.
145 *International Herald Tribune*, 9–10 Nov. 2002, p. 1.
146 Walzer (1992).
147 *Federalist* No. 2, 31 October 1787.
148 Walzer (1990/1992, p. 23).
149 *Ibid.*, p. 27.
150 Kallen (1924, p. 124).
151 Walzer (1990/1992, p. 38).
152 Walzer (1974/1992, p. 92).
153 Walzer (1994a).

154 Walzer (1994b), writing for the ethnic publication *Congress Monthly*, praised Jewish youth for maintaining the Jewish tradition of civil rights agitation on university campuses, while also calling for revision of some aspects of Judaism.
155 Walzer (1990/1992, p. 24).
156 Walzer (1995).
157 de Tocqueville (2001), quoted by Wood (2001, p. 49).
158 Scruton (1990, p. 304).
159 Canovan (1996, pp. 134).
160 Frank (1997); Hyatt (1990); MacDonald (1998); Svonkin (1997). Ethnic collaboration is a universal phenomenon in business (e.g. Bonacich 1973; Landa 1981; Light and Karageorgis 1994) as well as in the cultural industries and the social sciences (Greenwald and Schuh 1994; White 1966).
161 E. Kennedy, quoted from Auster (1990, pp. 10–26).
162 Q. Skinner (1998, p. 32).
163 *Ibid.*, fn. 103.
164 Connor (1972).
165 W. Connor (personal communication, 18.4.2002).
166 Cassese (1993).
167 D. Miller (1994) reviews some of the arguments for secession, though not from a biological perspective.
168 Salter (in press-c).
169 Richerson et al. (1996, pp. 164–5).
170 Corning (1983).

# 8. Fitness Portfolios for Individuals who are Intermarried, of Mixed-Descent, Childless, Adopting, Homosexual or Women in Modern Societies

*Summary*
While every human being has genetic interests, there are differences in availability of investment portfolios. In this chapter I assess the genetic portfolios of five classes of persons who find themselves in circumstances that constrain or skew optimal genetic investment in some way. The guiding question for the opening section is, how would an adaptively-minded individual distribute resources, for example between family and ethny, if he or she were married to someone from a different ethnic group or was descended from such a union? A similar question is posed in section 8.2 concerning strategies for individuals who find themselves without children. Section 8.3 assesses the genetic interests of individuals who adopt children, while section 8.4 assesses the interests of homosexuals, and 8.5 the interests of women who are torn between motherhood and careers outside the home.

No single pattern of investment is adaptive for everyone. In Chapter 6 I argued that circumstances could limit the types of fitness portfolios available to individuals. There might be a lack of collective goods in which to invest; some regimes punish ethnic nepotism. In this chapter I discuss circumstances associated with domestic life style rather than with politics. For each circumstance I try to answer the same question: which fitness portfolio is most adaptive for the individual in that circumstance? In the first section I discuss the portfolios of individuals who marry outside their ethnic group, as well as their children's portfolios. The second section concerns childless individuals. Then I discuss families that include adopted children, followed by the portfolios of homosexuals. Finally I consider the reproductive options open to parents, especially women, who seek nondomestic careers.

Note that my main goal in this chapter is not to describe how people actually behave. Rather, I explore how individuals *would* behave if they were attempting to preserve their genetic interests. This speculative intent is signalled by terms such as 'a rational fitness maximizer would do such and such'.

### *8.1 Intermarried and Mixed-Descent Individuals*

Intermarriage between ethnic groups has been accelerating over recent centuries with the advent of mass transportation and with international economic and political integration. What are the genetic interests of people in mixed marriages and those of their offspring? As we shall see, intermarriage lowers but does not eliminate the fitness payoff of investing in familial or ethnic genetic interests. Initially it mixes these interests and continued interbreeding between two groups blurs and finally erases differences of genetic interests at the group level. In so doing a new lineage is born, whether family or ethny, with genetic interests derived from its parent groups. Van den Berghe observes that ethnic fusion approaching panmixia heals ethnic conflict.[1] Interbreeding increases the average coefficient of kinship between groups and inevitably changes attitudes as familial bonds are forged across the previous group boundary and the two groups grow more alike.

Mixing can be asymmetrical, for example when a relatively small number of immigrants marry into a much larger society. Those immigrants and their spouses have distinctive genetic interests influenced by patterns of marriage choice made by their descendants; those interests are also influenced by the relationship between the new society and the previous one from which the immigrants came.

Interbreeding has produced the great majority of modern humans, whether between clans, tribes, or larger populations. The resulting gene flow has kept all the scattered populations members of the one species, such that matings between members of any two populations produce fertile offspring. Clearly interbreeding is something to which all humans can adapt. That does not mean, however, that it is uniformly adaptive, that there are never fitness costs involved. Each new generation has its own set of genetic interests that might differ from those of ancestors, so a descendant mixed population can be healthy and behaving adaptively regardless of its kinship to preceding generations. Yet a new generation, genetic facts on the ground, can represent a significant loss of fitness for its parent populations due to dilution of distinctive genes and hence lowered efficiency of parental and ethnic investment. In this section I consider the initial generations following interbreeding between ethnically distant populations. I take the perspective of two generations, the one that intermarries and the one produced as a result. Both generations experience changes to familial and ethnic genetic interests as a result of the ethnic mixing. Due to lack of research on the subject, this section consists of reasoning about optimum strategies based on theory drawn from sociobiology and population genetics.

*Advantages of intermarriage*

Peace is a major potential advantage of intermarriage, as we shall see. Fitness maximizers who are intermarried or of mixed descent will exert pressure to minimize conflict between ancestral groups. However, the resulting peace lobby incurs costs, also discussed below.

Intermarriage can also increase individual fitness by providing a mate when none is available in the home ethnic group. A wider choice of mates can result from any factor that increases social scale, as well as factors that reduce endogamy, including individual psychological differences in ethnocentrism and cultural change, for example the break down of endogamous social controls and the spread of non-discriminatory ideology.

Mate quality can in principle be accentuated for some males by hypergamy, the tendency of females from low-status groups to 'marry up'. This is due to the resource advantage possessed by males belonging to high-status groups. Because females find high status attractive in males, other factors being equal, males from high-status groups enjoy a wider choice of mates. A male who would otherwise not attract a female in his home group due to his age, health or status, might find a mate from a lower-status ethny who is attracted by the relatively high status and wealth of his ethny. Hypergamy operating between groups thus allows males belonging to a high status group to attract higher quality mates than they could find in their own group, which boosts male fitness but can have mixed fitness effects for the female. The latter should stand to have their individual fitness enhanced by their high-status mates' resource advantage, but there is likely to be a loss of genetic quality when the male's status is derived from his group identity despite individual low quality.

Hypergamy combined with monogamy tends to lower the fitness of women from the high status group and males from the low status group. In monogamous societies high status females can fail to find a mate due to competition from more attractive low-status females, while males from the low status group lose potential mates to males whose attractiveness is enhanced by their ethnic identity. Mate choice between the resulting pools of single men and women is inhibited by the same status factor.

*The cost of intermarriage*

Parent-offspring conflict is a well-known concept in evolutionary biology. The principle was first described by Trivers who argued that because parents and children (in sexually reproducing species) do not share all of their genes, conflicts of interests can arise.[2] One example is the timing of weaning. Trivers described children's temper tantrums and other attempts to prolong breast feeding as evidence of a conflict of interest. Weaning is necessary to increase the chance of conception and to provide milk for a new child. A given birth spacing can be optimal for the parent's inclusive fitness but suboptimal for an existing child which has to share parental investment with new siblings. Might parent-offspring conflict of genetic interests be exacerbated by intermarriage? In the following section I sketch out some of the factors likely to influence the behaviour of a hypothetical fitness maximizer. Note that this is a complex question and the following ideas should be taken as preliminary. Even the theoretical domain of parent-offspring conflict, uncomplicated by the added factor of ethnicity, is far from resolved.[3]

An important theory bearing on the fitness effects of inter-ethnic marriage is Rushton's genetic similarity theory (GST), initially discussed in Chapter 2 (p. 40).[4] GST states that assortative mating is adaptive because the children that result share more than 50 percent of their parent's distinctive alleles, allowing greater efficiency of parental investment. Rushton argues that it is adaptive for individuals to mate with or befriend those who are phenotypically similar, because this similarity has a genetic component and therefore signifies genetic similarity. Grafen convincingly argues that Rushton's theory fails in the case of ethnically homogeneous populations, since points of phenotypic similarity do not indicate the broad genomic similarity needed to boost inclusive fitness.[5] However, Rushton emphasizes the salience of ethnicity as a criterion for assortative mating. As we saw in Chapter 3, ethnic kinship can reach the levels of close family relatedness, and thus must consist of the same broad genetic similarity along the genome. Assortative mating along ethnic lines—endogamy—is a pronounced trend in all mixed societies.

Rushton argues that the lower frequency of shared genes in ethnically mixed families might result in less intense bonding, greater conflict and fewer children. Indirect support for this contention comes from studies indicating that identical twins cooperate more than non-identical twins,[6] that adopted children receive less investment than genetic children,[7] and that full siblings cooperate more than half siblings.[8] Direct evidence would require research on ethnically mixed families, but such research is weak. The question I wish to answer here is not what people actually do, but what they should do if they wish to conserve their genetic interests. Is the strong tendency observed towards endogamy adaptive? And is it adaptive for fitness maximizers to adopt different reproductive strategies when their spouses are from different ethnies, perhaps producing more children and investing less in each?

Answering this question will require a little mathematics from population genetics, but this abstractness should not lead readers to think that we are dealing with abstractions. Alleles are real, physical things—strands of DNA. Ethnic admixture can be detected by observing these strands. If the characteristic gene frequencies of the parent populations are known, an individual's ancestral ethnies can be determined with high reliability.[9] Forensic investigations now routinely use this fact, since geneticists can detect victims' and suspects' ethnic mix if blood, hair or other cellular traces are left at the scene of the crime.

Children receive on average half their alleles from each parent. When the parents are themselves related, calculating the kinship coefficient is more complicated because each parent is not the sole source of copies of his or her genes. The spouse also contributes to the child a share of the other parent's genes. In a population consisting of two ethnies X and Y with genetic distance $F_{ST}$, endogamous (within ethny) matings produce children with higher kinship to their parents. In Appendix 1,[10] Harpending derives that higher kinship as

$$f_{xx} = 0.25 + 3\,F_{ST}/4 \qquad \ldots 8.1$$

where $f_{xx}$ is the kinship coefficient between children and parents within ethny X. (As discussed by Harpending in Appendix 1, the kinship coefficient is typically half the coefficient of relatedness, so that parental kinship in outbred populations is one quarter, not the commonly cited *relatedness* of one half.) Exogamous (between ethny) matings produce children with lower than average kinship

$$f_{xy} = 0.25 - F_{ST}/4 \qquad \ldots 8.2$$

where $f_{xy}$ is the kinship coefficient between children and parents, when the latter are drawn from ethny X and ethny Y. The difference between these two quantities is the fitness payoff for parents who choose endogamy instead of exogamy. That payoff, in each child, is

$$f_{xx} - f_{xy} = F_{ST} \qquad \ldots 8.3$$

Note that this fitness advantage is sensitive to the ethnic comparisons being made. To illustrate this point, let us return to the examples used in Chapter 3, and consider the hypothetical case of an individual English man or woman choosing a spouse from different ethnic groups. Assume that whatever the choice, the number of children will be the same (I am aware of no evidence indicating that family size is affected by degree of outbreeding). For a person of English ethnicity, choosing an English spouse over a Dane gains less than one percent fitness.

But choosing an English spouse over a Bantu one yields a fitness gain of 92 percent (dividing the English-Bantu $F_{ST}$ of 0.2288 from Table 3.1 on p. 64 by the outbred parental kinship of 0.25). The same applies in reverse order, so that a Bantu who chooses another Bantu instead of someone of English ethnicity has 92 percent more of his or her genes in offspring as a result. It is almost equivalent to having twice the number of children with an English spouse. Thus assortative mating by ethnicity can have large fitness benefits, the largest derived from choosing mates within geographic races. The large kinship effect of inter-racial mate choice is shown in Table 8.1. The numbers are the percentage fitnesses gained through racially endogamous matings, discounting costs such as foregone hybrid vigour.

| | AFR | NEC | EUC | NEA | ANE | AME | SEA | PAI |
|---|---|---|---|---|---|---|---|---|
| Africans | | | | | | | | |
| Non-European Caucasoids | 54 | | | | | | | |
| European Caucasoids | 66 | 6 | | | | | | |
| Northeast Asians | 79 | 26 | 38 | | | | | |
| Arctic Northeast Asians | 80 | 28 | 30 | 18 | | | | |
| Amerindians | 90 | 38 | 42 | 30 | 23 | | | |
| Southeast Asians | 88 | 38 | 50 | 25 | 42 | 54 | | |
| Pacific Islanders | 100 | 38 | 54 | 29 | 47 | 70 | 17 | |
| New Guineans and Australians | 99 | 47 | 54 | 29 | 41 | 58 | 50 | 32 |
| | AFR | NEC | EUC | NEA | ANE | AME | SEA | PAI |

*Table 8.1. Percentage parental kinship gained by endogamous versus exogamous mate choice between nine races. For example, a Pacific Islander who chooses another Pacific Islander as a mate will have children with 70 percent greater kinship than if he or she had chosen an Amerindian mate. (Calculated from Table 3.2, p. 68.)*

What would a hypothetical fitness-conserver do? It depends on how much effort can be profitably expended on parenting and ethnic group strategies. An individual whose reproductive strategy involves less parental investment might be less choosy in selecting mates and make up any kinship lost in each child by producing more. Assuming low parental investment, some argue that positively *avoiding* mating with fellow ethnics might be adaptive because doing so would increase the number of ethnic kin.[11] Not mating with co-ethnics means that they also are more likely to mate outside the group, the net effect being to accelerate reproduction of one's distinctive genes. To take the previous example, endogamy by Bantu or English increases kinship 92 percent over exogamy. But exogamy would double the number of descendants for each group by doubling the number of mating within each group, yielding an 8 percent advantage over endogamy (assuming equal population size and no endogamous matings). Admittedly such a leap in fitness could only occur once. Still, intermarriage can be seen as a means of spreading one's genes into other populations.

One problem with this argument is that it fails to account for the substantial efficiencies of parental investment foregone by exogamy and the resulting impact on familial and ethnic competitiveness. Humans are extremely '*K* selected', meaning that they have relatively few offspring and carefully nurture each one. This large parental investment increases in efficiency when more parental genes are carried by offspring. For each exogamous individual, maintaining individual fitness would require having more children, in the English-Bantu case 92 percent more, and investing less in each child. Outbreeding constitutes, if it is to be adaptive, a partial reduction of a high-*K* strategy. Such a compromise would not be necessary if fellow ethnics could be relied upon to be exogamous, since then all would experience the same fall in fitness and noone would lose relative to any other. But can fellow ethnics be relied upon? This leads to a more subtle failing of the outbreeding argument. In effect, it recommends a group strategy—cooperative outbreeding—without allowing for free riders. An individual who urges exogamy for co-ethnics but is personally endogamic, will have higher individual fitness than those who take his advice. Such free-riding might be prevented by social controls that enforced outbreeding, but how could a tradition of such controls develop in a group that is perpetually dissolving into other traditions, unless the tradition were externally imposed? Endogamy not only preserves high levels of kinship but also helps reproduce the traditions needed to maintain group strategies of many kinds, including endogamy.

The outbreeding argument also fails from the universalist position, where one seeks to optimize everyone's fitness down the generations. An individual who mates outside the group not only reduces his own parental kinship, but that of his mate and his children. Panmixia of the global population would cause parental kinship to decline to 0.25, from an average endogamous level of about 0.37.[12] This would mean a universal loss of parental kinship and hence individual fitness of 32 percent, unless family size increased by that amount. If the number of children is held constant, which it must in a sustainable global society, universal endogamy yields the highest individual fitness for the greatest number down the generations. Relative fitness would not suffer if every population underwent panmixia, but this brings us back to the problem of countering free riders and ensuring uniform, thorough interbreeding. Wilful free-riding can be predicted due to peoples resisting loss of identity, while involuntary free-riding would be inevitable due to different rates of mixing around the globe. Since the process would take many centuries to complete (if it ever could be), free-riding, whatever its causes, would have time to produce radically uneven fitness outcomes.

In mammalian species male parental investment is more variable and typically less than that of females because producing offspring can cost nothing more than some spermatozoa; but at the minimum mothers must gestate and suckle offspring. Considering the extended period of dependency of human neonates, women are generally more constrained to follow a high-investment reproductive strategy than are men, and should therefore gain more fitness from endogamous matings. The same strategy will be adaptive for fathers who invest in their children.

Endogamy should be especially adaptive for the middle classes in modern societies where family size is small and parental investment high.[13] For them there seems to be a trade-off in modern societies between individual fitness and exogamy. A family of replacement size, which averages about 2.1 children in societies with low child mortality, is still necessary to maintain relative fitness within a stable homogeneous population. But outbred families would need to have more children to reach replacement level, because their children carry a lower concentration of the parents' genes. For families with ordinary means, more children could mean less investment in each child, due to limited resources of time, energy and funds. Lower investment would then reduce educational and other opportunities, and with it social mobility. Yet the exogamous family must either have more children and risk them slipping in social and economic status, or accept below-replacement fitness. If this is an exaggeration, if extra children are not a strain on middle class resources, exogamous families can make up lost fitness by having larger families. But note that they will not recover their fitness deficit from a secular increase in family size, because each extra endogamous child carries a higher concentration of its parents' genes. To catch up, exogamous families must be larger than the average endogamous family, and as we have seen, sometimes a great deal larger. These considerations point to endogamy as a strong facilitator of a sustainable middle class strategy, except for those wealthy families for whom parenting resources are effectively unlimited. Traditional injunctions against intermarriage may be expressions of prejudice but they also convey prescient wisdom, especially in the case of genetically distant populations.

Even for wealthy families endogamy is an added bonus, since it should help retain valued characteristics in the family lineage. Any characteristic with high heritability, such as racial characteristics, personality and intellectual ability, will be more reliably passed on when both parents come from the same stock. In sexual reproduction polygenetic traits regress to the population mean, so that children's ratings on those traits tend to fall half way between the parental average and the population average. But when a local inbreeding population, such as an ethnic group, has more (or fewer) alleles for a characteristic than is typical in the global population, it can maintain a higher (or lower) average frequency of that characteristic. By the same logic, exogamy can free a lineage from unwanted characteristics, though at the cost of reduced fitness.

Everyone has an interest in maintaining the family component of his or her fitness portfolio. No personal commitment is more intense than that of parents to children. Except in times of war, we devote far more personal resources to our families than to any other social category. Increasing the proportion of endogamous marriages would therefore raise the fitness of the majority of humans. The dramatic way in which endogamy increases individual fitness is further justification for the universal nationalism I advocated in Chapter 7. Raising children within national communities would increase the likelihood of them marrying fellow ethics.

*Three caveats*

This is an appropriate place to reiterate the caution I expressed at the beginning of this chapter, that much of this discussion is based on theory drawn from population genetics. It is a recently researched subject. The formulas that I have used appear correct, but there might be factors for which they do not account. Moreover, the population genetic data used in the calculations will almost certainly be revised as more genes are assayed. Consequently, I am more certain of the proposition that intermarriage reduces kinship between parent and child than I am about the magnitude of that reduction.

The second caveat is a reminder from the opening paragraph of this chapter, that unless specifically stated otherwise, I am not discussing actual behaviour but the hypothetical behaviour of a 'fitness maximizer' or a 'rationally minded individual'.

The third caveat is that none of the foregoing should be interpreted as implying that mixed-race children are inferior to pure bred children. The only difference is that they carry fewer of their parents' distinctive genes than do purebreds. Mixed-race children are likely to benefit from hybrid vigour, suffering less from physical and psychological disease. In some cases their mix of physical and behavioural qualities might make them more competitive than a child descended from one ethny.

*Mixed-descent individuals*

So far, I have only discussed the situation from the perspective of parents' genetic interests. The starting position for taking the perspective of the children's generation is that their genetic interests are as important as anyone else's. Whether a child carries a greater or lesser concentration of its parent's genes does not prevent it living an adaptive life. But if the child is to behave adaptively throughout its life, its decisions will be affected by its parents' mate-choice behaviour, because those earlier actions place the child in a particular matrix of extended kinship relations, not only within the family but within and between ethnies.

Harpending (Appendix 1) shows that an individual descended from two ethnic groups has zero genetic kinship with random members of both ancestral ethnies.[14] Zero does not sound like much, but it is actually intermediate, since in a population consisting of two ethnies with genetic distance $F_{ST}$, co-ethnics have kinship $F_{ST}$, while pairs drawn from different ethnies have kinship $-F_{ST}$. If the mixed-descent individual chooses a spouse from one ancestral ethny, his or her children will be more closely related to that ethny. Choosing a spouse from the mixed category will keep the children in that category. When the ancestral ethnies are within the same regional population or geographic race, the children inherit a more generalized set of interests consisting of the common regional or racial genetic interests of those larger sets. Those interests are quantified in Table 3.2 (p. 68) based on assay data provided by Cavalli-Sforza and colleagues.[15] Such individuals are simultaneously of mixed descent with regard to their ancestral ethnies and of single descent with regard to the global population, when the latter is conceived as an assemblage of such regions. Hence the child of a Japanese-Korean couple is of mixed descent viewed from a Japanese or Korean perspective, but of single descent when viewed from the global perspective. Closely related ethnies can be considered as single subdivisions of the species.

The result is that most mixed-descent individuals still have ethnic genetic interests. Moreover, like their parents, a mixed-descent individual's choice of mate affects the ethnic genetic interests of his or her children. Some of those effects follow in a straightforward manner from population genetics and inclusive fitness theory.

*Implications of intermarriage for ethnic altruism*

Mixed marriages produce differences of ethnic genetic interest between parents and between parents and children, though this need not lead to a conflict of interests. A woman of unmixed ethnic background married to a man from a different ethnic group has genetic interests in her children as well as in her ethnic group. Her husband's interests overlap with hers, since their genes share a common fate in offspring. However they have different ethnic genetic interests that come into opposition in circumstances of competition between the two ethnies or differential investment by one partner in his or her ethny. Their children have genetic interests in both their ancestral ethnies. Should those ethnies engage in competition the children will not have an interest in showing exclusive loyalty to either group. This contrasts with the parents, each of whom has a large interest in only one ethny. In times of ethnic conflict, aid by either parent for his or her ethny constitutes an assault on the other parent's interests, and a deviation from the neutral strategy that is adaptive for the children. Conversely, the children's neutral strategy represents a break with familial solidarity when the parents are engaged in aiding their ethnies. Thus there will be less genetic payoff from family solidarity, because familial and ethnic interests do not coincide; indeed, they call for opposing investments, or opposition between commitment and neutrality. The long term effect of optimal investment might be to reduce ethnic cohesion until continued intermarriage produces a new tight web of relatedness constituting a new descent group—a new ethny. This new ethny would then take precedence over its two forerunner ethnies, and familial and ethnic investment would once again be aligned.

Generational conflicts of interest need never arise if the parents' ethnies never come into competition. Thus, for the adaptively minded, intermarriage is a motive for preserving ethnic peace. More generally, it increases the stakes for preventing zero-sum competition, such as contests over territory and social status. That is not to say that group competition can be prevented, especially in mixed societies. Recall from Chapter 6 that harmony of ethnic interests is brittle in multi-ethnic societies, since it is vulnerable to the fluctuations of relations between the constituent ethnies wherever they reside, whether within or outside the state borders. A significant change in relative group numbers within the state will tend to dissolve any implicit compact based on maintenance of the status quo. The important point here is that the solidarity of ethnically mixed families is predicated on harmonious ethnic relations in the larger society, which is mostly outside the control of family members.

(I should re-emphasise that these speculations are meant to apply only to hypothetical fitness maximizers. Research on the actual behaviour of ethnically mixed families is thin.)

The solidarity of single-descent families is not dependent on ethnic relations. Endogamy sets up synergies between domains of genetic interest, increasing efficiency. An investment in the family is simultaneously an investment in part of the ethny, and vice versa. Exogamy weakens or actually breaks this synergy, especially between genetically distant ethnies. The splitting-off of families from ethnic genetic interests makes a balanced apportioning of investment between the two domains difficult if not impossible in times of competition between the ancestral ethnies. Fitness maximizers are faced with the invidious choice of maintaining family or ethnic solidarity, but not both simultaneously.

The splitting-off of families from parental ethnies need not be permanent on both sides. Subsequent intermarriage with either ethny quickly restores alignment between familial and ethnic genetic interests, if one counts 'quickly' to mean a handful of generations over the better part of a century. So exogamy is not a permanent sentence of genetic distance from all ancestral ethnies, although rejoining one necessarily increases the distance from the others.

This is where the 'peace dividend' comes in. In Chapter 6 I argued that in situations of uncertainty genetic interest is more prudently defended by favouring investment in the family over the ethny. And now we see that a mixed-ethnic individual has divided ethnic genetic interests when his or her ancestral ethnies compete one with the other. These two considerations taken together strengthen the priority of family over ethny for individuals of mixed descent, at least when the ancestral ethnies contribute roughly equal shares to such individuals' genome. Thus a rational fitness maximizer of mixed descent will show less ethnic solidarity to either ancestral ethny than will mono-ethnic individuals, though special conditions can make such solidarity rational.

Differences in population size is one special condition that might cause mixed-descent individuals to switch from a neutral stance to active support of one of his ancestral ethnies. Normally the predominant ethnic interests of an individual who is descended from two or more ethnic groups will depend on the proportion of ancestors from each group. When three grandparents are of ethnicity P and one of ethnicity Q, then the predominant ethnic interest lies with group P. This follows from the nature of sexual reproduction, in which a random-sample of half each parent's genes is inherited by children, and a quarter by grandchildren. The genetic interest residing in an ethnic group is proportional to the size of the group multiplied by the fraction of the individual's lineage derived from that group. A mixed-descent individual might have a much larger genetic interest in an ancestral ethny to which only one grandparent belongs, when that ethny is large. Strategic circumstances are also important. If one ancestral ethny is not faced with invasion but the other is, it would be adaptive to aid the threatened ethny. Similarly, if one ancestral ethny is wealthy and the other poor, a given amount of altruism would boost genetic interests more if it went to the latter.

Multiculturalism presents a special strategic environment for balancing family and ethnic investment. The individual belonging to the majority ethny in a multicultural state who marries into a minority ethnic group is confronted by competition between family and ethnic interests, because the multicultural state is predicated on majority passivity and minority mobilization. According to the multicultural rules of the game, minorities are most likely to thrive and avoid majority discrimination when the majority does not assert its interests but minorities remain relatively mobilized. For a minority parent with pure- or mixed-descent children multicultural doctrine presents no conflict of interests so long as the children continue to identify with the minority ethnic group. But a majority parent with mixed-descent children is faced with an invidious choice. Cooperating with the multicultural regime will maximize his children's individual fitness by ensuring them access to the minority benefits that multiculturalism affords. However, that entails the parent withdrawing investment from his own ethny to prevent it from asserting its group interests and thus destabilizing the regime. Prioritizing the large genetic interest in his ethny would automatically jeopardize his childrens special benefits. The invidiousness of this conflict of interest is reinforced by the fact that strong family feeling and patriotism tend to coincide. If love of tribe entails nepotistic feelings, then intense altruism towards close kin sits incongruously with indifference to ethny.

The advantage of endogamy in aligning family and ethnic genetic interests in ethnically mixed societies perhaps explains the emphasis placed on endogamy by long-lived minorities (e.g. overseas Chinese, Orthodox Jews, Gypsies, diaspora Armenians, German communities in Eastern Europe), by ethnies in segmented societies, and colonial peoples.[16]

### *8.2 Childless Individuals*

Individuals who remain childless are an interesting case because there are fewer close kin in whom they can invest. This makes the extended family and ethny more attractive investments. In allocating resources between these two interests, the extended family of nieces, nephews, cousins etc. will usually remain the core investment in a prudent portfolio. If the individual finds himself with little or no family of any description, the ethny can be more heavily weighted.

The evolutionary logic for childless individuals to direct a greater share of resources to the ethny follows from Chapter 5, in which I showed that Hamilton's Rule for adaptive altruism can apply to the ethnic family. Interestingly, it also applies to one form of altruism, in which the individual refrains from drawing on group resources. The argument is provided by Hamilton's analysis of the adaptiveness of receiving altruism. Hamilton's 1964 model (Part I) concerns receiving as well as giving altruism. It benefits an individual's inclusive fitness to receive more than he gives, even from close relatives, when doing so results in more copies of his distinctive genes. This will usually be the case, because an individual's children carry half of his genome, while nieces and nephews carry only one quarter and cousins one eighth. But there are adaptive limits to accepting largesse. When a resource would allow a sister to produce more than two offspring, but only one by ego, it is maladaptive for ego to receive that resource from the sister. Hamilton maintained that a genetic predisposition to take altruism would then select for genes that channel taking from distant kin rather than from close kin.[17] 'In the model world of genetically controlled behaviour we expect to find that sibs deprive one another of reproductive prerequisites provided they can themselves make use of at least half of what they take. . . . Clearly from a gene's point of view it is worthwhile to deprive a large number of distant relatives in order to extract a small reproductive advantage.'[18] It follows that when an individual is unable to produce any more offspring or otherwise benefit from further resources, it is maladaptive to receive anything from kin, no matter how distant, since they have a chance of deploying their resource to produce children with whom the individual shares some distinctive genes.

### *8.3 Individuals who Adopt*

Adoption is usually adaptive for the homeless child. Even if the typical adoptee receives somewhat inferior treatment from the foster parents,[19] most survive and thrive. However, unrequited child hunger, especially women's, can be maladaptive. Cultural anthropologist Marshall Sahlins thought that sociobiological theory implied that rearing unrelated children is maladaptive.[20] Anthropological data support this contention because in traditional societies most adoption occurs within extended families,[21] meaning that adoptive parents invest in kin. However, the many benefits of having children, even when not genetically related to the adopter, contradict Sahlins. It is impossible to formulate a simple rule for comparing fitness costs and benefits of adopting. Adopted children reciprocate, to some degree, the investment they receive from the adoptive family. Preindustrial economies were based in the home where adopted children assisted with work, including care for siblings. An unrelated child is often of great emotional and social benefit to a family.

When a choice is available, it is of course most adaptive to adopt close kin. Adopting a niece or nephew or grandchild will usually conserve familial genetic interests more than adopting a cousin. Adopting relatives will generally be more adaptive than adopting a random member of one's ethnic group. The asymmetries of adaptiveness continue out in concentric circles of kinship. Investing in a co-ethnic child will generally be more adaptive than investing in a child chosen at random from the world population.

Unless the reciprocity from adopted children is large, it will be maladaptive for an individual (or group of individuals) to adopt a large number of children from genetically distant populations. It is also maladaptive for fellow ethnics, the more so in a social-welfare state, since the effect is to induce taxpayers to subsidize their own genetic replacement. In this situation the adoptive parents volunteer more than their own resources and fitness to foster children; they also donate the fitness of their fellow ethnics. To return to the examples given in Chapter 3, and again assuming that every territory has some finite carrying capacity, when 100 English couples adopt 100 Bantu children, that reduces collective English fitness by 46 children. But adopting 100 Danish children reduces the collective loss to less than one child. When the adopted children are English only the adopting couple loses fitness, and that loss is zero if they are infertile and unable to adopt kin.

That leaves a great many needy children in the world. What is the optimal strategy for genuine altruists—those motivated by children's needs rather than their own child hunger? An optimal strategy will show some responsibility towards the home ethny, as urged by the early American sociologist Edward Ross: '[T]hose who are to come after us stretch forth beseeching hands as well as the masses on the other side of the globe.'[22] Cross-racial support for children would be much less maladaptive for donors and their ethnic groups if they sent support to the children, instead of bringing the children to the supporting society. Individuals who wish to help homeless children but also care for their own people could subsidize adopting couples from the child's own ethny, or kindred ethnies.

Aid to foreign children, indeed foreign aid in general, is rendered evolutionarily stable when the donating ethny's fitness is secured within its own borders. Societies can be expected to show greatest generosity to foreigners when their benevolent feelings are not compromised by ethnocentric reactions to growing numbers of foreigners within their territory. This might help explain the pronounced correlation between national homogeneity and foreign aid as a percentage of gross domestic production.[23]

### *8.4 Homosexuals*

This is a short section because there is little that is special to homosexual genetic interests. Many men and women of homosexual orientation have opposite-sex spouses and children, due to custom or religious pressure or the wish for children. For them it is possible to invest in all the nested circles of genetic interest, from family to ethny and mankind. Whether childless or not, homosexuals have the same genetic interests in their ethny as heterosexuals. A difference in distribution of genetic interests arises only when a person has few or no children due to homosexual orientation or for any other reason. In that case the extended family and the ethnic group become more important stores of distinctive genes than the nuclear family.

Naturally many homosexuals seek reforms to custom and law that protect their proximate interests, especially freedom from persecution for the expression of their sexuality. Homosexuals also have ultimate interests. A homosexual wishing to conserve his or her genetic interests should be favourably inclined towards the family because this is the basic unit of reproduction for their kin and ethny. Anything that undermines a society's pronatal values is a threat to its members' genetic interests, whatever their sexuality. When the homosexual is childless the threat is skewed towards his or her extended-familial and ethnic interests and away from nuclear–familial interests. But the threat remains. It is therefore prudent for homosexuals to promote cultural and legal outcomes that protect their rights in ways that do not subvert the family or ethny.

### *8.5 Women in Wealthy Modern Societies*

For many life is easy at the beginning of the twenty first-century. The great majority of individuals in modern wealthy societies are no longer faced with the age-old problem of finding enough food to support a replacement number of children. Food is plentiful. Similarly, modern medicine has banished much of the ancient scourge of disease, so that it is no longer necessary to bear many children to ensure the survival to adulthood of a few. Childbirth has never been safer, and achieving replacement fertility has never been more convenient. Birthing methods are still improving, for example as anthropological knowledge is applied to Western populations, somewhat reversing the medicalization that began in the late nineteenth century.[24] A modern woman can produce her replacement number of children (2.1 on average) and nurture them to school age within a decade, out of a typical life of seven or eight decades. Investment can decline thereafter until the children leave home and become established. Fatherhood is even less burdensome, since men do not gestate or suckle their babies. Men often bear the breadwinning role, but the entry of women into the non-domestic labour force in the latter half of the twentieth century has tended to reduce this responsibility. This loss of responsibility has corresponded to a loss of power and confidence, as argued by Lionel Tiger.[25] Contributing factors include women's access to efficient contraception, their freedom to compete with men in the corporate world of work, and feminist deployment of the rhetorics of grievance and ridicule to assault male pride. The same benefits and costs await all societies as wealth rises.

The family component of the modern fitness portfolio has never required less effort. Everyone's familial fitness is effectively subsidized by the accumulated innovation and work of preceding generations, who built the industrial and scientific infrastructure. They also filled the world with people, putting an end to large families as a viable mass strategy. Choice is still available, of course. Those for whom children are a joy can make up in numbers the reproductive shortfall of the infertile and those for whom children are not a priority.

The reality of modern societies is much more complicated than the idyll sketched above. Industrialization creates a demand for more skilled workers and an administrative and professional middle class, which in turn requires greater investment in education and training. The secular decline in fertility known as the demographic transition that began in Europe in the mid nineteenth century corresponded to the mature phase of the industrial revolution and the spread of government-sponsored universal education.[26] The middle class culture that developed was characterized by restrained fertility combined with relatively intense parental investment.[27] Despite the greater investment in children, large amounts of parents' time was freed up, especially mother's (men were not as burdened to begin with). If parenting requires less than half an adult's lifetime, what to do with the remainder?

Western societies are still adjusting to these multiple transformations. In recent decades the strategy of reducing fertility has gone beyond the optimal point, since the middle class no longer replaces itself, and is losing relative fitness.[28] This decline is probably inadvertent, a side product of modernity. The advent of efficient contraceptive technology in the 1960s reduced the frequency of chance conceptions, but mainly for the conscientious middle class. The same technology has allowed an unprecedented rise in casual sex, reducing the proximate value of marriage. The hedonistic quest for self-fulfilment leaves many individuals single and childless. Some women, intending to have children, find that when they stop contraception in their mid to late thirties they are unable to conceive. Those who conceive late are less likely to produce a healthy child. Declining religious authority has made way for alternate lifestyles that deemphasize child bearing, including a homosexual subculture, while the sheer scale of modern societies has expanded many other subcultures into enclosing social worlds. For example, students often maintain the pattern of single or communal living in proximity to a university into their late 20s and early 30s.

Male adjustment to low fertility has not been as difficult as female adjustment. The traditional male role of supporting the family with work outside the home is flexible in accommodating more or fewer children. But family life in modern industrial societies has crammed mothers into a most unnatural mould. The role of mother and housewife, which once took up much of women's adult life, can now be completed in as little as twenty years, while life span has increased to over seven decades. The doctrine that a woman's place is in the home was perceived, understandably, as oppressive by women trapped in empty houses in vast residential suburbs lacking social or work amenities. The modern capitalist economy had produced an environment for women very different to the hunter-gatherer milieu of life-long reciprocal support by friends and relatives. Perhaps the most profound difference is that observed by primatologist Sarah Hrdy, in which the isolated nuclear family inhibits the 'cooperative breeding' observed in traditional societies—women helping each other rear children and manage gender relations.[29]

One result of women's changing roles is that they have entered the non-domestic workforce in large numbers and begun competing directly with men for resources and status, naturally demanding the abolition of discriminatory practices and assumptions. Wage earning has given women more independence, and brought them the satisfactions of achievement and society outside the home. In the area of high pressure careers it has also helped depress their fertility well below replacement. One survey of female professionals in the United States, conducted by Sylvia Hewlett in the 1990s, found that 49 percent of those over 40 were childless, compared with only 19 percent of comparable male executives.[30] The main cause appeared to be that only 57 percent of these women were married, compared to 83 percent of the men. Hewlett found that marriage became more difficult as the women's status rose. As a rule, they either married young or not at all. Only 10 percent of the women were first married after age 30, only 1 percent after age 35. Interviews indicated that few men were interested in marrying, or even courting, high status women.

The strains of women's changing roles has created a demand for therapeutic ideologies. The ideologies on offer at any particular time reflect broader intellectual trends. Feminists in the early twentieth century such as Margaret Sanger sought to ennoble the mother role,[31] while the Marxist oriented theorists of the 1960s and 1970s often sought to abolish it. The 1994 *Concise Oxford Dictionary of Sociology* under the entry 'Motherhood', states: '. . . [S]ome feminist theorists have suggested that it is the biological fact of child bearing that is the key source of women's oppression . . . .'[32] More moderate views opposed the denigration of female reproduction. But radical feminism continued to influence curricula in Western universities with a doctrine that denied sex differences and railed against what was seen as the idiocy of much female life. In this guise feminism is a movement against not only female fitness but also the feminine component of human nature. Kate Millett stated that feminine women show 'passivity, ignorance, docility, virtue and ineffectuality'.[33]

The radical feminist political movement has been so far removed from biological reality that it has opposed scientific findings about child development when those findings tended to raise the status of motherhood. Science itself has been criticized as a patriarchal mode of thought. An example is the chorus of denunciation that met the distinguished developmental psychologist John Bowlby, whose insights into the critical need of babies for secure attachment to a mother figure have been amply confirmed by empirical studies.[34] Bowlby stated the following in an interview in 1989, commenting on the usual standard of female employment in Britain at the time, not high-flying professional careers:

> This whole business of mothers going to work, it's so bitterly controversial, but I do not think it's a good idea. I mean women go out to work and make some fiddly little bit of gadgetry which has no particular social value, and children are looked after in indifferent day nurseries. It's very difficult to get people to look after other people's children. Looking after your own children is hard work. But you get some rewards in that. Looking after other people's children is very hard work, and you don't get many rewards for it. I think that the role of parents has been grossly undervalued. . . . all the emphasis has been put on so-called economic prosperity.[35]

Bowlby's views were attacked even though they are consistent with small families and a relatively brief parental phase of life. Perhaps an ideology that rejects hormonal effects on behaviour, or *any* significant biological effect, cannot be expected to take seriously women's reproductive interests. Radical feminist theorists are exclusively concerned with proximate interests such as resources and status, giving no weight to the ultimate interest of reproduction.

Sarah Blaffer Hrdy is one feminist who is familiar with the modern neo-Darwinian synthesis.[36] She notes that Bowlby's analysis of babies' need for attachment is compatible with day care, so long as it resembles a family and provides the child with a sense of belonging. 'Acknowledging infant needs does not necessarily enslave mothers.'[37] However, economic realities, especially the shortage of reliable parent substitutes, means that mothers who care about their babies' well-being will often have to bear the responsibility of providing secure attachment, that is, of being close, for the first few years.[38] Hrdy recommends Bowlby's advice to fathers to get more involved in parenting their newborn babies and to continue this investment through childhood.[39]

The success of ideas that denigrate motherhood among professional women is well described by Hrdy.[40] Women torn between (out-of-home) vocation and children suffer from 'motherguilt' and are naturally attracted by ideas that offer release, for example by Simone de Beauvoir's talk about enslavement of mothers (expressed years before she came to regret her own childlessness). The emotional appeal of such ideas can be considerable. Mothers who must work (outside the home) or desperately want to do so are often afflicted by self doubts about leaving a child in someone else's care:

> Just who comforted her infant when she was otherwise occupied? Were they loving? What will be the long-term psychological consequences of repeated separations? For a mother who has just wrung herself free from the desperate grip of a toddler, frantic and unwilling to stay at daycare while Mom gets to her shift at work, such questions are sickening. . . . How . . . expedient to simply reject any theory that legitimizes infant needs.[41]

As a result there is market demand for polemical 'get the mother off the hook' books that 'distort and caricature what evolutionists and developmentalists were saying about infant needs . . .'.[42]

Hrdy argues that the general fall-off in fertility in modern societies is natural, that it is due to women choosing to forego childbearing in order to better themselves economically and in status. 'Wherever women have both control over their reproductive opportunities and a chance to better themselves, women opt for well-being and economic security over having more children.'[43] At first sight this trend contradicts the evolutionary definition of success as reproductive fitness. Hrdy calls this definition 'crass', but also calls it her 'world view'. Apparently she does not rate inclusive fitness as a high or noble objective, even though it is more important, more fundamental, than individual survival, and she considers the latter to be a worthwhile goal. How does Hrdy reconcile her neo-Darwinism and apparent promotion of maladaptive choices by women?

Western women's below-replacement fertility does not contradict evolutionary expectations, Hrdy believes, because 'mothers evolved not to produce as many children as they could but to trade off quantity for quality, or to achieve a secure status, and in that way increase the chance that at least a few offspring will survive and prosper'.[44] Before reaching this conclusion, Hrdy reviews data on the reproductive history of women in hunter-gatherer societies, the closest observable approximation to human life in the Pleistocene. As many as half of women in traditional Kalahari Bushmen society died with no surviving offspring. 'This is why quantity has rarely been the top priority for a mother. The well-being of her children and their quality of life, usually inseparable from her own, were primary.'[45] A relevant point here is that the absence of effective contraceptive technology meant a reliable supply of pregnancies to most women, obviating the need for a psychological motivation for quantity.

Hrdy's interpretation is persuasive as explanation. Without using the term 'adaptive', she argues plausibly that in hunter-gatherer societies it was more adaptive to seek quality rather than quantity of children, because this was more likely to yield genetic continuity. The implication is that genetic interest in the form of continuity is the ultimate criterion of what is adaptive. Genetic extinction is hardly adaptive, and neither is loss of fitness in relation to other women in society. Certainly women might not have been selected to heed the number of children they produce but instead to focus on quality of life. But a further evolutionary question might suggest itself to those who accept the view that reproductive fitness is fundamental to success: is women's lack of focus on number of offspring adaptive in the modern world? A brief historical period of below-replacement fertility is not catastrophic, especially if it is an antidote to over-population. Neither does it depress relative fitness when most other women in one's society have similar family sizes. As Hrdy points out, raising one child to maturity is better than raising none, since it puts one's genes in the next generation. But it will usually be maladaptive for an individual woman to lose relative fitness, whatever the society's average number of children. In which way is that a 'crass' assessment? That does not reduce to Herbert Spencer's view that women's function is to have children, that women are breeding machines. The point is that it is in a *woman's* interest to have children, at least the small number nowadays needed to continue her genes.

A similar point has been made by behavioural ecologist Bobbi Low.[46] Low and colleagues built a model that simulates the number of descendants produced by modern women who time their pregnancies differently. Women who delay childbirth risk genetic 'extinction', Low argues, because of the exponential advantage of early child bearers over several generations.[47] Even when family size is held constant and mortality is assumed to be zero, delayed childbirth causes the lineages to all but disappear. Late-reproducers are swamped genetically by early-reproducers. Low recognizes that the demographic transition changed the rules to reproduction, at least for middle class families, to high investment in few children. There is an optimal time in a woman's life to give birth, an age that depends on each woman's circumstances. It probably pays off for many women to delay childbearing until they can bring more skills and resources to their parenting and home-building roles. 'If you have your kids too early, it takes away from your own ability to develop and get what you need. And you may not have enough to invest in them. So they may not get enough to get a good start in life.'[48] But Low does not ignore genetic interests: '[I]f you wait too long, you lose out reproductively.' 'We can ignore Mother Nature at some level, but there are real costs to ignoring what we evolved to do.'[49]

Hrdy goes to lengths to quote some of the antiquated and often offensive opinions of male evolutionists towards women, taking aim at Herbert Spencer and Charles Darwin.[50] She points out many erroneous ideas, especially the notion that women are meant exclusively to breed, and should get on solely with their appointed function. Hrdy approvingly quotes Simone de Beauvoir's sarcastic comment made in 1949, when she wrote: 'Woman? Very simple, say the fanciers of simple formulas: she is a womb, an ovary; she is a female—this word is sufficient to define her.'[51] As in all other areas of knowledge there have been false assertions. But Hrdy never spells out what modern evolutionary thought has to say about women's (or anyone's) reproductive interests. While acknowledging that reproductive interest exists, she does not accord it priority over other aims:

> [T]here is little agreement about whose interests are to be maximized in a world where conflicting self-interests—between parents and offspring, between mothers and fathers, within families, between families—are endemic. What goal are we trying to achieve? Secure adults? Good citizens? Independent ones? Self-starters focused on fast tracks? Satisfied mothers? Reproductively successful family lines (in the recent past usually patrilines)? Maximized human potential? And if so, whose?[52]

The biological view represented by Low offers support for the emancipationist program of securing dignity and opportunities for women. The difference with radical feminist doctrine is that an approach based on natural science would never write-off child-bearing and -rearing. A trend towards greater male appreciation of genetic interests would give women considerable bargaining power in convincing men to treat parenting as a high calling and properly the work of a husband-wife team. Repudiating genetic interests would surely be an efficient method for freeing males of their family obligations.

The traditional female investment portfolio is conservative, emphasizing the highly concentrated store of distinctive genes found in each child. It is the strategy most proofed against free riders, appropriate for the large investments required in gestation, parturition, lactation, and years of nurture. It is a strategy relatively immune to the vagaries of investment in low-concentration genetic interests. The male penchant for speculating in ethnic interests is congruent with their lower expenditure on early child rearing, and their greater expendability. Perhaps the two strategies are complementary.

If adaptiveness is to be respected, the discussion of sex roles should incorporate knowledge of the genetic interests of women and men and how those interests can be brought into collaboration. Since the most important interest of both sexes is the well-being of children and social groups in whom both sexes have an equal stake, this is surely a promising basis for a fruitful partnership.

*Notes*

1 van den Berghe (1981).
2 Trivers (1974).
3 Godfray (1995).
4 Rushton (1989b, pp. 506–10); Rushton et al. (1984).
5 Grafen (1990).
6 Segal (1999).
7 Case et al. (2000; 2001).
8 Jankowiak and Diderich (2000).
9 Cavalli-Sforza and Bodmer (1971/1999, pp. 491–99).
10 Originally based on a personal communications from Harpending, 17.3.2002 and 16.4.2002.
11 Grafen (1990, p. 51); Waldman (1989).
12 Cavalli-Sforza et al. (1994, p. 118) state that the global $F_{ST}$ averaged over the 122 genes measured in 491 populations was about 12 percent. This accords with Lewontin's (1972) observation that the amount of genetic diversity existing between populations is between 10 and 15 percent.
13 Draper and Harpending (1982).
14 Harpending (personal communications, 27.3.2002). Harpending derived this result as follows. Consider a child of mixed descent from ethnies A and B, and random individual of either ethny, say B. Sample an allele at random from a certain locus of the child's genome, then another from the same locus in the random individual of B. With probability $^1/_2$ the random allele in the child derives from ethny A, with kinship to individual B of $-F_{ST}$. With probability $^1/_2$ it derives from ethny B, with kinship to individual B of $F_{ST}$. The two probabilities cancel, leaving a kinship of child to individual B of zero. See Appendix 1 for related proofs by Harpending.
15 Cavalli-Sforza et al. (1994, p. 80).
16 Bonacich (1973); Landa (1992); MacDonald (1994); Phillips (1999); Zenner (1982; 1991/1996).
17 Hamilton (1964/1996, p. 40).
18 *Ibid.*, pp. 45–6.
19 Case et al. (2000); Daly and Wilson (1999).
20 Sahlins (1976/1977).
21 Silk (1980).
22 Ross (1914, preface).
23 W. Masters (in press).
24 Schiefenhövel and Schiefenhövel (1999).
25 Tiger (1999).
26 Sanderson and Dubrow (2000).
27 van den Berghe (1979).
28 Vining (1986); Lynn (1996).
29 Hrdy (1999).
30 Hewlett (2002).
31 Chesler (1992).
32 Marshall (1994).
33 Millett (1971, p. 26).

34 Bowlby (1988); Hrdy (1999, pp. 485–510).
35 Quoted by Hrdy (1999, p. 495).
36 Hrdy (1981; 1999).
37 Hrdy (1999, p. 494).
38 *Ibid.*, p. 497.
39 *Ibid.*, pp. 509–10.
40 *Ibid.*, pp. 488–93.
41 *Ibid.*, p. 492.
42 *Ibid.*, p. 492.
43 *Ibid.*, p. 9.
44 *Ibid.*, p. 10.
45 *Ibid.*, p. 8.
46 Low (2000); Marano (2002).
47 Marano (2002).
48 Marano (2002); and see Low (2000, pp. 332–3).
49 Marano (2002).
50 Hrdy (1999, p. 20). Hrdy's criticism of Spencer extends to his preference in women. Spencer refused a proposal from the early feminist George Eliot to have a child together. Spencer refused on a eugenic ground, citing Eliot's plain looks: 'beauty is a signal of fertility and genetic quality.' Hrdy rejects Spencer's opinion on the authority of a linguist (Pinker 1997, p. 480), both apparently unaware of the literature confirming Spencer's insight (e.g. Buss 1994; Grammer 1993; Rikowski and Grammer 1999). Perhaps Spencer's decision was based on an evolved mate-choice preference. Hrdy is persuasive when she suggests that Spencer's reason for rejecting Eliot was not theoretical but aesthetic. She quotes his autobiography on the matter: '[T]he lack of physical attraction was fatal.' Spencer described Eliot as being of 'ordinary feminine height [but] strongly built' whose physique showed 'a trace of that masculinity characterizing her intellect' (Hrdy 1999, p. 545, fn. 24).
51 *Ibid.*, p. 17.
52 *Ibid.*, p. 493.

# Part III: Ethics

# 9. On the Ethics of Defending Genetic Interests

*Summary*
I formulate an ethic of 'adaptive utilitarianism' according to which a good act is one that increases or protects the fitness of the greater number. I apply this ethic in an attempt to answer three fundamental questions raised by the concept of genetic interest, especially the ethnic component (followed by short answers): (9a) Under which conditions if any does defending genetic interests justify frustrating other interests? Since genetic interests are shared according to degree of kinship, individuals have duties to family, ethny, and humanity ahead of strictly private needs. (9b) Should the ultimate interest of genetic fitness be accorded absolute priority over other interests? In principle 'yes', but in practice 'not always', since the effect of a behaviour on fitness is often unknown. (9c) What is the proper action when ultimate interests conflict? When ethnies conflict, adaptive utilitarianism is best satisfied by universal nationalism, since this ideology teaches respect for everyone's ethnic interests. Genetic continuity is compatible with peace between ethnies, with equality of opportunities within ethnies, but not with equality of fitness outcomes within ethnies, since a system that ensured such equality would be evolutionarily unstable. The ultimate form of liberty is the freedom to defend one's genetic interests.

## *Introduction: The limited intuitiveness of tribal ethics*

The discovery or clarification of an interest begs consideration of what is justified in its defence. In this chapter I raise and attempt to answer some basic questions of morality concerning the defence of genetic interests, especially in the domain of ethnic rivalry. I do so in the spirit of consilience, or unity of all knowledge, urged by E. O. Wilson. The Enlightenment will finally reach maturity, Wilson argues, when mankind deploys the knowledge gained from science to forge wiser, more humane policies.[1]

Some evolutionary thinkers have tried to explain morality, not only its content and behavioural underpinnings, but its social functions and evolutionary origins. Like other types of behaviour, the function and evolution of morality are closely linked, as explained by E. O. Wilson in his essay *On Human Nature*: 'The genes hold culture on a leash. The leash is very long, but inevitably values will be constrained in accordance with their effects on the human gene pool. Human behavior—like the deepest capacities for emotional response which drive and guide it—is the circuitous technique by which human genetic material has been and will be kept intact. Morality has no other demonstrable ultimate function.'[2]

In this chapter my aim is not to discuss how moral behaviour evolved but to evaluate genetic interests from the standpoint of human morality as it is now. Nevertheless, I try not to lose sight of the implications of Wilson's view that the moral instincts can change due to differential reproduction. From an evolutionary standpoint an ethical system that weeds out the genes or culture of those who practise it is a failure. The same evaluation surely follows from a humanitarian perspective. As Eibl-Eibesfeldt concludes his ethological analysis of ethics, 'An orientation toward survival value is important for the rational development of norms'.[3]

Readers will not find an appeal to religion in this chapter, though respect for religion is a necessary part of mature Enlightenment values. This claim might come as a surprise to those familiar with the contest between 'rationality' and religion that has marked the history of the Enlightenment since its emergence in the seventeenth century. But the proper role of science is to understand religion, not undermine it. As argued in earlier chapters, traditional religions have been overwhelmingly adaptive. Some evidence points towards an evolved psychological predisposition for producing religious commitment,[4] perhaps explaining why atheists tend to embrace ostensibly secular ideologies with religious fervour. In defending religious thought from the evolutionary perspective, D. S. Wilson puts rationality in its place thus:

> Adaptation is the gold standard against which rationality must be judged, along with all other forms of thought. Evolutionary biologists should be especially quick to grasp this point because they appreciate that the well-adapted mind is ultimately an organ of survival and reproduction. . . . It is the person who elevates factual truth above practical truth who must be accused of mental weakness from an evolutionary perspective.[5]

I make no apology for making this chapter an exercise in rational argumentation, not religious doctrine, but it is possible that I have erred in not suggesting ways to deploy religious behaviour to solve ethnic problems.[6]

One stepping-off point for working out the ethical implications of recognizing genetic interests within a scientific frame would be to document existing rights and customs pertaining to those interests. This would show, I think, that genetic interests are often accounted for in some ways, but indirectly. Examples include the legal and informal protection of life, liberty and property, socially imposed monogamy in the Western tradition, economic privileges accorded the family, parental right of access to and control over children, the incest prohibition, means of redress against adultery, and those situations where tribal and national mobilization happen to be adaptive. This was the thrust of my review in Chapter 6 of actual strategies used by individuals and nations to defend genetic interests.

The almost universal disquiet over abortion, felt also by those who practise it, is a special case because it helps isolate a concern for the survival of an individual genetic identity. Part of the debate concerns fertilized ova (zygotes) of one or a few cells, aborted using a 'morning after' pill, and embryos that have not yet attained recognizably human form, aborted surgically before the eighth week after conception. Those who have qualms about aborting zygotes behave as if they are respecting its genetic interests or as if they have a genetic interest in it. Certainly zygotes, which lack a nervous system, have no wishes, urges, or feelings that might define their interests. The argument against abortion that is based on the sanctity of life and which refers to a fertilized ovum as a 'human being' or a 'person' assumes the central importance of genetic identity. Thus the protection of individual life, even of individuals whose existence is defined only by their genetic codes, is a matter of grave concern in contemporary societies.

I take these examples as indicating that individual survival and family genetic interests are already largely provided for by law and custom. Westermarck had a plausible explanation for this confluence of law and biology. Sympathy for family life is widespread because everywhere this is the site of reproduction.[7] This universal mode of reproduction entails a shared interest in family life and, in an empathic species, generates widespread sympathy for the needs of parents and children. Consequently it requires no instruction in the arcane idea of inclusive fitness to convince most people that parents have a natural right to discriminate in favour of their offspring. However, ethnic genetic interests are usually more dilute than family interests, more dependent on culture for their recognition and defence, and thus rendered less recognizable by the loss of traditional social organization to the relentless march of modernity. An interest that is only partially discerned or invisible to many eyes, is unlikely to be the subject of adaptive moral feelings. Thus morality is likely to be more adaptive on the small scale, in matters of individual and familial relations, than in large scale ethnic affairs. While practical ethics should never drift too far from the anchor of sentiments, to think responsibly about ethnic affairs requires more deliberation and systematic fact-finding than is necessary in the personal realm.

Finding facts about gene frequencies entails the application of scientific methods. In modern societies scientific understanding is indeed indispensable for answering, and even comprehending, some of our moral dilemmas. But science alone cannot clarify the morality of ethnic affairs, or any other issue. Neo-Darwinism helps us understand the causes of cooperation and competition across all social species, but practical ethics must arrive at behavioural prescriptions or injunctions, which are qualitatively different to descriptive or explanatory statements. As Hamilton notes, '[T]he idea that such [genetically adaptive] behaviour is natural in man does not mean that it is right or even sensible under modern conditions'.[8] Science is limited to showing ways to optimize human values. The eighteenth century Scottish philosopher David Hume pointed out that an 'ought' statement cannot be derived logically from one or more 'is' statements. Values cannot be deduced from facts. Claiming otherwise is to commit what is known as the naturalistic fallacy. The deduction of a normative moral statement requires at least one normative premiss.

Developing Hume's position, some philosophers have argued that morality is emotive, deriving from feelings or attitudes.[9] 'Moral convictions . . . consist in norms for anger and for that first-person counterpart of anger, guilt.'[10] Emotivist ethics look to the various moral emotions to bridge the is-ought gap, to conscience and outrage, similar to what Kant called the 'moral law within'. I also make this assumption. Whatever rational frame is constructed for processing moral problems, the assertion of moral convictions is a necessary element of practical ethics. To fit that frame, to maintain a clear distinction between fact and value, moral assertions must be clearly identified as such.

Another limitation of science is that we are finite creatures with limited intelligence and knowledge; rationality is inherently 'bounded' or constrained. We cannot live by formal logic or elaborate formulas. Instead we get by with inductive inferences and rules of thumb.[11] We are also guided by aesthetic and religious sensibilities. Constrained as we are, we make moral judgements of great consequence, and must do so if we are to decide conflicts of interests. Choices are also forced in the game of life, every day genetic interests being won and squandered. A commentator who fails to advise people on how to defend their most precious assets is, by default, advocating the status quo, with its winners and losers. Ignorance is the only excuse for withholding advice on such a critical issue. Ignorance in this context cannot mean incomplete social knowledge, since knowledge of society is still rapidly accruing and the pace of social change is not slackening. Any commentator who is better informed than the public in some aspect of life's problems is obliged to offer an opinion on right conduct, building into that advice an assessment of his or her analytic certainty. That is my intent in this chapter.

### *Conflicts of Genetic Interest and a Utilitarian Frame*

Alexander argues from a sociobiological perspective, plausibly I think, that issues of morality only arise from conflicts of interest, that is, a conflict of non-moral values.[12] Three moral questions strike me as fundamental in thinking about conflicts of genetic interest. First, does defending these interests ever justify frustrating other interests? Secondly, if there are such conditions, should an individual's ultimate (genetic) interests be accorded absolute priority over others' proximate interests? Thirdly, what is the proper action when ultimate interests conflict? There are many moral issues concerning particular instances of clashing genetic interests, but these will usually entail answering one or more of the three fundamental questions. My opinions on these three questions, emphasizing the difficult ethnic dimension, will follow presently, after I have described the consequentialist, utilitarian ethical frame within which I formulate those opinions.

Consequentialism developed within the broad analytic philosophical tradition shaped by D. Hume, J. S. Mill, and most relevant to the present issue, by E. Westermarck's anthropological analysis of the moral emotions.[13] Within this broad tradition the ethical system of utilitarianism is particularly useful for thinking about conflicts of interests. The system was developed by Jeremy Bentham in the eighteenth century and most famously by John Stuart Mill in the nineteenth. There are other rational ethical systems, such as the contractarian approach represented by J. Rawls.[14] But utilitarianism looks especially promising as a moral heuristic for thinking about conflicts of interests because it pays special attention to the number of individuals affected by an act, and is thus readily turned to formulating ethical principles concerned with effects on populations. Utilitarianism has the added advantage of having been debated for over two centuries, with the result that its strengths and weaknesses are relatively well known.[15]

Utilitarianism is a teleological or consequentialist ethic, meaning that it evaluates an act as morally right if the act has desirable effects. Circularity is avoided by only considering effects that are non-moral, such as increasing well being of some kind.[16] This may be contrasted with deontological ethics in which an act is considered right or wrong in itself, whether in obedience to tradition, a deity, or intuition.[17] Teleological ethics have a deontological component, since they assert some *consequence* to be morally right in itself. Moral judgement must be inserted at some point in the analysis. The advantage of teleological ethics for present purposes is that they are attuned to consequences of acts, and therefore readily applied to problems of conflicting interests where actions are all important in determining which interest prevails.

Teleological morality is concerned with comparison of acts and their effects to determine which provides the greatest non-moral goodness. Goodness is often taken to be some feature of the 'good life' or good living. Plato thought this non-moral goodness was a state of mind, and agreed with Socrates that the unexamined life is not worth living. Aristotle held goodness to be an activity, while Bentham in *Introduction to the Principles of Morals and Legislation* (1789) and Mill in *Utilitarianism* (1863) maintained that a good act is one that produces happiness. Mill thought that the morally best act maximizes happiness for the greatest number.

Teleological ethics must also specify whose non-moral goodness is to be maximized. Egoistic utilitarianism argues for the rightness of maximizing personal happiness, while a group-oriented ethic places a premium on the happiness of the group. Another important distinction is made between act and rule utilitarianism. An act utilitarian judges each moral decision using the basic principle of utility, while a rule utilitarian refers to a set of rules, which are justified by that principle. Rule utilitarians argue that act utilitarianism does not account for aggregate effects of many individual decisions and that without rules the social and economic practices vital for civilized life would be impossible. This is a view also present in deontological ethics. Note that although rule and act utilitarianism have their differences, they agree on the utility principle: an act or rule is right only in so far as it increases happiness. In this sense utilitarianism is an expression of generalized benevolence and is intrinsically democratic in its emphasis on serving the majority interest.

Utilitarianism has its problems, especially with justice. There are realistic scenarios in which an act or rule justified by utilitarian ethics is repellent to moral intuition. One such scenario is as follows. A suspected murderer surrenders to the town sheriff and convinces the latter of his innocence. The murder is causing widespread anger and a mob outside the gaol house, constituting most of the town's adult population, calls for the suspect's lynching. Acceding to the mob's demand would bring an immediate rise in average happiness in the town, but doing so strikes us as dereliction of the sheriff's duty and also as immoral on the grounds of punishing an innocent person. Let us change the scenario a little. Now the sheriff discovers that not only is the suspect innocent but the real murderer is the mayor. The latter killed out of passion and is most unlikely to repeat the offence. Indeed, his conduct has been exemplary all his life, while the suspect is a vagabond petty thief who lowers the tone of the town, damaging its nascent tourist industry. Convicting the mayor will ravage the town's social order, produce widespread shame and embarrassment among his extended family, and effectively kill off the tourist industry, throwing dozens of breadwinners out of work. Only the sheriff is aware of the evidence needed to convict the mayor and absolve the suspect. What should he do? Utilitarianism dictates that he let the suspect hang and the mayor go free. Yet this strikes us as unjust. Mill considered objections from our intuitions about justice to be utilitarianism's worst problem.

A doctrine of rights brings justice into utilitarianism. Following Immanuel Kant's dictum, each individual is valued and his freedom and dignity are treated as ends, not means. This is a concession to deontological thinking and, as illustrated in the above example, is not derivable from the utility principle, especially from its maximization clause. However, it is only a partial break with emotivism. The break consists of the fact that people everywhere treat others as pure means with good or only slightly disturbed consciences when the effect of such treatment benefits them, meaning that Kant's dictum is not always intuitive. But it is also true that the force of the dictum originates in the moral rejection of suffering. As noted earlier, the utility principle itself originates in the intuition that it is morally good to increase happiness. Rights are thus consistent with the emotivist value-assumption of utilitarianism, though they come into conflict with its maximizing logic.

Injecting rights into utilitarianism reverses the evaluation of the previous example. Procedural justice requires that innocent individuals not be punished, even if this is in others' interests. Maximization of the happiness of the greatest number must be pursued within the constraint of inalienable rights. Doctrines of rights are also limited, especially by their deontological rigidity. In many circumstances rights alone offer no positive guide to individual or group action, while teleological ethics can suggest new directions for satisfying values. Utilitarianism's justice deficit will be revisited below in the context of genetic interests.

The absence of religion is a shortcoming of utilitarianism. It is incongruous that a rational ethic ignores an institution and associated psychological predisposition that have underpinned morality in all civilizations. For the reasons stated at the beginning of this chapter, I shall not attempt to join religion and consequentialist ethics, but suspect that the rift, though a defining element of utilitarianism, weakens the latter in all its guises. Synthesis is needed.

Another weakness of utilitarianism is its happiness criterion. Happiness is an emotion, and thus a proximate rather than an ultimate interest. As an indicator of ultimate interests it is better than nothing, but fallible. Individuals suffering from mania appear happy and claim to be so, but are prone to maladaptive behaviour. Drug addicts experience periods of intense happiness, and this can be maintained for a time if the supply of drugs is kept up. Yet drug addiction tends to be maladaptive. Humans strive for resources and status, that is clear, but achieving this goal does not increase happiness in any simple or predictable way. By contrast reproductive fitness is an objective measurable by number of offspring and continuity of one's familial and ethnic lineage.

The weakness of the happiness criterion is not fatal to the utilitarian enterprise because, as noted earlier, other criteria of non-moral goodness can be substituted for it. Bentham considered one such criterion to be living in a society governed by rational legal and political institutions.[18] G. E. Moore argued that consequences other than happiness, such as beauty, are important in evaluating an act's morality.[19]

### *Adaptiveness as utility*

In this section I argue that the structure of the utilitarian ethic can be retained while replacing criteria such as happiness or beauty with adaptiveness. From the perspective of modern biology the most important consequence of any act is how it affects genetic interests, how it affects adaptiveness. The consequence of ultimate import is not happiness of the greatest number but *adaptiveness of the greatest number*. This notion underpins a survival ethic—which I shall refer to as 'adaptive utilitarianism'—which has important advantages over happiness and other proximate criteria.

This ethic cannot be reduced to the social Darwinist doctrine of 'survival of the fittest'. Like the social Darwinists I shall argue that the freedom to compete within limits is a vital adaptive right, but the criterion of 'the greatest number' also leads to an emphasis on the need for cooperation and adoption of procedures for peacefully resolving conflicts. Placing a value on genetic survival is not the same as the belief that *laissez-faire* competition will always produce the most desirable social and economic outcomes. Those destined to lose in such a free-for-all have no interest in promoting it, and those destined to win might find a more pleasant path to victory. As Thomas Huxley argued, it is not true that what is in nature is necessarily desirable.[20] Suffering and extinction are commonplace in the wild; does that make them good? Theories of how species survive and evolve are only relevant to ethics by showing how to fulfil our values. Evolutionary biology informs a survivalist ethic in the same way that the science of nutrition helps decide what to eat. In both cases the driving value is the love of life, not particular means for securing it.

Adaptiveness has the advantage of corresponding to knowledge of the human condition, especially to observable states. We can observe individuals' (or groups') resources, the amount of control they have over their environment, their state of health, their fertility and life span, ability to defend themselves, and so on. Adaptive utilitarianism does not have a transient emotional state as its criterion of goodness, while retaining much of the intuitive appeal of classic utilitarianism. An actor concerned with maximizing his fitness will, of course, help his own fitness. He will also prefer to help more individuals rather than fewer, will prefer protection of equal rights for individuals and groups. Genetic interest starts with ego in a less intrinsic way than does the happiness interest; it is essentially social through being distributed across many individuals, since kin and fellow ethnics are, as Keith noted, 'members one of another'.[21] Classic utilitarianism might be more intuitive in the absence of biological knowledge because it holds up a standard of goodness that is itself a motivation. Whatever happiness and pleasure are, we want them; and we want to avoid pain. But the happiness criterion constitutes a radical ego-centricity that sits oddly beside the qualifying clause 'for the greatest number'. Emphasis on genetic interests automatically demotes ego to just one source of our vital interests, albeit the most concentrated. We know that our distinctive genes are contained in our families and ethnies. Adaptive utilitarianism is thus intrinsically social in scope before adding a number clause.

Separating the goodness criterion from the motivation for adopting it incurs a cost. That cost is reliance on scientific knowledge as a precondition for understanding the ethic in the first place, and dependence on cultural devices for motivating adherence to the ethic once adopted. On the other hand adaptive utilitarianism should be more sustainable in the long run because it is better for us. An adaptive utilitarian would condemn any practice that reduced fitness below replacement level, no matter how pleasurable. Drug-taking comes to mind, but also the sort of middle class culture common in developed societies that values consumption, comfort, and status over children.

More than classical utilitarianism, the adaptive version confronts the reality of competition. Political philosopher Hiram Caton criticizes Bentham's assumption that utilities are in harmony, that general benevolence can be achieved without treading on anyone's toes.[22] Bentham's assumption was most true in the realm of market economics. He agreed with Adam Smith that there was a commonality of economic interests. Bentham wrote: 'The more we become enlightened, the more benevolent shall we become; because we shall see that the interests of men coincide upon more points than they oppose each other.'[23] Caton remarks that Bentham's belief that utilities and human desires are in harmony caused him to 'neglect the great life conflicts that impart elevation and gravity to ethics'. In other realms utilitarians subsequently realized that interests do conflict, and advocated government sanctions sufficient to deter aggression against property or person. Government's main aim, they held, was to establish an artificial harmony of interests.[24] If Bentham had lived to take Darwin into account, his view of competition might have been more profound. In biological perspective the fundamental life conflict is acquisition of limited resources and mates in furtherance of reproductive fitness. It would be unfair to look back with more than two centuries' hindsight at Bentham's pre-Darwinian age and blame him for not considering fitness as his criterion for right action. Today there is less excuse.

Neo-Darwinism and the 'conflict sociology' tradition[25] are converging to describe the biological dimension of competing interests at the individual and group levels. The challenge for ethicists is to formulate workable rules of conduct that do not ignore the clash of ultimate interests.

Adaptive utilitarianism does not make the fundamental error of assuming a universal commonality of interests, despite scope for win-win outcomes in many circumstances. An extreme example of this error in utilitarianism is Peter Singer's argument in *Animal Liberation.* Singer believes that opposition to discrimination against women and other races is inconsistent with discrimination against non-human animals. [26] His argument begins with Bentham's starting assumption that in formulating rational ethics we must treat everyone's interests as equally important. He quotes Sidgwick's articulation of the same point: 'The good of any one individual is of no more importance, from the point of view (if I may say so) of the Universe, than the good of any other.'[27] Singer notes that 'the leading figures in contemporary moral philosophy' have made similarly egalitarian assumptions. From this position the argument can proceed in a straightforward manner:

> It is an implication of this principle of equality that our concern for others and our readiness to consider their interests ought not to depend on what they are like or on what abilities they may possess. . . . [T]he basic element—the taking into account of the interests of the being, whatever those interests may be—must, according to the principle of equality, be extended to all beings, black or white, masculine or feminine, human or non-human.[28]

Most will agree that animals should not be made to suffer. But Singer's proposition goes much further by asserting that we should give all humans and thinking animals equal consideration. This proposition is absurd on the face of it and is supported by flawed argument. Let me begin with Singer's proposition, just quoted. It is a trivial matter of observation and self reflection to know that we do not in fact give everyone equal consideration. Singer thinks this has to do with beliefs about what people are like or on their abilities. Nowhere does he mention relatedness or interpersonal ties. Yet it is largely kith and kin, not classes of ability, that receive our beneficence. It is kinship that generally predicts most-favoured treatment. Thus he misses the key moral question of discrimination due to familial and ethnic identity: is it right to discriminate in favour of those one loves? Singer does not mention love or affiliation or bonding or attachment. Doing so would have made it more difficult to treat discrimination as something done only against and not in favour of a category of people. It has usually been adaptive to discriminate in favour of friends, children, and in natural contexts, band and tribe. In the absence of perceived obligations of office or ideology that bar nepotism, people feel little or no moral compunction in doing so. And we discriminate in favour of our fellow human beings in preference to non-human animals. That is why most of us refuse to eat humans despite regularly eating non-human flesh.

As to Singer's argument that we *ought not* discriminate, either between humans or between humans and animals, it rests on an appeal to authority as a means of establishing the equality principle. But Sidgwick was wrong to say what he did, for in proper philosophical circles one does not claim that the Universe has a point of view. It has no such thing. To claim otherwise is bald anthropomorphism. Only thinking beings have points of view, and so far all thinking beings are evolved animals with an interest in propagating their genes, for example by discriminating in favour of offspring and tribe. Nowhere does Singer offer an argument as to why we should take an Olympian stance. It is hardly prudent or intuitive to take a position wholly detached from human concerns when attempting to devise humane ethics. When we think like the evolved creatures we are, in light of biological knowledge, we would not be committing any error to assume that genetic interest is the ultimate good (or at least an important one). Since inclusive fitness is local or particular, an adaptively minded philosopher should be suspicious of any doctrine that wholly rejects discrimination.

Notice that Singer uses the term 'interests' and thinks we should give equal weight to the interests of all beings. His only definition of interest is that it is the avoidance of pain. Singer quotes Bentham to the effect that the capacity to suffer gives any creature 'the right to equal consideration'.[29] He continues: 'The capacity for suffering and enjoyment is *a prerequisite for having interests at all* . . . .' This capacity is also sufficient evidence of interests, he maintains.[30] Thus a stone has no interests, but a mouse does, because of their different capacities to suffer. The more difficult case is organisms that have no feelings, no nervous system, such as plants and simple animals, but this is not discussed. A realistic description of interests—one that deployed evolutionary knowledge—would include the criterion of successful reproduction. In this light, a stone has no interests but an amoeba does—the same genetic interest as an organism possessing a nervous system. We might not give equal consideration to the interests of our planet's millions of species, but that is no reason to deny a universal interest in life; and neither does a creature's possession of interests morally shield it from predation. Singer implies that if an interest exists humans must respect it. Yet any evolutionarily-informed definition of interests will indicate ways in which they might conflict, for example an amoeba's interest in colonizing our bodies and our interest in avoiding disease; or a calf's interest in not being eaten and our interest in eating it. Such differences of interests are ubiquitous. As Alexander argued, it is when such differences exist between humans that we need ethics.[31] A viable ethic is hardly going to emerge from an analysis that begins by ignoring conflicts of interests, indeed, which fails to identify the most basic interests.

The emphasis on suffering as the sole criterion of equal rights leaves Singer with a cognitively biased means for differentiating which creatures deserve our greatest sympathy. Chimpanzees and other apes come up as most favoured species using both the suffering criterion and the criterion of human genetic interests, since apes are both the most intelligent non-human species and our closest relatives. However, the two standards can also indicate stark differences in sympathy. For example, Singer's cognitive bias leads him to favour the survival of a grown chimpanzee over that of an early-term human foetus, since chimpanzees can think and therefore suffer while foetuses cannot. Dawkins expresses the same opinion, justifying it by drawing a parallel between 'speciesism' and racism.[32] In contrast, the stance from genetic interests is that, exceptional circumstances aside, any human foetus should be given priority over any non-human animal. In contrast to adaptive utilitarianism, an overtly particularistic ethic that values the survival of humanity above all else, Singer has declared: 'Under certain conditions . . . it would be wrong, other things being equal, to continue the human race. This result does go against some of our most cherished convictions about the duty of preserving our race, and so on. I do not think we should be too dismayed about this.'[33]

A universal ethical system, one that was applicable to non-human animals, and to organisms anywhere, would emphasize the value that unites all life: reproduction. Adaptive utilitarianism would apply to all species anywhere because it deals with the fundamental reproductive interest of all life. This is a persuasive aspect of an evolutionary ethic, that it provides a moral frame for any organism able to think about such matters, a level of generality not provided by nonevolutionary ethics.[34] Previous ethical systems have sometimes shown concern for non-human species, especially in the religions of non-Middle Eastern origin. But no ethic offers other species a rationale for formulating moral rules.

Adaptive utilitarianism does not relegate individuals to homogeneous categories as egregiously as does hedonist utilitarianism. The reason for this is the nested particularity of genetic interests. As explained in Chapter 2, except for identical twins, every individual carries a unique concentration of his or her genetic interests. Each individual has a set of nuclear-family genetic interests that is only shared with other close family members. This is followed by wider interests shared only by members of one's clan, ethny, race, species and finally by other species in expanding circles of lower genetic relatedness. These ever larger aggregate genetic interests only retain their value when connected to a particular individual or group.

Another weakness of utilitarianism that a survival ethic corrects is the arbitrariness of the clause prescribing that happiness be maximized. Whether the criterion is happiness, pleasure or economic profit, Mill and the economists who adopted his approach thought that it was impossible to get too much of a good thing. This is an improbable view if proximate interests are not goals in themselves but means to adaptiveness. Even too much wealth or too many mates is bad if the monopoly diminishes the society bearing one's genetic interests. Too much happiness can diminish prudence and thus harm other interests, such as status or wealth, reducing fitness. Like other proximate interests, happiness necessarily exists in balance with other states, and is thus best optimized rather than maximized. Adaptiveness, in the sense of ability to survive and reproduce, is different. One cannot be too well adapted.

Neo-Darwinian theory holds that over long periods natural selection iterates towards phenotypes with maximal fitness, even if a maximal state is never reached. In this theory any concession to other values results in replacement of the responsible genes (if genes were the cause) by genes better able to exploit the niche. The possibility of over-reproduction leading to ecological degradation and a population crash does not diminish survival as the ultimate value. Environmental protection is a means for securing long-term survival, not vice versa. In the long run constraint of population growth is only sustainable if it is somehow adaptive for the population that undertakes such a strategy. This can be the case, since continuity (survival) takes precedence over expansion. The notion of 'over adaptiveness' (surely an oxymoron) also fails to dislodge genetic interests as the ultimate interest. Some species become over specialized, adapted for one niche and thus vulnerable to rapid environmental change. Is this a case of over adaptiveness? No. Over-specialization is a type of maladaptation. Generalization can also be a losing strategy when a stable environment allows specialized competitors to more efficiently exploit resources. Adaptation to various contingencies serves fitness, not vice versa, because differential fitness is the selective agenda that dictates which phenotypes are adaptive. The only constraint on this theory is time. Natural selection of a fine-tuned adaptation can take a considerable amount of time, and maximization might never be reached if the environment changes faster than the rate of genetic evolution. Nevertheless, the direction of differential survival is generally towards better survivors and reproducers for a given environment.

'Preference utilitarianism' proposed by R. M. Hare also parallels adaptive utilitarianism while falling short of it.[35] Preference utilitarianism seeks to maximize not happiness but the satisfaction of preferences. Hare argued that an act is good to the extent that it satisfies the preferences of all those affected by the act. Hare's system is agnostic about what is good, instead taking as its criterion of goodness actual conscious preferences, whatever they might be. This avoids the assumption of mindless hedonism, but puts the ethicist in the hands of potentially ignorant and irrational minds. To avoid mob morality preference utilitarians usually seek to distinguish between desires that are informed and rational from those that are misinformed and irrational. The result can be a responsible paternalism directed towards children, but it can also be a license for the ethicist to impose his values on the ethical system, bypassing the need to consult preferences at all. Lenin's doctrine of a revolutionary vanguard offers an example of where this can lead.

From a biological perspective, the major weakness of preference utilitarianism is its dependence on conscious desires. Interests need not be the explicit objects of preferences. In Chapter 6 I argued that peoples living in anonymous mass societies are often more protective of their individual than of their ethnic fitness. Despite being outfitted with the potential for both family and ethnic feelings, humans are not as instinctively equipped to identify and defend ethnic genetic interests in the evolutionarily novel world of mass anonymous societies. Once adopted as a criterion of right action, adaptiveness has the advantage over subjective feelings of being associated with a method for discovering which preferences are desirable. Moreover, that method is scientific and thus itself testable. The result is an objective criterion that helps protect individuals from undue influence from transient emotions and cultural fashions, and from the personal values of the ethicist.

### *Justice, a mixed adaptive ethic, and bounded rationality*

I noted earlier that utilitarianism has a weakness in not meeting intuitive standards of justice. Hamilton noted something similar in his criticism of the amoral nature of reproductive behaviour. He wrote:

> I am doubtful whether the findings from natural selection throw any light on the problem of how it is rational to act when the desirabilities of outcomes are really in the pattern of a Prisoner's Dilemma. But natural selection, the process which has made us almost all that we are, seems to give one clear warning about situations of this general kind. When payoffs are connected with fitness, the animal part of our nature is expected to be more concerned with getting 'more than the average' than with getting 'the maximum possible'. Little encouragement, I think, can be drawn from the fact that this may, in some cases, imply less than maximum population densities; it implies concurrently a complete disregard for any values, either of the individual or of groups, which do not serve competitive breeding. This being so, the animal in our nature cannot be regarded as a fit custodian for the values of humanity.[36]

There are problems with this passage,[37] but it emphasizes the potential hazards of a purely adaptationist ethic. Adding rights to adaptive utilitarianism to form a mixed ethic would temper the pure ethic with some non-reproductive values. An effective mixed ethic would state that ends do not always justify means, which must meet certain ethical standards independently of their effect on fitness outcomes. The mixed ethic would retain the dynamic life force of adaptive utilitarianism, the core principle that, other factors being equal, it is morally better to act so as to increase the fitness of the greatest number. But the social environment would be shielded from the core's aggressive surges by an enveloping mantle of rights that dampen irregularities, a senate of humane wisdom that lacks the creative leadership powers of the lower house but which holds veto power over excessive legislation. For example, in some situations population exchange might enhance the genetic interests of all participating ethnies (or of the greater number of individuals), and thus qualify as ethical according to pure adaptive utilitarianism. But if the process entailed violation of rights, the mixed ethic would disqualify it.

It is a simple matter to add rights clauses but this threatens incoherence. Can coherence be retained? I noted earlier that the sense of justice is, like the utility criterion, an emotivist ethical concept. If happiness and other proximate values can be derived from the adaptiveness criterion, perhaps rights also can be similarly derived (and therefore be human rights). A derivation might be feasible along the following lines. Everyone shares an interest in the rule of law as a counter to arbitrary rule, since the latter threatens the freedom to raise a family and acquire resources. This implies a universal interest in maintaining due legal process and the right to individual life, liberty and property. Whatever other interests exist, it is in the general interest that these be pursued within the constraint of basic human rights. An ethical view that parallels this argument is 'negative utilitarianism', which asserts that avoiding the loss of utility takes moral priority over advancing it. One might formulate a type of rule utilitarianism in which rights are derived from the core principle of adaptiveness as utility. An emphasis on procedural rights would have the added advantage of allowing evolution to continue. Evolution entails competition and attendant inequality of individual fitness. Evolution is possible with the enforcement of equal opportunities but not with equal outcomes.

'Bounded rationality' is a final weakness of classic utilitarianism that carries over to the adaptive type. The great complexity of society renders social analysis difficult. Even the study of small group dynamics is unfinished business in social psychology. It follows that no one is intellectually equipped to apply a criterion of goodness to each act to predict all the many ramifications, making act utilitarianism impracticable. The same epistemological problem afflicts adaptive utilitarianism: how can anyone be expected to estimate all the ramifications of his or her acts? A life spent in contemplation would partially solve this dilemma by reducing the number of acts and thus ramifications of all kinds, except that it would also depress fitness. A more viable ethic is to adopt rules of conduct that generally yield a tolerable happiness outcome in the case of rule utilitarianism or fitness outcome in the case of adaptive rule utilitarianism. In the latter case, rules of conduct must balance investment in family, ethny and humanity in different proportions as circumstances warrant. That presupposes complex analysis and inevitable approximateness of results. Uncertainty of analysis weakens the act-utility logic of the pure ethic and gives extra credibility to the rule-utility approach that resembles the mixed ethic.

In formulating a prudential ethic, expected error in analysis calls for a conservative approach that gives precedence to higher concentration of genetic interests. This supports individual rights as a bulwark against the tyranny of majorities, and indicates again that a mix adaptive ethic might be derived as a corollary of the pure ethic. For the same reason, ethnic groups should be accorded rights against world opinion. Bounded rationality also indicates continuity over expansion and evolution over revolution. Adaptive rule utilitarianism would consist of rights, laws and practices that protected individuals' and groups' continuity and freedom to defend genetic interests.

Modern theories of justice are moving away from the notion that some perfect harmony of interests can be achieved, but rather 'resides in the recognition and methodical management of expected and necessary disagreements'.[38] This view, that 'justice is conflict', is an additional reason to suppose that an adaptive rule utilitarianism derived strictly from the pure ethic could take the place of an arbitrarily mixed ethic. This accords with Alexander's argument that sexually reproducing species will always have differences of genetic interests.[39] Ethics cannot hope to produce perfectly harmonious societies. Instead, ethics is limited to being a form of conflict resolution. Though the utilitarian aspiration to consensus ethics can never be realized, the objective of minimizing mutually destructive conflict is attractive and can be approached by resolving disputes in ways that minimize harm and maximize mutual benefits.

### *Pure, mixed, and rights-centred ethics*

To this point I have defined a pure form of adaptive utilitarianism ('what is adaptive for the greatest number is good'), as well as a mixed form (the pure form plus individual rights). I shall contrast these two ethics with a 'rights-focused' ethic consisting only of individual rights, the utilitarian clause being removed altogether. This last ethic is agnostic about the moral goodness of genetic (or any other) interest. Only the means used to pursue interests count in evaluating actions. Since being adaptive is morally uncompelling, it is deemed unacceptable if any harm whatsoever is committed in its name. In other words, it is neither right nor wrong to pursue genetic interests; but the means used in that pursuit do vary in moral quality.

The rights-centred ethic is coherent when deontological, that is, when it takes the form of absolute rules against certain means, perhaps derived from religion or intuition. It is incoherent when expressed teleologically, that is when means are rejected for their harmful effects. Avoiding harm implies that weight is being given to an interest, and the ethic would become a selective application of teleological principle. My reasons for considering this ethic are first that its extreme underemphasis of genetic interests provides a useful contrast with the first two ethics and, secondly, that it is realistic with respect to ethnic affairs. For decades in the West it has been highly unfashionable to apply prudential thinking to ethnic relations for avoiding harm to majority populations.

The first two ethics share the assumption that ethnies are stores of genetic interests for their members, and the rights-centred ethic is unaffected by consideration of interests of any kind. As noted in Chapter 2 this is a view rejected by some analysts. If ethnic genetic interests do not exist the implications for our three candidate ethics are simple to deduce: the pure ethic loses all force; the mixed ethic reduces to a set of individual and minority rights; and the rights-centred ethic remains unchanged. The three ethics are summarized in Table 9.1. In the following sections I answer the three questions raised earlier in the chapter from the standpoints of these three ethics. My conclusions are summarized in Table 9.2.

| Ethic | Is EGI morally good? | Do means justify ends? |
|---|---|---|
| 1. Pure adaptive utilitarianism | Yes | Yes |
| 2. Mixed adaptive ultilitarianism | Yes | Yes, but constrained by rights |
| 3. Rights-centred ethic | No, nor bad | No; rightness of means unrelated to consequences |

*Table 9.1. Three ethics for evaluating defence of ethnic genetic interests (EGI).*

| Question<br>Ethic | a. Can it be moral for EGI to frustrate other interests? | b. Schould genetic interests, including EGI, have absolute priority? | c. What is the right action when genetic interests conflict? |
|---|---|---|---|
| 1. Pure adaptive utilitarianism | Yes | Yes, when EGI clearly at stake. | Compete within adaptive limits. |
| 2. Mixed adaptive utilitarianism | Yes, in defence and expansion that preserves competitor. | No when EGI conflicts with individual rights. | Compete but respect rights. |
| 3. Rights-centred ethic | No, since this entails causing harm. | No, since only means matter. | Stop competing, since it entails harm. |

*Table 9.2 Right action in conflicts of ethnic genetic interests (EGI) according to three ethics.*

### *Three Ethical Questions*

#### *(9a) Does defending ethnic genetic interests ever justify frustrating other interests?*

This question is a core ethical issue raised by the concept of genetic interests. There is no ethical issue when an individual chooses to survive and reproduce without infringing others' interests. However, prioritizing genetic continuity can involve favouring fellow ethnics over members of other ethnies. Some feel that it is morally good to care for one's family, and thus for subjective and objective analogies of the family, including the ethny. This amounts to the loyalty to band and tribe that has characterized humanity for most of its history. In Western societies it is now normal to oppose this view by asserting the unfairness of discrimination and by pointing to the misery of ethnic conflict that has afflicted humanity from primordial times to the industrialized warfare and genocide of the twentieth century. What do our three candidate ethics recommend, and how credibly?

Pure adaptive utilitarianism implies that it can be moral to harm minority interests in defence or expansion of majority ethnic genetic interests. This follows from the unqualified statement that it is positively good to behave adaptively and better to pursue the adaptiveness of the greatest number. The mixed ethic allows frustration of others' interests but only in defence. Since an aggressing group violates others' rights to life, liberty and property and accompanying procedural rights in order to expand its genetic interests, its victims are justified in harming the aggressor in acts of defence. The rights-centred ethic holds that pursuing ethnic genetic interest is not bad, so long as no harm is done in the process. This disallows aggression because it violates rights, but it also disallows defence when the aggression only threatens genetic interests, since these are given nil weight.

While I find the mixed ethic most intuitive, it would have us believe that for most of human history humanity has been behaving immorally in the continuing saga of competition between individuals and groups. Disallowing aggression might appeal in a crowded world of high-tech weaponry where economic effort is more productive than territorial conquest, but that is surely no reason to relegate our ancestors (some quite recent) to moral purgatory. At least when evaluating the human past, some ethical space must be left for manoeuvre, for the cut and thrust of competition.

Analogy between family and ethny is a more fundamental approach to getting a grip on the moral issue at stake, one that warrants adding a right to the mixed ethic (see p. 306). From Chapter 4, genetic interest is not obvious as an emotional priority unless expressed in phenotypic categories. Emotional tagging of genetic interests is most reliably accomplished using the family analogy.[40] The analogy is compelling because there is only a quantitative difference between kin and ethnic genetic interests. What can be said to criticize a person who claims a genetic interest in kin and does the sort of things a loving parent, aunt or uncle would normally do? Is parental care administered in awareness of the genetic interests it serves less moral than the naïve variety? Is there anything wrong with awareness of an empirical truth? If not, then criticizing the informed variety is tantamount to criticizing parental love. More interestingly, if being genetically informed *increases* parental motivation, is such extra care or the belief that causes it immoral? Parents who are more caring than other parents are praised, not condemned. I cannot detect anything intrinsically wrong with being motivated by knowledge of genetics. While utopian socialists such as the early Bolsheviks and the Israeli kibbutz movement have tried to abolish the nuclear family and the parental favouritism it involves,[41] attempts to redirect parental investment to nonkin have usually given way to popular demand; parental favouritism remains almost universally accepted. This inevitably contributes to inequality among children but nevertheless the parental bond is so strong that this primary form of discrimination is considered morally unexceptional. Discrimination in favour of one's family is generally considered a weak basis for alleging immorality except as an abuse of public office. Indeed, *failure* to care disproportionately for one's close kin is widely condemned.

The analogy between family and ethny is instructive. If there are nationalist acts that defend ethnic genetic interests (and surely many types do) *and* that are considered morally acceptable, then on what basis can we morally condemn a genetically informed nationalism that resembles the naïve type? If defensive nationalism is justified, how does it matter that the defenders know they are protecting their genetic interests? And even if the defenders' awareness of their shared genetic interests changes their behaviour, for example by inspiring greater self sacrifice and greater vigour in killing the aggressors, does that make genetically informed defensiveness less justified than the naïve type? The same argument applies to ethnic minorities struggling to escape discrimination. Would their cause be any less just if minority activists or their majority supporters understood that they were defending minority genetic interests? The preceding discussion supports the mixed ethic.

If acceptance of discrimination in favour of kin in private affairs is widespread, then perhaps knowledge of the familial nature of ethnicity will lead to greater understanding and tolerance of ethnocentrism. This would be appropriate only in the case of favouritism directed towards co-ethnics, not positive aggression directed at other ethnies. Ethnic altruism can, of course, have negative side effects on other groups, but this is analogous to the inequality among children that is a side effect of parental altruism. If it is in most individual's interests to be allowed to practise ethnic nepotism (except where public office calls for disinterestedness), then most should tolerate ethnic favouritism in altruism. This is the unambiguous implication of the pure ethic, which favours the fitness of the greatest number. It is only supported by the mixed ethic if individual and minority rights are guaranteed. The rights-centred ethic would disallow freedom to discriminate because this would harm some individuals, though consistency requires that this ethic also oppose parent's right to favour their own children.

Perhaps the family analogy can be taken further. It is parents' duty to care for their children. Do we have a similar duty to nurture our ethnies? Such a duty would imply that it is morally right to defend one's ethny. It would also mean that fellow ethnics could be held accountable for their actions towards the ethnic family, similarly to family members being considered to have acted improperly when they failed to aid kin. The logic for asserting both family and ethnic duties is the same, and derives from the nature of genetic interests. An individual who fails to help a family member in time of need or who directs scarce resources towards nonkin harms the jointly-held genetic interests of the whole family. The same is true with ethnies. Failing to favour a fellow ethnic over a member of a closely related ethny has a very slight effect on ethnic genetic interests, and so might not be judged a breach of duty. But when ethnic kinship is high, as is the case in competition between members of different races, failure to show ethnic loyalty is the genetic equivalent of betraying a child or a grandchild. The individual stakes need not be so high for serious damage to be done when collective goods are undermined. Examples include betraying sensitive information to the enemy in time of war or using the influence of high office or wealth to confuse national identity or the perception of national interests. Any action that weakens an ethny's capacity to mobilize, for example to resist mass immigration of distantly related ethnies, must be considered a blow to ethnic fitness. As we saw in Chapter 3, in such cases the damage done to genetic interests can be orders of magnitude greater than an act of family betrayal, so that the duty to ethny might be greater than to family in circumstances of threat to the group's collective interests. Though individual transgressions have slight effect, general transgression harms a precious collective good. The adaptive utilitarian injunction against individual transgressions must reflect the aggregate effect, not that of an individual act. The situation is similar to fines used to reinforce the rule 'Keep off the grass!'. Any one person walking on the grass does little damage, but in order to prevent cumulative damage from a large number of trespassers, the fine must be steep enough to deter each person. That is a feature of rule adaptive utilitarianism, and it is compatible with the pure and mixed ethics, though not with the rights-centred ethic because the means requires punishment to protect an interest.

The plausibility of the analogy between duty to family and tribe is greatly strengthened by the fact that tribal duty has been the norm throughout human history and prehistory. Only those who accept a secular doctrine of original sin could contemplate the notion that humans have been evil for all of their existence and have only become pure in the last decades in a few benighted countries. It is more rational to assume that the absence of ethnic duty is a bold experiment, possibly an immoral one. Tribal fealty should be taken as still applying unless modern ethnies or their members' relationship to them have changed so much over the last few decades that this ancient duty no longer applies. Have they?

A duty to ethnic patriotism based on an analogy with family duty would be most plausible if individuals feel a duty towards anonymous offspring. (I could find no systematic studies on the subject.) Most members of an ethny are strangers and therefore knowledge of their genetic commonality is abstract. An individual who feels a duty towards offspring who are strangers, namely children adopted off and raised by another family, might also feel an obligation to the ethny as a whole or a large number of co-ethnics, which constitute an equal or larger share of genes as a single offspring. If this is not true, the analogy between familial and ethnic duty relies on either of two preconditions, one sufficient, the other necessary. The first is tribal feeling analogous to parental motivation. Anthropology and psychology confirm that it is easy to trigger ethnocentrism in humans, though historical circumstance and education introduce considerable variability in intensity.[42] But in this situation a sense of duty would follow automatically from the tribal feeling.

The second condition, identification, is more fundamental, but only a necessary condition for duty to ethny. A parent usually knows his or her children and duty towards them is contingent on that awareness. We would not condemn someone for failing to care for children he did not know existed (though we might be critical of the behaviour that led to such ignorance). In the past it was similarly unlikely for anyone to be unaware of his tribal identity, but the deracination common in the modern world sometimes confuses identity. In fact most people are aware of their ethnic roots through knowledge of their family tree. However, ethnic identity is merely a necessary condition for it to be obligatory to support one's ethny.

Either of two conditions, when added to ethnic identification, might produce feelings of ethnic obligation. They might even produce in observers the conviction that the subject has such an obligation. First, if a person comes to believe that his ethny is his extended family, he might feel obliged to aid it even though he has no tribal feeling. Alternatively, his belief might lead to tribal feeling that then releases a sense of obligation. Secondly, when members of an ethny do in fact belong to an extended genetic family, as is often the case, the average member is a store of genetic interests for every other member. Understanding the nature of this mutual interest might produce a feeling of responsibility to preserve that interest, for example by making oneself a productive member of society and by acting to maintain group cohesion. Would an observer feel that such an individual was obliged to favour his ethny in either condition? Arguably yes, if the observer felt that family members have a special duty towards one another, or that it is wrong to harm other's vital interests. Note that in this discussion I have deployed emotivist ethics in the tradition of Westermarck.

To conclude this part of the discussion, I return to the question of whether social conditions have changed such that duty to ethny no longer holds. Conditions have changed, and no further proof is needed than the fact that tribal duty is felt much less intensely, at least in societies with developed economies, than over the preceding millennia of the human experience. Sentiment has declined, but genetic interests remain, and due to globalisation are larger than ever. Nevertheless, individuals cannot be held accountable for not defending their ethnic interests unless the conditions discussed above apply. Note, however, that those conditions are variable, both due to blind cultural evolution and human design.

Another approach for evaluating the moral status of an interest and the means used to defend it is to consider the ethics of frustrating pursuit of that interest. The rights attached to membership in various groups are often acknowledged due to apprehension of abuses suffered by one group at the hands of another. A right or at least a presumption in favour of defending ethnic genetic interests can become apparent by considering the ethics of blocking defence of such interests. Judging such blocking behaviour to be unethical implicitly acknowledges a right or presumption to defend ethnic genetic interests.

As discussed above in Chapter 5, strategies for frustrating other groups' pursuit of their genetic interest include directing investment away from it by masking or misidentifying a group's identity to its members. This can be achieved by constructing fictive ethnicity using the tools of ideology and the mass media. Fictive non-ethnicity can also be constructed, by denying the existence of a group that does in fact exist. What is the propriety of doing so? The intuitive answer, I think, is contingent on context. Demobilizing an aggressor nation appeals because it reduces the likelihood of an unjust war; demobilizing a defensive nation does not appeal for the converse reason. The family analogy again helps clarify the issue. Recall from Chapters 2, 3, and 5 that an ethny is usually a large store of genetic interest homologous to that residing in families, though often at a lower concentration. Both groups are large stores of genetic interest for their members. Genetically, directing members' loyalty away from their ethnies is equivalent to directing parents' loyalties away from their children. Discrediting, confusing or leading an individual's altruism away from his or her family, whether nuclear or ethnic, will usually be tantamount to reducing an individual's inclusive fitness. Sentimental interests are also damaged since such acts tend to break or prevent the formation of meaningful prosocial relationships. The contemporary intellectual emphasis on the evils produced by ethnic sentiment should not blind us to the positive role of ethnicity in providing a sense of belonging, social cohesion, and continuity. Whether ultimate or proximate interests are considered, turning individuals away from their ethnies is aggressive in effect, and if done with knowledge of genetic or sentimental interests, also in intent. It would seem to be immoral on that basis.

The family analogy and the wrongness of blocking familial and ethnic interests, both indicate that an additional right is derivable from the mixed ethic. That right is, approximately, the freedom to aid ones' family and ethny which, biologically expressed, is the right to protect one's genetic survival. Note that this right brings the mixed ethic closer to pure adaptive utilitarianism by emphasizing the right to compete, rather than to achieve some minimal outcome. If competition is to be legitimate, which conditions should pertain?

There are many interests the pursuit of which entails engaging in competition for scarce resources, such as high-status jobs, a house with a view, and (more directly connected to genetic interests) mates. Every society has procedures for legitimately advancing individual, and thus family, interests ahead of others. Even in egalitarian hunter-gatherer societies where hunters do not control the distribution of their kills, successful hunters appear to benefit by a rise in status among other men.[43] In non-hierarchical societies, popularity is a precious resource since it facilitates exchange and mate choice. In contemporary Western societies winning exams and undercutting another retailer's prices are accepted methods for gaining more resources. Until the demographic transition, social and economic success generally led to higher fitness. Unless it can be shown that ethnic genetic interests, unlike all other interests, possess no imperative weight at all, there is no reason why defending these interests ought not compete with others, even if this has differential fitness effects.

This reinforces the presumption of the right to strive for the advancement, not just the defence, of one's family and ethny. The pure and mixed ethics contrast starkly in the means they allow. The bald pure ethic, stripped of its corollaries presented earlier, retains an element of 'nature red in tooth and claw'. Its guiding criterion, adaptiveness of the greatest number, gives maximum protection to majorities, leaving small ethnies exposed. Except in the special case of nuclear and perhaps biological warfare, the risk of mutual destruction is small, and pure adaptive utilitarianism would often leave winners free to mistreat their weaker opponents. Prudence on the part of winning groups does not guarantee losers' rights. It is true that ethnies that are successfully expanding have less to gain from risking their social and economic environment and so are wise to limit their gains before the point at which the losers use desperate measures. But if such measures are not available (losing groups tend to have limited resources) the cost of total victory can be small and the losing group safely displaced (killed, driven off, or forcibly assimilated). It is in the face of the pure ethic that Alexander's caution about unrestrained interest seeking rings true: '[Ethical] rules consist of *restraints* on particular methods of seeking self-interests, specifically on activities that affect deleteriously the efforts of others to seek their own interests.'[44] The mixed ethic has such restraints in the right to survival of the losing group with reduced but viable resources. But it balances these rights against the right to strive for one's ethny, which permits asymmetrical outcomes. Recall Hamilton's view that some competition must be allowed to continue.[45] On the international scene limited national wars are tolerable, he thought, but not genocide or wars of mass destruction. This view is moving in the direction of a mixed adaptive utilitarian ethic, though in his strictly biological analysis Hamilton did not discuss rights.

Note that losers accept the legitimacy of the outcome when the competition is regulated such that the loss is not total and there is much left to gain from remaining within the society that sanctioned the contest. By the same conditions, losers are less constrained to protect the overall social system as their losses mount. Threatened extinction of the group would still not justify strategies that destroyed the biosphere, but certainly would allow strategies that destroyed society—namely revolution, revolt, and secession.

Finally in this section I want to discuss the ethics of caring for family. Much of the foregoing argument has relied on an analogy between family and ethny, yet some reject any assumption that nepotism is morally good. This assumption should be examined. Defending genetic interests might be criticized as immoral or amoral if good conduct were defined as an act of assistance that *reduces the inclusive fitness of the giver*. This form of helping behaviour, sometimes called genetic altruism, is maladaptive in its effect on the carer, in contrast to kin or ethnic altruism that stand a chance of raising the carer's inclusive fitness.[46] Westermarck's concept of 'promiscuous alms giving', which he attributed to primitive Christianity, approximates genetic altruism. Individuals serving in official capacities in business and government are under contractual obligation not to discriminate in favour of family or ethny. Such discrimination is frowned upon, quite accurately and properly, as 'nepotism'. As already noted in this chapter, early Bolshevism and other utopian communist experiments went further and attempted, unsuccessfully after much tragedy, to suppress the nuclear family as well as ethnic loyalties. The ethical intuition behind this was that favouring family and ethny are forms of selfishness that compete with broader solidarity (of the proletarian class or humanity in general).

Extreme universalism still lives in the idea that any type of altruism that is delimited, in which love and assistance are granted to some categories more than others, amounts to xenophobia.[47] The evolutionary mechanism credited with causing this aggressive denial of sympathy is kin selection. As explained in Chapter 5 (p. 123), the significance of kin selection (and inclusive fitness processes in general) is that it allows altruism to be adaptive and thus spread throughout a species. Kin favouritism protects the genes that code for altruism from being weeded out of the gene pool. Evaluating all discrimination, even acts of supreme love, as xenophobia, is thus tantamount to asserting that altruism must be maladaptive to be moral. One leading evolutionary theorist goes beyond the tantamount by contending that only non-fitness-enhancing behaviour can be moral.[48] Alexander suggests that this is an unconsciously self-serving moral sentiment that, when expressed, influences some susceptible individuals to show indiscriminate altruism that benefits the moralist.[49] By definition such behaviour will tend to reduce the relative fitness of the genetic altruist. Unfortunately, Alexander offers no argument as to why this sentiment would not influence the 'moralist's' own kin, and offers no evidence that such advocates preach a different morality to their families.

Alexander implies that universalism is often a competitive tactic, or can be, and concludes thus: 'If morality means true sacrifice of one's own interests, and those of his family, then it seems to me that we could not have *evolved* to be moral.' Darwin considered moral behaviour to have evolved to be universalistic within the tribe but particularistic between tribes. In his view self sacrifice was a losing strategy for isolated individuals but a winning one when put to the service of inter-tribal competition:

> It must not be forgotten that although a high standard of morality gives but a slight or no advantage to each individual man and his children of the same tribe, yet an increase in the number of well-endowed men and advancement in the standard of morality will certainly give an immense advantage to one tribe over another. A tribe including many members who, possessing in a high degree the spirit of patriotism, fidelity, obedience, courage, and sympathy, who were always ready to aid one another, and to sacrifice themselves for the common good, would be victorious over most other tribes; and this would be natural selection.[50]

As an evolutionist Westermarck treated promiscuous alms-giving with some contempt.[51] He too considered the universalist elements of moral behaviour to be not always genuine, describing the moral emotions as possessing the qualities of 'disinterestedness, *apparent* impartiality, and *flavour* of generality'.[52]

Westermarck was in two minds about universalism as a criterion of moral rightness. On the one hand he favoured the impartial application of standards, but on the other he advocated a descriptive approach to discerning human morality. He sought to discern moral principles in cross-cultural comparison of moral behaviour. Maintaining strict universalism as a moral criterion risks abandoning the descriptive approach that led Westermarck to develop his ideas about moral relativism. According to Westermarck, the universalism of the moral emotions applies most strongly within delimited groups that resemble the bands and tribes in which humans evolved, corresponding in some respects to present day ethnies and their political form of nations. Not to allow for this empirical fact would be a large concession to the abstract universalism proposed by Kant. Even universalism applied only within a society is unnatural when too abstracted and strict, as in the ethics of John Rawls.[53] The latter defined social justice according to a highly abstract and denatured conception of fairness in which actors must choose general rules while blind to their own interests, behind a 'veil of ignorance' that prevents discrimination in favour of particular others, even self. Rawls named this imaginary situation the 'original position', an ironic choice of words considering the wealth of information about human origins in clan based societies that existed by the 1970s. Westermarck is sympathetic to impartiality, as are all Enlightenment thinkers, but his evolutionary empiricism provides an escape route from the inhumanness of Rawls's system. Biology also militates against the extreme universalisability maxim of Kant, according to which a rule of conduct only qualifies as a moral rule if it applies equally to everyone, unimpeded by familial or other group boundaries.

Biological realism qualifies all abstract ethics. One does not find in Westermarck's writings approval of nineteenth century social Darwinism, the program of coopting the descriptive statement 'survival of the fittest' as a moral guide to public policy without regard for prosocial values of compassion and cooperation. Westermarck sought to describe and understand the moral sentiments and how they gave rise to institutions. He was not concerned with metaphysics, especially the derivation of inhuman cosmic laws from the equally inhuman mechanics of natural selection. This same empirical view of humans as an evolved species would, I think, have kept him from the other metaphysical extreme of asserting that only non-fitness-enhancing behaviour can be moral.[54] His approach is compatible with Allan Gibbard's theory of normative judgements that analyses moral emotions from a Darwinian perspective. Like Walzer, Gibbard argues that it is unrealistic to expect humans to be strictly impartial and universal in their compassion:

> It does make sense to be upset and angered when faraway deeds are specially heinous. It even makes sense to engage one's feelings in fictions from time to time. It makes no sense, though, to go through life with utterly impartial feelings; each of us needs feelings specially engaged in himself and in special other people. Only small portions of human life can claim our fullest emotions.[55]

Or as Walzer summarizes—in a maxim that partly reconciles Kant and Rawls with human nature—'[t]he crucial commonality of the human race is particularism'.[56]

Perhaps the way to square the empirical reality of ethnic nepotism with universalism, both elements of Westermarck's ethics, is to define a moral emotion as an emotion that is disinterested *to a significant degree*, rather than absolutely. It should not need to be perfectly disinterested, perfectly impartial, and perfectly general, to be judged moral. These conditions can be satisfied by appearances, perhaps facilitated by self-deception,[57] as much as by objective conditions.

Whichever solution is adopted, a realistic, humane ethics must set up criteria of goodness that do not frustrate all adaptive behaviour or condemn core adaptive elements of human nature. That is an error adaptive utilitarianism does not make. It makes pragmatic sense to designate certain aspects of our nature as original and indelible sins. The large scale cooperation on which modern societies depends must be partly enforced against free riders. Humans are flexible strategizers capable of free riding in creative, ingenious and relentless ways when circumstances make this adaptive. But it is absurd to extrapolate this to the view that what is adaptive cannot be morally good conduct. Ethics informed by knowledge of evolution would set achievable goals, and goals that do not genetically eliminate the moral person. Indiscriminate altruism is theoretically possible, but it is only sustainable across many generations if it sustains the genetic interest of the altruist.

Let me summarize my answer to the first question, 'Does defending ethnic genetic interests ever justify frustrating other interests?'. Pure adaptive utilitarianism approves of ethnic genetic interests frustrating other interests so long as this maximizes the adaptiveness of the greater number. The mixed ethic also approves but affords basic rights to all parties, including minorities. The rights-centred ethic rejects frustrating other interests because this causes harm. I have argued in favour of the mixed ethic. The analogy between family and ethny helps clarify the moral issues at stake. Ethnic identity combined with a feeling or belief that the ethny constitutes an extended family is likely to induce a sense of duty towards it. An observer might agree that such a duty exists. The analogy indicates that if it is wrong to divert a parent's nurture from his or her children, then it is also wrong to divert individuals from identifying with or aiding their ethnies. The family analogy would lose force if it were plausibly argued that adaptive behaviour cannot be moral because it is discriminatory. This is Kant's classic argument that moral rules must be universalisable. However, this position is vulnerable to a *reductio ad nauseam*, since it implies that mother love is not morally good and that moral behaviour is impossible in the long-term future. It is also vulnerable to *reductio ad absurdum*, because it implies that moral behaviour could not have evolved.

### *(9b) Should an individual's ultimate (genetic) interests be accorded absolute priority over others' proximate interests?*

The message of modern biology is that genetic fitness is the ultimate interest, meaning precisely that it is of absolute importance. This is surely the starting position of any ethical discussion of the choice between genetic and other interests. In circumstances where the choice is clear, failing to put genetic interests first can be as maladaptive as throwing away one's own life or letting one's children drown. Failing to discriminate in favour of one's own children is likely to depress fitness, as is failing to defend the interests of one's ethny when a threat to group fitness is reasonably clear cut, as in territorial displacement (ethnic cleansing), subordination, and genocide.

Fortunately for those who hold proximate values dear, whether one gives greater emphasis to genetic interests or to other values will rarely be an either/or choice. Most humans are evolved to value adaptive proximate interests such as bonds of kith and kin, status, wealth, and health because they are adaptive. More accurately, striving for the things we hold dear is adaptive or was adaptive for much of our evolutionary past. So our lives are unlikely to be turned upside down if we act to increase or secure our genetic interests. This will amount to nothing more than shuffling existing priorities.

But choice of lifestyle is generally not a moral problem unless it affects others. The more pressing question is, should my genetic interests take priority over others' non-genetic interests? Despite the overriding importance of genetic interests, I shall conclude that an affirmative answer is only ethical under special circumstances.

Asserting that survival is of overriding importance implies that person P's genetic interests should have priority over person Q's non-genetic interests. It is easy to imagine or recall circumstances where such priority is right. Borrowing someone's vehicle without permission is usually immoral, but few would object if the purpose were to rush an injured person to hospital. Also, in emergencies it is expected in some societies that strangers at the scene of an accident will render assistance, even if this causes them some risk. Several individuals were prosecuted in the German city of Munich in the 1990s for failing to make efforts to rescue boys who had fallen through ice in a lake, and drowned. Presumably the attempt would have involved some risk. Public outrage at the indifference shown by the bystanders indicates an intuitive ranking of interests. The boys' near certain loss of life was accorded greater weight than the bystanders' inconvenience or slight risk of life had they attempted rescue. In effect greater genetic interests were given moral priority over lesser- or non-genetic interests, at least between members of the one nation.

The problem with using such examples as the basis for formulating general ethics is that they are rare. Even if we assume that genetic interests should be accorded moral precedence, there is the problem of distinguishing ultimate from proximate interests. Which proximate interests do *not* have ultimate payoffs? Some proximate interests have obviously strong impacts on ultimate interests, such as resources, personal and group survival, and rank. But others, such as reputation, freedom from stress, and freedom of expression, have more subtle ultimate impacts. The analytic problem facing the adaptive utilitarian is that ultimate interests can usually be advanced only by promoting proximate interests, tending to confound attempts to distinguish between the two.

Some proximate interests are redundant for securing genetic interests, for example an extra penny on the fortune of a wealthy person. Arguably progressive income taxes are usually acceded to without armed revolt partly because they target wealth that is not critical to survival. However, withdrawing even a superfluous interest can detract from the precious freedom to advance interests in general. The uncertainty of distinguishing proximate from ultimate interests argues against giving the latter priority over the former, except when the distinction is glaring, such as in the examples of individual fitness described above, or the ethny's independence,[58] status, relative numbers or territory.

When the distinction between ultimate and proximate interests is uncertain, the effect is to favour the latter, approximating respect for individual rights. This is further support for the mixed ethic as a corollary of the pure ethic and parallels the argument from act to rule utilitarianism. To use the analogy of classical utilitarianism, when it is difficult in principle to estimate effects of individual acts, rules such as rights are indicated.

Despite the blurring of ultimate and proximate interests, the distinction can sometimes be made with reasonable certainty. In resolving disputes, genetic interests should take precedence over proximate interests. Such resolution is likely to be considered legitimate by affected parties and thus contribute to a lasting peace that facilitates everyone's interests. In cases where the distinction is not clear, precedence should be given to those interests more closely connected with inclusive fitness. For example, in territorial disputes geographically large nations usually experience less threat to their genetic interests than do small nations. A transfer of territory to the large nation can be catastrophic for the small nation while gaining little for the larger. Adaptive utilitarianism presumes in favour of smaller nations' interests. Another example concerns inequality. Within a nation state adaptive utilitarianism supports redistribution from rich to poor when survival or fitness is at stake, for example in the support of medical services and education.

To summarize the answer to the second question, the pure ethic designates it moral to give ethnic genetic interests absolute priority over other interests when the choice is clear cut. However, since genetic interests are largely advanced through proximate interests, this absolute priority applies only in special cases of dire or clear risk to the ethny, such as invasion, territorial displacement, replacement migration, or genocide. The mixed ethic, while giving moral weight to ethnic genetic interests, does not grant them absolute priority because of the individual and minority rights it contains. The rights-centred ethic rules out absolute priority for ethnic interests because it apportions it no weight in comparison to harm done in attaining it.

*(9c) What is the proper action when ethnic genetic interests conflict?*

> KING HARRY
> Therefore take heed how you impawn our person,
> How you awake our sleeping sword of war;
> We charge you in the name of God take heed.
> For never two such kingdoms did contend
> Without much fall of blood, whose guiltless drops
> Are every one a woe, a sore complaint
> 'Gainst him whose wrongs gives edge unto the swords
> That makes such waste in brief mortality.
> . . .
> May I with right and conscience make this claim?
>
> Shakespeare, *Henry V*, 1600, Act 1, Scene 1

First a preliminary issue. Is the reduction of an individual's or group's genes necessarily a moral issue? One might argue that in the absence of a motivation to aggress, as is the case in all those animal and plant species that lack consciousness, genetic replacement presents no moral problem. And since humans generally are not aware of genetic interests, perhaps there is nothing immoral about one person or group's genes replacing others'. Morality would only become involved if the process of gene replacement entailed aggressive motives or fear or perceived loss on the part of the losers, and it would be these motives and fears that would raise moral issues, not the genetic effects. This argument leaves considerable room for ethical considerations since gene replacement is often achieved by aggressive acts motivated by intention to harm or callousness towards victims, including violent conquest, territorial displacement, and deception. Even if genetic conquest is unaccompanied by aggressive intent, an ethical issue still arises if the conquered suffer physically or emotionally in the process. Also, it is not true that humans are completely unaware of genetic interests. As noted in Chapter 6, all societies have cultural substitutes, often kin metaphors that serve emotionally to tag family, clan, and ethny. As knowledge of genetic interests spreads there will be fewer excuses for aggressing against those interests and perhaps greater resistance by victims, making the process of genetic replacement ever more ethically problematic. Furthermore, it is not absolutely clear that morality resides only in motives, since ethics can involve evaluating situations as well as apportioning blame. An observer might not blame a winning group for replacing another group's genes, but still consider the fact of replacement morally objectionable and thus something to prevent or reverse if possible.

Those who do not consider peaceful genetic replacement to be a moral issue will have no moral objection to their own painless genetic extinction. But adaptively minded individuals will strenuously object, so that it would be unrealistic to depend on consensus to determine right and wrong in matters of conflicting genetic interests. Evolved organisms are unlikely to accept for very long, the legitimacy of a social order that weeds out their lineages. This points to another flaw in the classical utilitarian ethic that the adaptive version implicitly corrects. Bentham and Mill thought that happiness (or any other utility) of the greater number could be the basis of a rational ethic. Much ethical store was invested in this rationality. In the Newtonian age, rationality was equated with a syllogism consisting of valid reasoning from true premisses. Rationality was considered something one cannot disagree with, in the sense that one cannot disagree with the deductive truths of logic and arithmetic. The utilitarian reliance on general benevolence in the form of the happiness of the *greatest number*, was assumed capable of generating moral and political consensus. Conflicts of interests and competition were things to be pushed aside and conquered, not incorporated into the ethical system. Utilitarianism is an ideology of harmony formulated by members of the dominant class. While any utilitarian ethic will favour the majority, an ethic informed by evolutionary theory must take into account minority interests, even if it overrides them in some circumstances.

As noted earlier in the chapter, everyday observation and neo-Darwinian theory tell that conflict of interests is endemic to the human condition. As Alexander and other sociobiological theorists have pointed out, the members of sexually reproducing species will always have differences of genetic interests.[59] Hence ethics cannot hope to produce perfectly harmonious societies but is limited to being a system for resolving inevitable conflict. The classical utilitarian aspiration to consensus is an impossible dream, but some degree of consensus can be maintained by resolving disputes in ways that minimize harm and maximize mutual benefit. This is hardly a new idea. Milton wrote in 1644:

> For this is not the liberty which we can hope, that no grievance ever should arise in the Commonwealth, that let no man in this world expect; but when complaints are freely heard, deeply considered, and speedily reformed, then is the utmost bound of civil liberty attained that wise men look for.[60]

Western legal traditions have become sophisticated systems for resolving disputes in ways that protect rights and, in so doing, retain legitimacy for the social structure. Thus a realistic appraisal of human fractiousness need not crush the hope for harmony through justice. That hope is most realizable when justice consists of equal treatment in process, not equal outcomes, between competing parties.

The Darwinian version of conflict resolution is consistent with existing methods, and like them allows for asymmetrical outcomes. In economics and many small scale social situations, conflict resolution can have a Darwinian flavour when competitors are left to resolve their disputes using means sanctioned by law or custom. The adaptive utilitarian ethic favours procedures leading to benefits for the greater number. The pure ethic would tolerate methods that harmed individual and minority rights in the interest of the majority, while the mixed ethic would protect those rights.

Conceptualising ethics as a form of conflict resolution does not tell us whether conflicts of genetic interest can ever be peacefully resolved. How can legitimate order be retained when one side of a dispute loses in a struggle for life, which is the case when genetic interests conflict? The rights-centred ethic, with its 'peace at any price' value, is also no help when neither party is willing to pay the price of extinction. Can there be any compromise over ultimate interests? Impelled by such overriding concerns, will not any rule book be torn up that does not guarantee a draw or a victory? If in clear-cut contests only ultimate interests can justify opposing other ultimate interests, and if the least acceptable outcome is the status quo ante, then do not opposing genetic interests cancel each other, so that neither side can morally advance its interests at the expense of the other? This is the idea advanced by M. Walzer, quoted previously in Chapter 7 (p. 192): '[Tribalism] cannot be overcome; it has to be accommodated, and therefore the crucial universal principle is that it must always be accommodated; not only my parochialism but yours as well, and his and hers in their turn.'[61]

Perhaps examples of the legitimate resolution of conflicts of genetic interests can point the way to principles of ethical resolution. 'Live and let live' is an excellent adaptationist maxim enshrined in all legal systems. Personal fitness is protected by the prohibition on taking another's life except in self-defence, and by property rights.

An important example of the resolution of conflicts of genetic interests is the prohibition of polygyny in modern societies. Monogamy became the norm in Medieval Christian Europe, a unique development among hierarchical societies. Monogamy is ecologically imposed in hunter-gatherer societies. No one man can accumulate the resources or authority needed to attract and keep more than one woman. But wherever societies become stratified, polygyny arises because powerful males take several wives.[62] In Medieval Europe this pattern was broken by an alliance forged between the Church and commoners against the aristocracy.[63] In analysing Medieval marriage patterns, Macdonald adopts Alexander's concept of socially imposed monogamy, meaning monogamy enforced by a system of social controls. So powerful was the Church-commoner alliance that kings were unable to make heirs of their favourite bastards, and could not easily divorce their wives. English Protestantism received the imprimatur of King Henry VIII as a means of avoiding the Pope's ban on his, in effect, serial monogamy. For years the Church prevented Henry from annulling his marriage to Catherine of Aragon and marrying Anne Boleyn, but this ban was circumvented when Henry appointed his own 'pope' as the head of the Protestant England Church. Socially imposed monogamy spread throughout Europe and continued after the Church began to lose authority from the early nineteenth century. The spread of Western social patterns in recent history has also spread socially imposed monogamy, for example to East Asia and westernising Islamic states.

Socially imposed monogamy is instructive for present purposes because women are a critical reproductive resource for men: no woman, no offspring. Polygyny deprives some men of this resource and reduces their stake in defending the society. The monogamy imposed on the Medieval aristocracy reduced their fertility but did not eliminate it, while the fertility of many commoners was raised from zero. The overall effect was to increase or maintain the fitness of the greater number. Socially imposed monogamy is thus a corollary of pure adaptive utilitarianism. This again indicates that at least some individual rights are inherent to the pure ethic, consistent with it being an expression of general benevolence like its classic utilitarian predecessor.

One analogy of socially imposed monogamy would be universal nationalism in the form of sovereign territory and genetic continuity for every ethny, but only if this defended the genetic interests of the greatest number. As argued in Chapter 7, this would seem to be true, since (1) the world population is largely composed of ethnies, (2) each is a large store of its members' genes, and (3) nation states are the most effective territorially-based ethnic group strategy yet devised. Thus, as a generally applied principle, universal nationalism would seem to serve most ethnic interests and therefore the interests of most human beings. The rule-utilitarian character of this principle indicates that elements of the mixed ethic flow rationally from the pure ethic, without arbitrary clauses.

A further corollary supports this conclusion. Any one ethny might have its interests sacrificed while not much diminishing the global interest. Since all ethnies are minorities within the global population, all have an interest in adopting the mixed ethic, or a derivative of the pure ethic that resembles it. Unless a compelling case is found for delimiting adaptive utilitarianism to a particular state territory, pure adaptive utilitarianism implies that all small ethnies are obliged to give up their territories to large ethnies, and large ethnies give up theirs to the rest of the world, since this benefits the greater number. But applying the ethic globally implies the nationalist principle of the ethnic right to monopolize a territory. It is thus in the general interest to defend the concept of national sovereignty.

As with monogamy, universal nationalism would need to be socially imposed by alliances and other international treaties. These would provide collective security in return for the right to coordinate defence policies and punish defectors. This might seem a strange recommendation to be based on a utilitarian ethic, since Bentham and Mill opposed the Leviathan state.[64] But in the modern context individuals are defenceless unless they participate in powerful group strategies, foremost being the state. As Q. Skinner has forcefully argued, individual freedom is dependent on the freedom of the society as whole within the international arena.[65] Otherwise all citizens are enslaved to foreign interests. The power of the state, properly deployed, defends the autonomy of its citizens. This is the classical view originating in Roman thought and continued through such thinkers as Machiavelli. If the nation state itself is not to be a free rider on its citizen's altruism, further social controls are needed to constrain its conduct in domestic affairs. One such control discussed in Chapter 7 is the ethnic constitution that explicitly dedicates the state as an instrument for the protection of a particular nation.

There are profound ethical implications in making the state simultaneously the champion of a nation's interests in the international scene and a disinterested arbitrator of family interests within the nation. Such an arrangement is fair from the biological perspective, by regulating conflicts between and within nations. Also, the more a state succeeds in making itself ethnically homogeneous, or by preserving that status, by fair means, the more it resolves the ethical problem of conflicting genetic interests between majority and minority. Universal nationalism is biological justice, and this is bound to produce legitimacy. I argued in Chapter 7 that the nation state is a putative ethnic group strategy, since it advertises itself as, in effect, a primordial tribe. Its legitimacy derives from its claim, explicit or implicit, to fulfil the basic tribal functions of defending members' individual and shared genetic interests. When states that pose as nation states abrogate their tribal promise their claim on citizens' altruism loses legitimacy, or deserves to lose it. Citizens would be justified, based on adaptive utilitarian ethics, to reform or tear down their states and build new ones whose ethnic composition and constitution better serve their genetic survival.

Just solutions to social conflicts of interests are more stable than solutions based on force or subterfuge. Bismarck was right in his belief that ethnic autonomy helps underpin stability within and therefore between states. The principle became popular towards the end of the nineteenth century, and President Woodrow Wilson's policy for reorganizing Europe after the First World War operated on this principle. Wilson's policy has been accused of contributing to the instability of the post-War period. In fact that instability can be attributed to the compromise of Wilsonian principles, especially the harsh terms forced on the Central powers by the victors, including the dismemberment of Germany, Austria and Hungary. Another cause was reaction to the terror spread by the communist revolution in Russia. The fact is that in an age of national awareness and popular will, instability in a region continues until empires have been broken down into states that can pass themselves off to the population as nations. That is why democracy often leads to secession, a decline in ethnic diversity, and a rise in fortunes.[66]

*Kinship overlap between ethnies*

Kinship overlap between ethnies has implications for the ethics of group strategies. Recall from Chapter 3 that some members of an ethny might be closer in kinship to another ethny than to their own. This overlap has not been quantified, and might occur at significant levels only between closely related ethnies. Assuming that it does exist, what does this mean for the ethics of pursuing ethnic genetic interests?

The mixed ethic guarantees minority rights, and thus opposes any group strategy (including state policy) that causes harm to minority interests, wherever they might be located. That bars aggressive policies towards ethnies with whom the home ethny has a kinship overlap. It does not bar competitive group strategies directed at ethnies with negligible overlap. Neither does it prevent the imposition of social controls on an entire society aimed at maintaining majority interests, for example conserving its relative fitness within its territory, so long as this does not reduce minority interests. A mixed ethic that guarantees only individual rights, not protection of minority genetic interests, offers greater latitude for majority-centric strategies, for example immigration policies that increase the majority's representation within the state territory.

The pure ethic justifies overriding the interests of minorities when this clearly benefits majority interests. Thus in circumstances where majority and minority interests are mutually exclusive, genetic overlap would not rule out group strategies that harmed minority interests, wherever they were located. Strategies that would make multiple sets of interests compatible are preferable, according to the pure ethic. When overlap is so large that the minority interest approaches that of the majority, the pure ethic dictates mutually beneficial strategies. Since overlap is greatest between closely related ethnies, the adaptive utilitarian ethic offers its strongest injunction against aggression between ethnies belonging to the same regional population or race.

*Freedom and adaptiveness*

To this point I have been arguing that conflicts of genetic interest can be resolved by the mixed ethic combined with social controls, in a manner analogous to the way that Western socially imposed monogamy resolves conflicts of interest caused by polygyny. I also noted that maximizing the genetic interests of the greater number is compatible with unequal outcomes. Free societies leave some scope for competitors to resolve their own differences within established law and custom, guaranteeing mainly procedural justice. I want to end this essay by discussing the connection between freedom and adaptiveness.

Alexander's contractual ethic leads him to conclude that it is immoral to infringe someone else's genetic interests.[67] However, as argued in section 9a, no society absolutely condemns competitive advantage, and probably could not do so and remain a viable society. The modern system of capitalism and the social Darwinism it expresses acknowledge the need to let unsuccessful types of enterprise fail. Hamilton believed the same with regard to defective genes, but thought these could be weeded out in humanitarian fashion through organized eugenics.[68] His main point was that social systems must leave room for adaptive genetic replacement. Hamilton bluntly stated that equality of fitness is impossible because of the accumulating mutation load that would result.[69] If true, this rules out a strictly contractual ethic, at least one that would equalize individual reproductive interests. If Hamilton was right, evolution must continue in the form of differential fitness at the individual level unless and until germ-line mutations are corrected as a routine medical procedure. Continued evolution entails processes of differential reproduction. This argues for acceptance of a substantial degree of individual competition impinging on inclusive fitness. Individuals should have the right to pursue their genetic interests within the law but not possess the right to equal fitness outcomes. The foregoing favours the pure ethic or mixed ethic with rights based on equal process rather than equal outcomes.

Hamilton's argument that continued individual competition is needed to limit a growing mutation load leaves room for a contractual approach to protecting ethnic group interests. While biology rules out a strict equal-outcomes approach to individual genetic interests, it does not rule out protection of ethnic genetic interests. It might be possible to protect the continuity of almost everyone's ethnic genetic interests by preventing genocide and aggressive war and associated military measures, as well as mass migration between genetically distant populations. Eibl-Eibesfeldt has pointed out that in a world of limited space, mutual recognition of every ethny's right to continuity places moral limits on population growth.[70] In such a world any society's sustained failure to control birth rates would constitute an act of international aggression, implying some powerful alliance system or a world government able to control population growth. This is the logical extension of contractual ethics when applied to ethnic genetic interests. The alternative, as implied by Hamilton, is to leave some room for national competition, including autonomous control of population, immigration, and defence policies.[71]

The latter is surely preferable to a system of rigid, permanent social control. A global system that prevented all conflict would be prone to the abuses I discussed in Chapter 7. A warrant to police conflict might be turned instead to eliminate freedom. In a heavily populated world ethnic interests will often conflict due to competition for limited and unevenly distributed resources such as living space, raw materials, and favoured economic niches. Even peaceful tactics for advancing relative fitness, such as accelerated reproduction and immigration, will infringe others' ultimate interests and are thus inherently aggressive. This means that international controls would have to be draconian if they were to suppress all conflict. The most basic freedoms would be lost.

After discussing the possibility of ending conflict by subordinating all humans to the status of cells in a giant super-organism, Hamilton wished for something different. He wished for a solution that 'kept us all potentially free-living'.[72] The solution he recommended is liberal eugenics. I think that Hamilton could have gone further and plausibly argued that free-living itself is adaptive. Reliable defence of any interest will, ultimately, only come from those with a stake in that interest. From the adaptationist perspective, some autonomy should be left to individuals and societies to compete.

A caged eagle might be better fed than its wild cousin, its plumage sleeker. But it is cut off from nature, from the freedom to advance or defend its fitness. Like an eagle soaring above its domain, humans are really free only when at liberty to compete, even if inadvertently, if only in defence, even if risks and opportunities are socially buffered. For a naturally evolved organism the ultimate form of liberty is the freedom to defend its genetic interests.

*Notes*

1 E. O. Wilson (1998).
2 E. O. Wilson (1978a, p. 167).
3 Eibl-Eibesfeldt (1989, p. 718).
4 Irons (2001).
5 D. S. Wilson (2002, p. 228).
6 In this chapter I take a rational approach for several reasons, the overarching one being that only rational thought equipped with the knowledge provided by science can find solutions to the pressing problems of Man's own making, in particular the loss of

human genetic and cultural diversity with concomitant loss of local kinship. No religion is fully equipped to describe the moral problem posed by conflicting genetic interests. Although genetic interests are as old as life, the scientific idea of inclusive fitness is not explicitly discussed in any religious tradition. Applying religious doctrine to genetic interests will entail rational, including scientific, thought. Secondly, there are many religions but only one standard of scientific rationality. Only the scientific method has proven capable of ending the incessant debates that characterized philosophical (let alone religious) differences before empirical and mathematical studies began their exponential growth. Thirdly, as implied by D. S. Wilson, while evolutionary biologists should acknowledge the adaptiveness of some religions, this recognition itself does not and should not violate the norm of empirical truth-seeking. Solutions, including resolution of ethical dilemmas, might necessitate religion, but discerning that fact and choosing adaptive forms requires analysis. Secondary questions arise, such as whether the benefits of religious belief and ritual can be gained if the religion is instrumentally chosen. If so, is the religiously committed mind nevertheless blind to the terrain of biological interests; and is that terrain only illuminated by the cold light of biological science?

7 Westermarck (1971/1912).
8 Hamilton (1975, p. 142).
9 Westermarck (1971/1912); Stevenson (1944).
10 Gibbard (1990, p. 126).
11 Gigerenzer et al. (1999); Simon (1965); Stove (1982).
12 Alexander (1987).
13 Salter (in press-a).
14 Rawls (1971; 1999). While carrying much that is intuitive about fair relations between ethnic groups, in *The Law of Peoples* (1999) Rawls fails to identify any ethnic interests, let alone distinguish between proximate and ultimate ones. Contractarian ethics are, of course, compatible with weight being given to interests, but Rawls's example indicates that they are less interest-bound than utilitarianism.
15 My discussion of utilitarianism is based largely on J. J. C. Smart (1967).
16 The circularity to be avoided is: 'An act is morally good if its effects are morally good.' Such a proposition does not answer the fundamental question of what is morally good.
17 Though not rationally derivable, systems of deontological rules can be rationally ordered. For example, G. E. Moore observed that identical situations should receive the same ethical judgement, and some deontological ethicists set up rules intended to categorize and judge situations. Immanuel Kant proposed a principle for identifying higher deontological rules: can the rule be universalised? So deontological ethical systems need not be irrational, though based on one or more absolute rules received from intuition, tradition, or a deity.
18 Bentham (1789, chapter 1, section IV).
19 Moore (1903).
20 Huxley (1894).
21 Keith (1968/1947, p. 316).
22 Caton (1988, p. 558).
23 Quoted in Caton (1988, p. 558).
24 Raphael (1981/1994, p. 40).
25 Collins (1975); Sanderson (1999, p. 406).

26 Singer (1975/1991, pp. 2–3).
27 Sidgwick (1907, p. 382), quoted in Singer (1975/1994, p. 5).
28 Singer (1975/1991, p. 5).
29 *Ibid.*, p. 7.
30 Singer (1975/1991, p. 8).
31 Alexander (1987).
32 Dawkins (1989, p. 10).
33 Singer (1976, p. 97).
34 S. Pinker has suggested the universal potential of an ethic based on genetic interests: '[I]t might be that the best ethical theory really does embrace termites, Martians, and humans, but is couched in terms of "well-being" or "interests", which happen to differ among termites, Martians, and humans, and which therefore draw out different specialized implications of that single moral system. And perhaps the differing interests in turn can ultimately be explained in terms of the genetic interests of the agents in question' (Pinker 1998).
35 Hare (1981).
36 Hamilton (1971/1996, p. 219).
37 Most of this passage, which concluded his chapter in the *Man and Beast* symposium of 1971, is reasonable. Human nature evolved locally, but economic and other social decisions now have impacts across large societies. We are thus tempted to seek relative advantage even when this reduces everyone's absolute welfare. The unreasonable part comes at the end, where Hamilton refers disparagingly to the 'animal part' of human nature, as if this is inferior to the culturally-based part. Is it credible, within the adaptationist paradigm that Hamilton helped establish, to suppose that any part of human nature (and therefore a human universal) did not evolve precisely because it 'served competitive breeding', however indirectly, or did not interfere with it?
38 Hampshire (2001, p. 43); see Hampshire (2000).
39 Alexander (1987).
40 Thinking about phenotypes must be what drives protective feelings for unborn children, even when they are little more than genetic information stored in the ovum. Genetic information alone is emotionally meaningless, unless associated with images of human beings (usually babies) via knowledge of human development. That association is not fictional, since genomes do in fact grow to become babies if left undisturbed in the womb.
41 Heller (1988); Shepher (1969); Spiro (1995).
42 Salter (1998/2002); and see Chapter 6.
43 Hawkes et al. (2001).
44 Alexander (1987, p. 81).
45 Hamilton (1975/1996, p. 344).
46 Helping nonkin can also be adaptive if it leads to reciprocity (Trivers 1971). Thus gifts to friends do not count as genetic altruism, since reciprocity is likely.
47 Buck and Ginsburg (2000).
48 Williams (1988). See critiques of Williams's view by Hrdy and Cobb following Williams's article.
49 Alexander (1987, p. 177).
50 Darwin (1871, p. 203).
51 Salter (in press-a).
52 Westermarck (1912, II, p. 739, emphases added).

53 Rawls (1971).
54 Williams (1988).
55 Gibbard (1990, p. 127).
56 Walzer (1994a, p. 200).
57 Alexander (1979, pp. 134–6); Lockard and Paulhus (1988).
58 There are degrees of ethnic independence, the most basic and least assertive being continuity of self identification down the generation, followed by degrees of economic and political cooperation within the ethny, then the capacity to act cohesively as a group, rising to possession of a state apparatus with some degree of political autonomy over internal affairs, culminating in the status of a full nation state with recognized borders and sovereignty over internal and external affairs.
59 Alexander (1979); Trivers (1974).
60 Milton, *Areopagitica*, 1644, p. 1.
61 Walzer (1994a, pp. 199–200).
62 Betzig (1986).
63 MacDonald (1995). Betzig claims that the Medieval aristocracy also practised polygyny, but in an exchange with MacDonald following his 1995 paper failed to provide convincing evidence for this contention.
64 Q. Skinner (1998).
65 *Ibid.*
66 Alesina and Spolaore (1997).
67 Alexander (1987, e.g. p. 81).
68 Hamilton (1996, p. 194).
69 Hamilton (1969/1996).
70 Eibl-Eibesfeldt (in press).
71 Hamilton (1975, p. 148).
72 Hamilton (1996, p. 194).

# 10. Afterword

This essay has ranged across several fields of knowledge, including genetics, evolutionary theory, ethology, ecology, various policy areas, the political theory of the state, and ethics. Since mastery of any of these fields is the work of a lifetime, the unavoidable conclusion is that I am not competent to write this essay. Readers should thus approach the arguments presented in this book with a critical attitude. I recommend that you look on it as a stimulus to debate, rather than a statement of final wisdom. I have done my best to get the analysis right, but errors probably remain.

In this book I have argued that an overlooked interest possessed by all individuals is genetic reproduction. This has implications not only for self preservation and personal reproduction, but for the distribution of altruism between family, ethny and humanity. My primary aim has not been to explain human behaviour, but rather to offer social and political theory about what individuals *should* do if they want to behave adaptively. I have suggested strategies for defending genetic interests in a sustainable manner under various circumstances, and offered some thoughts on policy and ethical dimensions. Much of the argument is built on empirical and analytic assumptions that can be tested by: (a) the continued clarification of ultimate interests, including the relative importance of genes and culture; (b) the identification of kin, including ethnic kin, through genetic assays; and (c) the efficacy of strategies for defending genetic interests.

The philosophical component can also be tested, though by debate rather than science. That debate should centre around the ethical status of genetic interests and what is justified in their pursuit. Should growing knowledge of human biology confirm that genetic continuity is the ultimate, or even *an* ultimate, interest, it is difficult to imagine an interest so fundamental being irrelevant to politics. Surely humans' genetic interests can be afforded respect, not in an unobtainable perfect harmony but protected as best we can against the dangers of the modern world with all the compromises and adjustments that entails.

One claim for which I offer no qualification is the centrality of biology in understanding the human condition. Until the discovery of our evolutionary origins, we were like blind men in pursuit of the adaptive life, the wise advancing falteringly, arms outstretched with senses and instincts strained to give some warning of danger. Sustainable altruism was necessarily limited to an intimate circle of kith and kin and sometimes tribe whose similar features, words and smiles were palpable. The Golden Rule was blurred by incomplete information about the ultimate interests at stake in inter-populational exchange. Now the Enlightenment is illuminating our fundamental interest, that which, ultimately, causes us to strive, suffer, and rejoice. The known human universe is enlarged by ever more complete genetic maps, and theoretical biology provides powerful beacons with which to navigate them. We can discover kin in far off lands and reach out across continents and oceans. We can also map others' interests, reducing the chance of inadvertent collisions in life's voyage. Conflict will always be a possibility while men are free, but we now know the fundamental kinship of all mankind, and this is a reason to limit the methods used to settle squabbles that will always occur within the human family.

# Appendix 1: Kinship and Population Subdivision by Henry Harpending

*Summary*

The coefficient of kinship between two diploid organisms describes their overall genetic similarity to each other relative to some base population. For example, kinship between parent and offspring of *1/4* describes gene sharing in excess of random sharing in a random mating population. In a subdivided population the statistic $F_{ST}$ describes gene sharing within subdivisions in the same way. Since $F_{ST}$ among human populations on a world scale is reliably 10 to 15 percent, kinship between two individuals of the same human population is equivalent to kinship between grandparent and grandchild or between half siblings. The widespread assertion that this is small and insignificant should be reexamined.

## *Coefficient of Kinship*

It is easy to understand why parental care evolved in many lineages: parents and offspring share genes so that parental effort devoted to offspring is in fact effort devoted to the parent's own genes. Hamilton (1964) formalized this insight and extended it to arbitrary degrees of relationship. When Hamilton and others described the theory they often spoke in terms of gene identity by descent, thinking for example of the one half of the nuclear genes in a diploid offspring that are identical to those in the parent. Many authors also spoke of shared genes. Neither of these descriptions is completely accurate. I may share many genes with, say, an onion, but this gene sharing is not relevant to the evolution of social behaviour within humans.

A better way to think of kinship, relationship, and Hamilton's theory is think of gene sharing in excess of and in deficit of random gene sharing. A parent shares many more than half his genes with an offspring, but in a random mating population half those genes are surely identical because they came from the parent, while gene sharing with the other half of the child's genome is just what is shared with any random member of the population.

While Hamilton wrote his theory in terms of the coefficient of relationship, most population geneticists reason instead with the coefficient of kinship. Once kinship is known, relationship follows immediately from a simple formula (Bulmer 1994).

Here is the definition of kinship between person $x$ and person $y$: pick a random gene at a locus from $x$ and let the population frequency of this gene be $p$. Now pick a gene from the same locus from $y$. The probability that the gene in y is the same as the gene picked from $x$, $p_y$ is

$$p_y = F_{xy} + (1 - F_{xy})p$$

An interpretation of this is that with probability $F$ the genes are the same, with probability $1 - F$ they are different, in which case the probability of identity is just the population frequency $p$ (Harpending 1979). Rearrangement gives the definition of the coefficient of kinship:

$$F_{xy} = (p_y - p)/(1 - p) \qquad (1)$$

Kinship coefficients in a random mating diploid population are simple and well known. For example, pick a gene from me, then pick another gene from the same locus from me. With probability 1/2 we picked the same gene, while with probability 1/2 we picked the other gene at that locus. Therefore the probability that the second gene is the same as the first is just $1/2 + p/2$, and substitution of this conditional frequency in the formula for kinship shows that my kinship with myself is just 1/2. The same reasoning leads to the well known values of 1/4 with my child, 1/8 with my grandchild, my half-sib, or my nephew, and so on.

It is very important that the coefficient of kinship not be confused with the coefficient of relationship. These are conceptually and numerically different creatures. The coefficient of relationship can be thought of as 'fraction of shared genes' between two organisms. Relationship is familiar to many biologists since W.D. Hamilton developed his famous theory of kin selection in terms of the coefficient of relationship. However most subsequent development of the theory has been in terms of kinship coefficients.

In a random mating diploid population the relationship between the two coefficients is simple: the coefficient of relationship is just twice the coefficient of kinship. This simple rule of thumb breaks down as soon as any complications like inbreeding or population structure are introduced. The best general definition of the coefficient of relation $R_{xy}$ between individuals $x$ and $y$ is (Bulmer 1994):

$$R_{xy} = F_{xy}/F_{xx}$$

where $F_{xy}$ is the kinship between $x$ and $y$ and $F_{xx}$ is the kinship of x with himself. This has the interesting property that it is not necessarily symmetric: $R_{xy}$ is not in general equal to $R_{yx}$.

*Population Subdivision*

Most of the applications of Hamilton's theory in biology have used kinship and relationship derived from genealogical relationships. For example parental care evolves, we think, because parents and offspring share genes. But gene sharing (in excess of random gene sharing, always) can arise in other situations. In a subdivided population, individuals share genes with other members of the same deme, and these shared genes are fuel for evolution by inclusive fitness effects in exactly the same way that pedigree relationships, like that between parent and child, are fuel for evolution by inclusive fitness effects.

I derive here the relationship between population subdivision and kinship in a very simple case, but the formulae apply much more generally than our simple derivation implies. At this point I must mention that our derivations apply to large populations. In the case of small groups ('trait groups', as D. S. Wilson calls them) I would have to consider that if we pick a gene from an individual, the frequency of that gene in the rest of the deme gene pool is slightly reduced. An exact treatment of small demes leads to annoying algebraic terms of order 1/n where n is the deme size. I am concerned with large groups and I ignore these terms.

Consider a population made up of two demes of exactly the same size and a genetic locus with exactly two alleles. The conclusion of the algebra below is that the familiar statistic that describes population subdivision, $F_{ST}$, is precisely kinship between members of the same deme. In other words genetic differences between demes imply genetic similarity within demes, and $F_{ST}$ is just the coefficient of kinship between members of the same deme due to the population structure. For example $F_{ST}$ among human populations is about 1/8, and this is just the coefficient of kinship in a single population between grandparent and grandchild, uncle and nephew, or two half-sibs. In a diverse world, members of the same population are related to each other to the same degree that grandparents and grandchildren are related to each other in a single population.

There are two demes of equal size labelled A and B. At a locus the frequency of a gene is $p_A$ in deme $A$ and $p_B$ in deme B. The frequencies in the two demes of the alternate allele are $q_A$ and $q_B$. The overall mean frequencies are simply $p$ and $q$. It is convenient to use a slightly different notation to describe the gene frequencies:

$p_A = p + \delta$
$p_B = p - \delta$

so of course

$q_A = q - \delta$
$q_B = q + \delta$

Now imagine that we pick a gene at random from the population, then pick another gene from the same locus from the same deme. What is the coefficient of kinship within demes? In order to find this we use the formula (1) above.

With probability 1/2 we pick someone from population A initially, and with probability $p_A$ we pick the allele whose frequency is $p_A$. With probability $q_A = 1 - p_A$ we pick the alternate allele. Putting these possibilities into equation (1) we have

$$F = (1/2)p_A(p_A - p)/q + (1/2)p_B(p_B - p)/q + (1/2)q_A(q_A - q)/p + (1/2)q_B(q_B - q)/p$$

Using the substitutions listed above, this becomes

$$F = \{(p + \delta)(\delta) + (p - \delta)(-\delta)\}/2q + \{q - \delta)(-\delta) + (q + \delta)(\delta)\}/2p$$

$$= 2\delta^2/2q + 2\delta^2/2p$$

and since $p + q = 1$

$$F = 4\delta^2/4pq$$

$$= \delta^2/pq$$

This is simply the $F_{ST}$ genetic distance between the two populations—the variance of the gene frequency divided by the mean gene frequency multiplied by its complement. When $F_{ST}$ is reported for a collection of populations, it is in effect an average of all the pair wise population $F_{ST}$ statistics. The statistic is computed for each allele at each locus, then averaged over all loci.

Many studies agree that $F_{ST}$ in world samples of human populations is between ten and fifteen percent. If small long-isolated populations are included, the figure is usually somewhat higher. A conservative general figure for our species is $F_{ST} \approx 0.125 = 1/8$. This number was given by Cavalli-Sforza in 1966, and a widely cited paper by Lewontin (1972) argued at length that this is a small number, implying that human population differences are trivial. An alternative perspective is that kinship between grandparent and grandchild, equivalent to kinship within human populations, is not so trivial. For further discussion see Klein and Takahata (2002, pp. 387–390).

*Kinship in a Subdivided Population*

Equation (1) and its derivation shows that if we pick a gene at random from a population of two demes we find that its overall frequency is $p$, then the frequency of that gene in the same deme is on average

$$p_{same} = p + (1 - p)F_{ST}$$

while the frequency of that gene in the other deme is on average

$$p_{other} = p - (1 - p)F_{ST}.$$

Using equation 1 and these relations we can derive kinship and relationship coefficients within and between demes easily.

An individual's coefficient of kinship with someone from his own deme is just $F_{ST}$ while his kinship with someone from the other deme is $-F_{ST}$. What about kinship with oneself in a subdivided population? Pick a gene from an individual, then pick another at random from the same individual: with probability 1/2 we picked the same gene and with probability 1/2 we picked the other one, in which case the probability it is the same is $p + (1 - p)F_{ST}$. Therefore

$$p_{self} = (1/2)(1 + p + (1 - p)F_{ST})$$

Using equation 1, we find that

$$F_{self} = (1 + F_{ST}) / 2$$

rather than the simple 1/2 kinship with self in a single random mating population. It is simple to derive familiar family kinship coefficients in the same way: for example kinship with a child when the other parent is from the same deme is

$$F_{child} = 1/4 + 3\ F_{ST}/4$$

and so on. In general, if the kinship in a random mating population with a relative is 1/x, then in a subdivided population the kinship with that same relative is

$$F_{relative\ of\ degree\ x} = 1/x + (1-x)F_{ST}/x \qquad (2)$$

What about kinship with a relative who is a hybrid between the populations? Consider, for example, a child whose other parent is from the other deme. Pick a gene from the parent: the probability of picking the same gene from the child is 1/4, the probability of picking a gene from the child not identical to the first but from the same deme as the parent is 1/4, and the probability of picking a gene from the other deme is 1/2. Putting these together, the probability of the picking the same gene is

$$p_{hybrid\ offspring} = 1/4 + 1/4(p + (1-p)F_{ST}) + 1/2(p - (1-p)F_{ST}).$$

Using equation (1), this becomes

$$F_{hybrid\ offspring} = 1/4 - F_{ST}/4.$$

In general the same derivations shows that kinship with a hybrid relative of degree $x$, meaning a relative with whom kinship in a random mating population would be $x$, is

$$F_{hybrid\ relative\ of\ degree\ x} = 1/x - F_{ST}/x. \qquad (3)$$

The difference between equations (2) and (3) is just $F_{ST}$, the difference between kinship with an intra-demic relative and a hybrid relative. Notice also that as $x$ becomes large, equation (2) shows that kinship with a random member of the same deme is $F_{ST}$ and kinship with an otherwise unrelated hybrid offspring is 0.

## References

Bulmer, M. (1994). *Theoretical evolutionary ecology*. Sinauer, Sunderland, Massachusetts.

Cavalli-Sforza, L. L. (1966). Population structure and human evolution. *Proceedings of the Royal Society of London B*, 164, 362–79.

Hamilton, W. D. (1964). The genetic evolution of social behavior, parts 1 and 2. *Journal of Theoretical Biology*, 7, 1–51.

Harpending, H. (1979). The population genetics of interactions. *American Naturalist*, 113, 622–30.

Klein, J. and Takahata, N. (2002). *Where do we come from? The molecular evidence for human descent*. Springer, Berlin.

Lewontin, R. C. (1972). The apportionment of human diversity. *Evolutionary Biology*, 6, 381–98.

# Appendix 2: Glossary

|  |  |
| --- | --- |
| adaptively minded | Consciously striving to behave adaptively. Also see 'rational fitness maximizer'. |
| adaptive | Behaving in such a way as to maintain or increase genetic representation in future generations, i.e. to conserve or expand genetic interests. |
| allele | An alternative form of a gene at a locus on the genome. |
| altruism | Helping behaviour that carries a net cost for the helper. The cost can be in resources, stress to the organism, the risk of injury and death, or individual fitness. The capacity for psychological altruism in the form of selfless affiliation towards kin or community members evolved because it increased inclusive fitness. For further discussion see p. 110, n. 33. |
| altruism, genetic | Helping behaviour that lowers the inclusive fitness of the helper. See 'altruism, promiscuous'. Because genetic altruism is self-eliminating, it occurs at low frequencies. |
| altruism, kin | Nepotism, being helping behaviour directed towards kin. This is the most intense form of altruism and is found in all social species. |
| altruism, promiscuous | Coined by Westermarck to mean altruism that is directed indiscriminately, i.e. not selectively towards kin or reciprocators. Promiscuous altruism aids free riders and is therefore self-eliminating over evolutionary time. |
| altruism, reciprocal | Coined by Trivers to mean helping behaviour predicated on return of the fa- |

| | |
|---|---|
| | vour.[1] Reciprocity constitutes a form of psychological altruism when it is motivated by selfless affiliation. |
| autochthonous | A people is autochthonous that has lived in a region for a long time. |
| carrying capacity | For a given level of technological and economic development, a territory's carrying capacity is the population beyond which further population growth results in some value being lost. Lost values include privacy, access to open space, and sustenance. When the last value is lost, exceeding a territory's carrying capacity results in die-offs. |
| causes, proximate and ultimate | A basic distinction in evolutionary biology between short-term and long-term processes, but more fundamentally between non-evolutionary and evolutionary causes. For example, behaviour and physiology have proximate causes in psychological and developmental processes that unfold relatively quickly, but those processes themselves are the outcome of evolutionary (ultimate) processes, especially natural selection in the EEA, working over many thousands or millions of years. |
| cline | A slight change in gene frequencies. Genetic differences between neighbouring autochthonous populations mainly take the form of clines, though often there are some genes uniquely associated with a population. Clines accumulate to form relatively abrupt genetic differences between geographic races. |
| co-ethnic | A member of the same ethny. |
| collective goods | See 'goods'. |
| competition, genetic | This occurs when an individual or group tends, advertently or inadvertently, to replace another individual's or group's |

| | |
|---|---|
| | genes. No competitive feelings or intentions are necessary for genetic competition to occur; indeed they are usually absent. |
| concept nation | A nation defined by adherence to a set of ideals or constitutional principles rather than by membership of a cultural or ethnically defined population. |
| constitutional patriotism | Patriotism directed at a set of ideals rather than towards a country or ethny. |
| domain general | A mental ability is 'domain general' if it helps solve a wide range of adaptive problems in the EEA. General intelligence is one such ability. |
| domain specific | A mental ability is 'domain specific' if it is specialized to deal with one or a few adaptation problems in the EEA. This type of ability is often referred to as a 'mental module'. Language acquisition and memory for faces are domain specific abilities. |
| Environment of Evolutionary Adaptedness (EEA) | The range of environments in which a species evolved to its present state. The EEA is an ultimate cause. |
| ethic, deontological | An ethical rule that focuses on an action's intrinsic characteristics rather than on its consequences. Thus an act is held to be good or bad due to law or intuition or religion. It is contrasted with teleological ethics. |
| ethic, pure | The version of the adaptive utilitarian ethic that contains no clause protecting individual rights, so that it evaluates acts purely on the basis of their effects on the fitness of the greatest number. |
| ethic, mixed | The version of the adaptive utilitarian ethic that contains a clause protecting individual rights. Although the mixed ethic defines a good act as one that maximizes the fitness of the greatest number, the |

| | |
|---|---|
| | rights clause condemns acts that violate individual human rights. |
| ethic, teleological | An ethical rule that evaluates actions according to their outcomes, not according to whether they are intrinsically good or bad. |
| ethnic cleansing | The (usually forced) removal of individuals from a territory on the basis of their ethnic identity. |
| ethnic constitution | A constitution that defines a country in ethnic terms and explicitly provides for the protection of the country's ethnic gentic interest or interests. As the supreme law of a country, an ethnic constitution is sufficient to define a country an ethnic state. |
| ethny | A population sharing common descent. 'Ethny' is a preferable term to 'ethnic group' because members of such a category rarely form a group. Ethnies are usually concentric clusters of encompassing populations, such as tribe, regional population, and geographic race. The term 'ethny' used in this book usually means 'a named human population with myths of common ancestry, shared historical memories, one or more elements of common culture, a link with a homeland and a sense of solidarity among at least some of its members'.[2] However, it sometimes has a more general meaning, and thus corresponds most closely to the biological concept of the population. |
| Evolutionary Stable Strategy (ESS) | A behaviour pattern that does not reduce the inclusive fitness of the actor or actors, and which remains adaptive when the majority of the population adopts that pattern. An ESS can remain in place for many generations. |

| | |
|---|---|
| fitness, absolute and relative | A lineage's distinctive genes (its genetic interests) face extinction unless the lineage reproduces at or above the rate of other lineages forming the population. This is true whatever the lineage's absolute fitness, the rate at which its numbers are growing, since depressed relative fitness leads to progressively lower genetic representation within the population. |
| fitness, individual | The genetic contribution by one genotype to the next generation relative to the contribution of other genotypes of the same species. Individual fitness depends on number of offspring. |
| fitness, inclusive | A genotype's individual fitness plus the individual's effect on the fitness of all copies of its distinctive alleles. Note that fitness is a dynamic concept, the effect of a genotype on the gene frequencies of subsequent generations, while genetic interest is aggregate kinship, a static gene count. A behaviour boosts fitness when it tends to preserve or increase genetic interests. |
| fitness portfolio | The allocation of life effort across genetic interests—self, family, ethny, and humanity. Also referred to as 'investment portfolio'. |
| gene pool | All the genes contained by a breeding population. |
| genome | An individual's complete set of genes. |
| genotype | An organism's complete set of genetic information which, through interaction with the environment, results in the phenotype. |
| genetic interest | The number of copies of an individual's distinctive genes. These are most concentrated in the individual, then in first degree relatives, thence in decreasing concentration to clan, tribe or ethny, geographic race, and species. In terms of |

| | |
|---|---|
| | population genetics, genetic interest can be quantified as aggregate kinship, as an equivalent number of children or other close kin. |
| genetic interest, ethnic | The number of copies of a random individual's distinctive genes in his or her ethny, not counting the copies in kin. The size of ethnic genetic interest is relative to the kinship of genetic competitors. When competitors are closely related ethnies, the interest can be relatively small. When competitors are distantly related, especially from different geographical races, ethnic genetic interest can be many orders of magnitude greater than familial genetic interests. |
| genetic interest, familial | The number of copies of an individual's distinctive genes in family members, including the subject. In an outbred population, copies of half an individual's genes are found in parents, full siblings, and children. A quarter are found in nieces and nephews and grandchildren, an eighth in great grandchildren and first cousins, and so on. An individual's familial genetic interest is the sum of these fractions, with the individual counted as one complete genome. |
| genetic stake | Genetic interest. |
| genocide | The killing of part or all of an ethny, on the basis of ethnic identity. |
| goods, public and collective | Some benefit possessing two properties, jointness of supply and nonexcludability. Jointness of supply means that the benefit is not diminished by consumption, so that any number of people can benefit. Nonexcludability means that no-one can be prevented from benefiting from the good. A lighthouse satisfies both conditions. In this book a distinction is made between public |

goods, which benefit everybody, and collective goods, which benefit a particular group.

**group selection** — A process in which a characteristic evolves due to competition between groups rather than (or in addition to) competition between individuals. Kin selection was initially seen as an alternative to group selection. Indeed, Hamilton's 1964 formulation belonged to the tradition initiated by R. A. Fisher which criticized the idea that altruism can evolve due to self sacrifice benefiting a group or species. However, Hamilton's 1975 reformulation of inclusive fitness theory indicates that group selection can increase the frequency of altruism, if altruism serves inclusive fitness. D. S. Wilson and Sober reintroduced group selection on this basis.

**interest, proximate** — A benefit to which we are guided by appetites and other motivational states. Proximate interests include food, status, sex, social bonds, children, and property. Motivations to acquire proximate interests evolved because they increased genetic fitness.

**interest, ultimate** — The ultimate interest is reproduction, the goal towards which all life is shaped through natural selection. Adaptive information carried in genes is transmitted between generations, and is therefore an ultimate interest. For humans culture might also be an ultimate interest, though this is less certain (see Chapter 4, section g).

**investment** — The allocation of some limited resource such as time, energy, or money to aid reproduction. Most organisms invest in such fitness-enhancing activities as their own

| | |
|---|---|
| | growth, self maintenance, and mating guided solely by innate physiological and psychological mechanisms. Human investment is also influenced by culture and conscious decision. |
| kinship | (1) The sharing of an ancestor. (2) The coefficient of kinship *f* is the probability that an allele chosen at random from a locus in one individual (or population) is identical to an allele chosen from the same locus in another individual (or population). |
| kinship, coefficient of | Usually denoted by *f*. Between two individuals A and B, *f* is the probability that an allele drawn from a particular locus in A is identical to an allele drawn from the same locus in B. *f* can be measured by gene assay. |
| kinship overlap | The kinship of two groups overlaps when some inter-group pairs have greater kinship than some intra-group pairs. |
| multi-ethnic society | Any society in which no single ethny comprises the overwhelming majority of the population. Such societies can be pluralist or multicultural, but they can also be national if one ethny or closely related group of ethnies dominates. |
| multicultural society | A multi-ethnic society in which all ethnies officially are accorded equal status according to the pluralist ideology of multiculturalism. At the end of the twentieth century, this system had been applied mainly to English speaking countries plus the Netherlands. In practice multiculturalism entails, and is probably only feasible with, the tolerance or ascendancy of minority ethnic solidarity combined with the loss of majority solidarity . |

natural selection — The differential weeding out of less adaptive forms under changing environmental pressures.

naturalistic fallacy — The assertion that a value statement can be deduced (derived logically) from statements of fact. The Scottish philosopher David Hume was the first to point out that an 'ought' statement cannot be deduced from any number of 'is' statements.

panmixia — The thorough interbreeding of two or more populations.

phenotypes — A complete organism that develops using information carried in the genotype and the environment.

pluralism — An ideology or doctrine according to which society can safely dispense with political, cultural or ethnic homogeneity. Pluralists hold that individuals achieve full autonomy only if institutions nested within the larger society are allowed autonomy, such as trade unions, political parties, religious groups, and ethnic minorities. Pluralism positively evaluates tolerance and other behaviours that promote harmony within diversity. Ideas of pluralism dating from the early twentieth century were developed by such theorists as Horace Kallen into the idea of multiculturalism.[3]

population — A set of organisms of the same species living in or recently migrated from a defined territory, for example one bounded by natural features of rivers, mountains and seas and, in the case of humans, by cultural and political boundaries. Interbreeding is or has been freer within a population than between it and adjacent populations. While all ethnies are or were

| | |
|---|---|
| | recently populations, not all populations are ethnies. |
| portfolio | See 'fitness portfolio'. |
| public goods | See 'goods'. |
| race | Two populations constitute different races when they are sufficiently genetically distant from one another that they are physically distinct—especially on inspection of external characteristics such as colour, hair form, and physiognomy. Since such differences are most visible in populations resident for millennia on different continents, the race concept is usually tied to a continental name, such as 'African', 'East Asian', 'Australian', etc. |
| rational fitness maximizer | A hypothetical individual who consciously seeks to maximize his or her inclusive fitness. The concept is similar to the model 'rational utility maximizer' posited in econometrics. It can be useful to imagine that animals are fitness maximizers, an assumption not made in this book. See Chapter 6 for multiple examples of humans falling short of fitness maximization. |
| relatedness | Hamilton developed inclusive fitness theory using the coefficient of relatedness *r*, being the proportion of genes identical by descent shared by two individuals.[4] Subsequently quantitative genetic theorists replaced *r* with the coefficient of kinship *f* because of the latter's greater precision. In most cases $2f = r$, such that kinship with self is 0.5, parental kinship is 0.25, and sib-sib kinship is 0.25 (in the context of an outbred population). |
| replicator | Any entity of which identical copies are made. In neo-Darwinian theory, the gene is the basic replicator. Culture is thought to be a replicator by some theorists. |

Dawkins contrasts replicators with vehicles.

social identity theory — The theory, based on data from psychological experiments, is that individuals are predisposed to categorize the social world into various kinds of groups, identify with one or more of these, and positively evaluate those with which they identify.

state — An assemblage of institutions that administers individuals living with a demarcated territory. The state apparatus monopolizes the use of coercion within its territory. A separate meaning of 'state' is an administrative unit within a federal system of government.

state, ethnic — A type of nation state whose laws or traditions unambiguously give special protection to one or more ethnies. Typically these laws or traditions assume or assert that the country belongs to one or more founding ethnies. An ethnic state constitutes a territorial ethnic group strategy, by reserving a demarcated territory for a particular ethny.

state, nation — A nation state is a state that administers an ethny. All nation states originated with a founding ethny, and mobilized mass support by emulating aspects of the primordial ethnic group strategy. However, nationality can be redefined in terms of cultural criteria, degrading the efficacy of some types of nation state as ethnic group strategies.

strategy, group — Any coordination of a group to achieve a group goal, such as mutual defence or contribution to collective goods such as government and welfare. Any set of individuals can embark on a group strategy. When that set is an ethny, the strategy can be referred to as an ethnic group strategy.

| | |
|---|---|
| strategy, territorial | Monopoly of a demarcated territory has been a core element of most ethnic group strategies, evident in hunter-gatherer societies, tribal people, and nations. Such monopoly aids a group strategy in several ways, for example by helping identify the group, securing resources such as water, game and arable land, and focussing security on border defence. A major benefit is that territorial monopoly reduces the sensitivity of group continuity to relative fitness—decline in numbers relative to the species. |
| teleology | Explanation by reference to some purpose, typically the assumption that a behaviour is directed towards a goal, from the Greek root *telos* for end or purpose. Teleology is fallacious when applied to understanding evolution. Vitalists assumed that life strives to achieve some preordained final state, while modern evolutionary theory states that genetic change is due to random drift and natural selection. |
| tribal genetic interest | The original ethnic genetic interest. |
| ultimate/proximate | See separate entries under 'causes' and 'interest'. |
| universal nationalism | The idea that ethnic self rule is advantageous for optimizing the general good. The universal nationalist puts his or her own ethny first, but also respects the autonomy of other peoples. It contrasts with chauvinistic nationalism which is thoroughly ethnocentric. |
| utilitarianism, classic | An ethical principle according to which an action is good if it maximizes happiness for the greater number. |
| utilitarianism, act adaptive | The version of adaptive utilitarianism according to which each individual is re- |

sponsible for estimating the adaptive effects of his or her behaviour.

utilitarianism, adaptive — A survivalist ethic modelled on classical utilitarianism, but which substitutes 'adaptiveness' for 'happiness'. The ethical principle is as follows: An act is right to the extent that it maximizes the adaptiveness of the greatest number.

utilitarianism, rule adaptive — The version of adaptive utilitarianism according to which rules are formulated which, if generally obeyed, increase the adaptiveness of the greatest number. The rule version is more compatible with the mixed ethic than is the act version.

vehicle — Any relatively discrete entity which carries replicators and whose characteristics are influenced by those replicators to advance their own preservation and propagation. The least controversial vehicle is the individual organism, the most controversial is the large group, such as the ethny. In this book I argue that ethnies are relatively distinct entities that carry concentrations of their members' distinctive genes and are therefore potential vehicles of members' genetic interests.

*Notes*

1 Trivers (1971).
2 Hutchinson and Smith (1996, p. 6).
3 Kallen (1916/1956).
4 Hamilton (1964).

# References


Adorno, T. H., Frenkel-Brunswik, E., Levinson, D. J., and Sanford, R. N. (1950). *The authoritarian personality.* Harper & Row, New York.

Akenson, D. H. (1992). *God's peoples: Covenant and land in South Africa, Israel, and Ulster.* Cornell University Press, Ithaca, N.Y.

Alba, R. D. (1985). *Ethnicity and race in the U.S.A.* Routledge, Chapman & Hall, New York.

Alcock, J. (2001). *The triumph of sociobiology.* Oxford University Press, New York.

Alesina, A., Baqir, R., and Easterly, W. (1999). Public goods and ethnic divisions. *Quarterly Journal of Economics,* 114(November), 1243–84.

Alesina, A., and Ferrara, E. L. (2000). Participation in heterogeneous communities. *Quarterly Journal of Economics,* 115(3), 847–904.

Alesina, A., and Spolaore, E. (1997). On the number and size of nations. *The Quarterly Journal of Economics,* 112(November), 1027–56.

Alesina, A., and Wacziarg, R. (1998). Little countries: Small but perfectly formed. *The Economist.* 3 January, 63–5.

Alexander, R. D. (1971). The search for an evolutionary philosophy of man. *Proceedings of the Royal Society of Victoria,* 84(1), 99–120.

Alexander, R. D. (1979). *Darwinism and human affairs.* University of Washington Press, Seattle.

Alexander, R. D. (1985/1995). A biological interpretation of moral systems. In *Issues in evolutionary ethics,* (ed. P. Thompson), pp. 179–202. State University of New York Press, Albany, NY.

Alexander, R. D. (1987). *The biology of moral systems.* Aldine de Gruyter, New York.

Anderson, B. R. O. G. (1993). Nationalism. In *The Oxford companion to politics of the world,* (ed. J. Krieger, W. A. Joseph, M. Kahler, G. Nzongola-Ntalaja, B. B. Stallings and M. Wier), pp. 614–19. Oxford University Press, New York.

Appiah, K. A. (1996). Race, culture and identity: Misunderstood connections. In *Color conscious. The political morality of race,* (ed. K. A. Appiah and A. Gutman) , pp. 3–74. Princeton University Press, Princeton, NJ.

Armstrong, J. (1982). *Nations before nationalism.* University of North Carolina Press, Chapel Hill.

Arnhart, L. (1995). The new Darwinian naturalism in political theory. *American Political Science Review*, 89(2), 389–400.

Ashburn-Nardo, L., Voils, C. I., and Monteith, M. J. (2001). Implicit associations as the seeds of intergroup bias: How easily do they take root? *Journal of Personality and Social Psychology*, 81(5), 789–99.

Auster, L. (1990). *The path to national suicide: An essay on immigration and multiculturalism*. AICF, Monterey, Va.

Banfield, E. C. (1967). *The moral basis of a backward society*. Free Press, New York.

Barbujani, G., and Sokal, R. R. (1990). Zones of sharp genetic change in Europe are also linguistic boundaries. *Proceedings of the National Academy of Sciences USA*, 87(5), 1816–19.

Barry, B. (1996). Nationalism, intervention and redistribution. Insitut für Interkulturelle und Internationale Studien (INIIS), Universität Bremen. Working paper 3/96.

Bateson, P. P. G. (1978). Book review: *The Selfish Gene*. *Animal Behaviour*, 26, 316–18.

Beaumont, P. (2002). Israel fears invasion of immigrants. *The Observer*, London, 16 June.

Beckstrom, J. H. (1993). *Darwinism applied: Evolutionary paths to social goals*. Greenwood Publishing Group, Westport.

Bentham, J. (1789). *Introduction to the principles of morals and legislation*. London.

Betts, K. (2002). Boatpeople and the 2001 election. *People and Place*, 10(3), 36–54.

Betzig, L. (1986). *Despotism and differential reproduction: A Darwinian view of history*. Aldine, Chicago.

Betzig, L. (1992). Roman monogamy. *Ethology and Sociobiology*, 13, 351–83.

Bigelow, R. S. (1969). *The dawn warriors: Man's evolution towards peace*. Little Brown, Boston.

Birrell, R. (1995). *A nation of our own. Citizenship and nation-building in Federation Australia*. Longman Australia, Melbourne.

Blackmore, S. (1999). *The meme machine*. Oxford University Press, New York.

Blalock, H. (1967). *Toward a theory of minority-group relations*. Wiley, New York.

Blalock, H. M. J. and Wilken, P. H. (1979). *Intergroup processes: A micro-macro problem*. The Free Press, New York.

Blok, A. (2002). Mafia and blood symbolism. In *Risky transactions. Kinship, ethnicity, and trust*, (ed. F. K. Salter), pp. 109–128. Berghahn, Oxford and New York.

Bodmer, W. F. and Cavalli-Sforza, L. L. (1976). *Genetics, evolution, and man.* W. H. Freeman and Company, San Francisco.
Boehm, C. (1993). Egalitarian behavior and reverse dominance hierarchy [with peer commentary]. *Current Anthropology*, 34(3), 227–54.
Bonacich, E. (1973). A theory of middleman minorities. *American Sociological Review*, 38, 583–94.
Bookman, M. (1997). *The demographic struggle for power: The political economy of demographic engineering in the modern world.* Frank Cass, London/Portland, Oregon.
Borjas, G. J. (1999). Immigration, the issue-in-waiting. *New York Times*, New York, 2 April.
Borjas, G. J. (2001). *Heaven's door: Immigration policy and the American economy.* Princeton University Press, Princeton.
Bouchard, T. J. (1994). Genes, environment, and personality. *Science*, 264 (17 June), 1700–1.
Bouchard, T. J. (1996). IQ similarity in twins reared apart: Finding and responses to critics. In *Intelligence: Heredity and environment*, (ed. R. Sternberg and C. Grigorenko), Cambridge University Press, New York.
Bouchard, T. J. (1998). Genetic and environmental influences on adult intelligence and special mental abilities. *Human Biology*, 70, 257–79.
Bowlby, J. (1988). *A secure base: Parent-child attachment and healthy human development.* Basic Books, New York.
Boyd, R. and Richerson, P. J. (1985). *Culture and the evolutionary process.* University of Chicago Press, Chicago.
Boyd, R. and Richerson, P. J. (1987). The evolution of ethnic markers. *Cultural Anthropology*, 2, 65–79.
Boyd, R. and Richerson, P. J. (1992). Punishment allows the evolution of cooperation (or anything else) in sizable groups. *Ethology and Sociobiology*, 13, 171–195.
Boyd, R. and Silk, J. (1997). *How humans evolved.* Norton, New York.
Brimelow, P. (1995). *Alien nation. Common sense about America's immigration disaster.* Random House, New York.
Browne, A. (2000). Race and politics. The suicide of the English nation? UK whites will be minority by 2100. *The Observer*, London, 3 September.
Brubacker, R. (1992). *Citizenship and nationshood in France and Germany.* Cambridge University Press, Cambridge, MA.
Buchanan, P. (1998). Our Christian dispossession. *New York Post*, Washington, DC, Saturday, 28 November.

Buchanan, P. J. (2002). *The death of the West. How dying populations and immigrant invasions imperil our country and civilization.* St. Martin's Press, New York.

Buck, R., and Ginsburg, B. E. (1991). Emotional communication and altruism: The communicative gene hypothesis. In *Altruism. Review of personality and social psychology.* (Ed. M. Clark), pp. 149–75. Sage, Newbury Park, California.

Buck, R., and Ginsburg, B. E. (2000). Communicative genes in the evolution of behavioral and social systems: Empathy and altruism. Paper presented at The American Political Science Association, 1 September. Washington, D.C.

Burke, E. (1790/1968). *Reflections on the revolution in France.* Penguin, Harmondsworth.

Burnham, J. (1941). *The managerial revolution: What is happening in the world.* John Day Company, New York.

Burnham, J. (1964/1975). *Suicide of the West: An essay on the meaning and destiny of liberalism.* Arlington house, New Rochelle, N.Y.

*Business Week* Executive Programs (1996–1997). Europe Challenged: 1996 Business Week Europe Conference of Chief Executives. *Business Week, International Edition.* Dec.30–Jan.6, Special advertising section.

Buss, D. M. (1994). *The evolution of desire: Strategies of human mating.* Basic Books, New York.

Buss, D. M., *et al.* (1999). Jealousy and the nature of beliefs about infidelity: Tests of competing hypotheses about sex differences in the United States, Korea, and Japan. *Personal Relationships*, 6(1), 125–50.

Butovskaya, M., Diakonov, I., and Salter, F. (2002). Effects of sex, status and ethnic similarity on willingness to give to beggars: A questionnaire study. Unpublished manuscript.

Butovskaya, M., Salter, F., Diakonov, I., and Smirnov, A. (2000). Urban begging and ethnic nepotism in Russia: An ethological pilot study. *Human Nature*, 11(2), 157–82.

Campbell, D. T. (1983). The two distinct routes beyond kin selection to ultrasociality: Implications for the humanities and social sciences. In *The nature of prosocial development: Theories and strategies*, (ed. D. Bridgeman), pp. 11–41. Academic Press, New York.

Canovan, M. (1996). *Nationhood and political theory.* Edward Elgar, Cheltenham, UK.

Case, A., Lin, I.-F., and McLanahan, S. (2000). How hungry is the selfish gene? *Economic Journal* 110(466), 781–804.

Case, A., Lin, I.-F., and McLanahan, S. (2001). Educational attainment of siblings in stepfamilies. *Evolution and Human Behavior*, 22, 269–89.

Cassese, A. (1993). Self-determination. In *The Oxford companion to politics of the world*, (ed. J. Krieger, W. A. Joseph, M. Kahler, G. Nzongola-Ntalaja, B. B. Stallings and M. Wier), pp. 822–3. Oxford University Press, New York.

Caton, H. P. (1983/1994). Descriptive political ethology. Griffith University. Unpublished manuscript, 5 pp.

Caton, H. P. (1988). *The politics of progress: The origins and development of the commercial republic, 1600–1835*. University of Florida Press, Gainesville.

Caton, H. P. (1998/2002). Reinvent yourself: Labile psychosocial identity and the lifestyle marketplace. In *Ethnic conflict and indoctrination: Altruism and identity in evolutionary perspective*, (ed. I. Eibl-Eibesfeldt and F. K. Salter), pp. 325–43. Berghahn, Oxford and New York.

Cattell, R. B. (1937). *The fight for our national intelligence*. P. S. King, London.

Cavalli-Sforza, L. L. (1986). Cultural evolution and genetics. In *Human genetics: Proceedings of the 7th International Congress Berlin 1986*, (ed. F. Vogel and K. Sperling), pp. 24–33. Springer, Berlin.

Cavalli-Sforza, L. L. (1991). Genes, peoples and languages. *Scientific American*, 264 (Nov. 91), 72–78.

Cavalli-Sforza, L. L. (2000). *Genes, peoples, and languages*. (trans. Seielstad, Mark). North Point Press, New York.

Cavalli-Sforza, L. L. and Bodmer, W. F. (1971/1999). *The genetics of human populations*. Dover, Mineola, NY.

Cavalli-Sforza, L. L. and Cavalli-Sforza, F. (1995). *The great human diasporas: The history of diversity and evolution*. (trans. Thorne, Sarah). Addison-Wesley, Reading, MA.

Cavalli-Sforza, L. L. and Feldman, M. W. (1981). *Cultural transmission and evolution: A quantitative approach*. Princeton University Press, Princeton, NJ.

Cavalli-Sforza, L. L., Menozzi, P. and Piazza, A. (1994). *The history and geography of human genes*. Princeton University Press, Princeton, NJ.

Chagnon, N. A. (1980). Mate competition, favoring close kin, and village fissioning among the Yanomama Indians. In *Evolutionary Biology and Human Social Behavior*, (ed. N. A. Chagnon and W. Irons), pp. 86–131. Duxbury, North Scituate, MA.

Chesler, E. (1992). *Woman of valor. Margaret Sanger and the birth control movement in America*. Schuster, New York.

Clarke, A. L. (2002). Assessing the carrying capacity of the Florida Keys. *Population and Environment*, 23(4), 405–18.

Cliquet, R. and Thienpont, K. (1999). In-group/out-group relations and evolutionary biology: Implications for social science and social policy. In *In-group/out-group behaviour in modern societies. An evolutionary perspective*, (ed. K. Thienpont and R. Cliquet), pp. 281–301. NIDI CBGS Publications, Brussels.

Coase, R. H. (1937). The nature of the firm. *Economica*, 4, 386–405.

Cohen, J. (1995). *How many people can the earth support?* Norton, New York.

Collins, R. (1975). *Conflict sociology: Toward an explanatory science*. Academic Press, New York.

Collins, F. S. and Jegalian, K. G. (1999). Deciphering the code of life. *Scientific American*, 281(6), 50–55.

Collinson, S. (1993). *Beyond borders: West European migration policy towards the 21st century*. Royal Institute of International Affairs, London.

Connor, W. (1972). Nation-building or nation destroying. *World Politics*, 24 (April), 319–55.

Connor, W. (1985). The impact of homelands upon diasporas. In *Modern diasporas in international politics*, (ed. G. Sheffer), pp. 16–46. Croom Helm, London.

Connor, W. (1993). Beyond reason: The nature of the ethnonational bond. *Ethnic and Racial Studies*, 16(3), 373–89.

Corning, P. A. (1983). *The synergism hypothesis: A theory of progressive evolution*. McGraw-Hill, New York.

Corning, P. A. (2000). Biological adaptation in human societies: A 'basic needs' approach. *Journal of Bioeconomics*, 2(1), 41–86.

Cosmides, L. and Tooby, J. (1992). Cognitive adaptations for social exchange. In *The adapted mind: Evolutionary psychology and the generation of culture*, (eds. J. H. Barkow, L. Cosmides and J. Tooby), pp. 163–228. Oxford University Press USA, New York.

Courtois, S., Paczkowski, P. A., Bartosek, K., Werth, N., Panne, J.-L. (1999). *The black book of Communism: Crimes, terror, repression*. (trans. M. Kramer and J. Murphy). Harvard University Press, Cambridge, MA.

Crawford, M. H., Koertvelyessy, T., Pap, M., Szilagyi, K., and Duggirala, R. (1999). The effects of a new political border on the migration patterns and predicted kinship (PHI) in a subdivided Hungarian agricultural population: Tiszahat. *Homo*, 50(3), 201–210.

Daily, G., and Ehrlich, P. (1992). Population, sustainability, and Earth's carrying capacity. *BioScience*, 42, 761–71.

Daly, M., and Wilson, M., ed. (1999). *Stepparental investment*. Special issue of *Evolution and Human Behavior*, 20(6). Elsevier. New York.

Daly, M., Salmon, C., and Wilson, M. (1997). Kinship: The conceptual hole in psychological studies of social cognition and close relationships. In *Evolutionary social psychology*, (ed. J. Simpson and D. T. Kenrick), Erlbaum, Mahwah, NJ.

Daly, M., Wilson, M., and Weghorst, S. J. (1982). Male sexual jealousy. *Ethology and Sociobiology*, 3, 11–27.

Darnton, R. (2002). A €uro state of mind. *New York Review of Books*, 49(3), 30–32.

Darwin, C. (1859). *On the origin of species by means of natural selection, or The preservation of favoured races in the struggle for life*. Murray, London.

Darwin, C. (1871). *The descent of man, and selection in relation to sex*. (1913 edn). Murray, London.

Dawkins, R. (1976). *The selfish gene*. Oxford University Press, Oxford.

Dawkins, R. (1978). Replicator selection and the extended phenotype. *Zeitschrift für Tierpsychologie*, 47, 61–76.

Dawkins, R. (1979). Twelve misunderstandings of kin selection. *Zeitschrift für Tierpsychologie*, 51, 184–200.

Dawkins, R. (1981). Selfish genes in race or politics. *Nature*, 289(5798), 528.

Dawkins, R. (1982a). *The extended phenotype: The gene as the unit of selection*. Freeman, Oxford and San Francisco.

Dawkins, R. (1982b). *Replicators and vehicles*. Cambridge University Press, Cambridge.

Dawkins, R. (1989). *The selfish gene*. (2nd edn). Oxford University Press, Oxford.

Degler, C. (1991). *In search of human nature: The decline and revival of Darwinism in American social thought*. Oxford University Press, Oxford.

de Tocqueville, A. (2001). *Democracy in America*. (Translated, edited and with an introduction by H. C. Mansfield and D. Winthrop). University of Chicago Press, Chicago.

Dio, C. (1925). *History*. (trans. E. Cary). Putnam's Sons, New York.

Draper, P., and Harpending, H. (1982). Father absence and reproductive strategy: An evolutionary perspective. *Journal of Anthropological Research*, 38, 255–73.

Durham, W. H. (1991). *Coevolution: Genes, culture and human diversity*. Stanford University Press, Stanford, CA.

Dyson, F. (2000). Science, guided by ethics, can lift up the poor. *International Herald Tribune*, New York, 29 May, 10.

Easterly, W. (2000). Can institutions resolve ethnic conflict? Unpublished manuscript, World Bank, New York.

Easterly, W. and Levine, R. (1997). Africa's growth tragedy: Policies and ethnic divisions. *Quarterly Journal of Economics*, 112 (November), 1203–50.

Eco, U. (2002). *Five moral pieces*. Secker and Warburg, New York.

Eibl-Eibesfeldt, I. (1970). *Liebe und Hass. Zur Naturgeschichte elementarer Verhaltensweisen.* Piper, München. [1972 English edition: *Love and hate. The natural history of behaviour patterns.*]

Eibl-Eibesfeldt, I. (1982). Warfare, man's indoctrinability and group selection. *Ethology (Zeitschrift für Tierpsychologie)*, 60, 177–98.

Eibl-Eibesfeldt, I. (1989). *Human ethology*. Aldine de Gruyter, New York.

Eibl-Eibesfeldt, I. (1991). Deutschlands Zukunft: Nationalstaat oder multikulturelle Gesellschaft? In *Nachdenken über Deutschland*, Vol. 1. (ed. D. Keller), pp. 38–63. Verlag der Nation, Berlin.

Eibl-Eibesfeldt, I. (1994). *Wider die Mißtrauensgesellschaft. Streitschrift für eine bessere Zukunft [Against the society of mistrust. Prospects for a better future].* Piper, Zürich.

Eibl-Eibesfeldt, I. (1998). *In der Falle des Kurzzeitdenkens [The trap of short-term thinking].* Piper, Zürich.

Eibl-Eibesfeldt, I. (1998/2002). Us and the others: The familial roots of ethnonationalism. In *Ethnic conflict and indoctrination: Altruism and identity in evolutionary perspective,* (ed. I. Eibl-Eibesfeldt and F. K. Salter), pp. 21–53. Berghahn, Oxford and New York.

Eibl-Eibesfeldt, I. (in press 2003). Ethnicity, the problem of differential altruism, and international multiculturalism. In *Welfare, ethnicity, and altruism. New data and evolutionary theory,* (ed. F. K. Salter), Frank Cass, London.

Eibl-Eibesfeldt, I., and Salter, F. K., ed. (1998/2002). *Ethnic conflict and indoctrination: Altruism and identity in evolutionary perspective.* Berghahn. Oxford and New York.

Engardio, P. (2002). The chance of a lifetime: Poor nations get a "demographic dividend". *Business Week.* 25 March, 34.

Entine, J. (1999). *Taboo: Why black athletes dominate sports and why we're afraid to talk about it.* Public Affairs, New York.

Erasmus, C. (1977). *In search of the common good. Utopian experiments past and future.* Free Press, New York.

Erikson, E. H. (1966). Ontogeny of ritualization in man. *Philosophical Transactions of the Royal Society of London*, B251, 337–49.

Esses, V. M., Dovidio, J. F., Jackson, L. M., and Armstrong, T. L. (2001). The immigration dilemma: The role of perceived group competition, ethnic prejudice, and national identity. *Journal of Social Issues*, 57(3), 389–412.

Faist, T. (1995). Ethnicization and racialization of welfare-state politics in Germany and the USA. *Ethnic and Racial Studies*, 18(2), 219–50.

Fehr, E., and Gachter, S. (2002). Altruistic punishment in humans. *Nature*, 415, 137–40.

Fisher, R. A. (1930/1999). *The genetical theory of natural selection. A complete variorum edition.* (2nd edn). Oxford University Press, Oxford.

Flohr, H., and Tönnesmann, W., ed. (1983). *Politik und Biologie. Beiträge zur Life-Sciences-Orientierung der Sozial Wissenshaften.* Parey. Berlin/Hamburg.

Francis, S. (1984/1999). *James Burnham.* Claridge Press, London.

Frank, G. (1997). Jews, multiculturalism, and Boasian anthropology. *American Anthropologist*, 99(4), 731–45.

Frank, S. A. (1995). Mutual policing and repression of competition in the evolution of cooperative groups. *Nature*, 377 (12 October), 520–22.

Freedman, D. G. (1974). *Human infancy.* Halsted, New York.

Freedman, D. G. and Freedman, N. C. (1969). Behavioral differences between Chinese-American and European newborns. *Nature*, 224, 1227.

Geiger, G. (1988). On the evolutionary origins and function of political power. *Journal of Social and Biological Structures*, 11, 235–50.

Gellner, E. (1983). *Nations and nationalism.* Cornell University Press, London.

Gibbard, A. (1990). *Wise choices, apt feelings: A theory of normative judgment.* Clarendon, Oxford.

Gigerenzer, G., Todd, P. M., and the ABC [Centre for Adaptive Behavior and Cognition] (1999). *Simple heuristics that make us smart.* Oxford University Press, New York.

Gilens, M. (1999). *Why Americans hate welfare: Race, media, and the politics of antipoverty policy.* University of Chicago Press, Chicago.

Glantz, D. M., and House, J. M. (1995). *When titans clashed: how the Red Army stopped Hitler.* University of Kansas, Lawrence, Kansas.

Godfray, H. C. J. (1995). Evolutionary theory of parent-offspring conflict. *Nature*, 376, 133–8.

Goetze, D. (1998). Evolution, mobility, and ethnic group formation. *Politics and the Life Sciences*, 17(1), 59–71.

Goetze, D. (1999). Resource subtractibility and the evolution of groups. 49 page Manuscript.

Golby, A. J., Gabrieli, J. D. E., Chiao, J. Y., and Eberhardt, J. L. (2001). Differential responses in the fusiform region to same-race and other-race faces. *Nature Neuroscience*, 4(8), 845–50.

Gottfredson, L. S. (1997). Mainstream science on intelligence: An editorial with 52 signatories, history, and bibliography. *Intelligence*, 24(1), 13–23.

Gottfredson, L. S. (1997). Why *g* matters: The complexity of everyday life. *Intelligence*, 24(1), 79–132.

Gould, S. J. (1981/1996). *The mismeasure of man.* (2nd edn). Norton, New York.

Grafen, A. (1990). Do animals really recognize kin? *Animal Behaviour*, 39, 42–54.

Graham, H. D. (2002). *Collision course: The strange convergence of affirmative action and immigration policy in America.* Oxford University Press, New York.

Grammer, K. (1988). *Biologische Grundlagen des Sozialverhaltens.* Wissenschaftliche Buchgesellschaft, Darmstadt.

Grammer, K. (1993). *Signale der Liebe. Die biologischen Gesetze der Partnerschaft.* Hoffman und Campe, Hamburg.

Greenfeld, L. (1992). *Nationalism: Five roads to modernity.* Harvard University Press, Cambridge, MA.

Greenwald, A. G., and Schuh, E. S. (1994). An ethnic bias in scientific citations. *European Journal of Social Psychology*, 24 (Nov-Dec), 623–39.

Greider, W. (1996). *One world, ready or not. The manic logic of global capitalism.* Simon & Schuster, New York.

Gudmundsson, H., Gudbjartsson, D. F., Frigge, M., Gulcher, J. R., and Stefansson, K. (2000). Inheritance of human longevity in Iceland. *European Journal of Human Genetics*, 8(10), 743–9.

Guillén, M. F. (2001). Is globalization civilizing, destructive or feeble? A critique of five key debates in the social science literature. *Annual Review of Sociology*, 27, 235–60.

Gulcher, J., Helgason, A., and Stefánsson, K. (2000). Genetic homogeneity of Icelanders. *Nature Genetics*, 26(4), 395–6.

Gurr, T. R. and Harff, B. (1994). *Ethnic conflict in world politics.* Westview Press, Boulder.

Habermas, J. (1998). *The inclusion of the other: Studies in political theory.* MIT Press, Cambridge, MA.

Haldane, J. B. S. (1955). Population genetics. *New Biology (London)*, 18, 34–51.

Hall, J. A. (1993). State. In *The Oxford companion to politics of the world*, (ed. J. Krieger, W. A. Joseph, M. Kahler, G. Nzongola-Ntalaja, B. B. Stallings and M. Wier), pp. 878–83. Oxford University Press, New York.

Hamilton, W. D. (1963). The evolution of altruistic behavior. *American Naturalist*, 97, 354–56.

Hamilton, W. D. (1963/1996). The evolution of altruistic behavior. In *Narrow roads of gene land. Vol. 1: Evolution of social behaviour*, Vol. 97. (ed. W. D. Hamilton), pp. 6–8. W. H. Freeman, Oxford.

Hamilton, W. D. (1964). The genetic evolution of social behavior, parts 1 and 2. *Journal of Theoretical Biology*, 7, 1–51.

Hamilton, W. D. (1964/1996). The genetic evolution of social behavior, parts 1 and 2. In *Narrow roads of gene land. Vol. 1: Evolution of social behaviour*, Vol. 97. (ed. W. D. Hamilton), pp. 11–82. W. H. Freeman, Oxford.

Hamilton, W. D. (1969/1996). Population control. In *Narrow roads of gene land. Vol. 1: Evolution of social behaviour*, (ed. W. D. Hamilton), p. 196. W. H. Freeman, Oxford. [originally in *New Scientist* 44 (20 Oct.) 1969, pp. 260–1.]

Hamilton, W. D. (1970). Selfish and spiteful behaviour in an evolutionary model. *Nature*, 228, 1218–20.

Hamilton, W. D. (1971). Selection of selfish and altruistic behavior in some extreme models. In *Man and beast: Comparative social behavior*, (ed. J. F. Eisenberg and W. S. Dillon), pp. 59–91. Smithsonian Institute Press, Washington, DC.

Hamilton, W. D. (1971/1996). Selection of selfish and altruistic behavior in some extreme models. In *Narrow roads of gene land. Vol. 1: Evolution of social behaviour*, Vol. 97. (ed. W. D. Hamilton), pp. 198–227. W. H. Freeman, Oxford.

Hamilton, W. D. (1972). Altruism and related phenomena, mainly in social insects. *Annual Review of Ecological Systems*, 3, 193–232.

Hamilton, W. D. (1975). Innate social aptitudes of man: An approach from evolutionary genetics. In *Biosocial anthropology*, (ed. R. Fox), pp. 133–55. Malaby Press, London.

Hamilton, W. D. (1975/1996). Innate social aptitudes of man: An approach from evolutionary genetics. In *Narrow roads of gene land. Vol. 1: Evolution of social behaviour*, (ed. W. D. Hamilton), pp. 315–52. W. H. Freeman, Oxford.

Hamilton, W. D. (1996). *Narrow roads of gene land. The collected papers of W. D. Hamilton*. W. H. Freeman, Oxford.

Hampshire, S. (2000). *Justice is conflict*. Princeton University Press, Princeton, NJ.

Hampshire, S. (2001). He had his ups and downs. *New York Review of Books*, 48(8), 42–4.

Hardin, G. (1968). The tragedy of the commons. *Science*, 162, 1243–8.

Hardin, G. (1974/2001). Living on a lifeboat. *The Social Contract*, 12(1), 36–47. [Reprinted from *BioScience*, October 1974.]

Hardin, G. (1993). *Living within limits. Ecology, economics, and population taboos*. Oxford University Press, Oxford.

Hardin, G. (1999). *The ostrich factor: Our population myopia*. Oxford University Press, New York.

Hare, R. M. (1981). *Moral thinking: Its levels, method, and points*. Oxford Clarendon Press, Oxford.

Harpending, H. (1979). The population genetics of interactions. *American Naturalist*, 113, 622–30.

Hawkes, K., O'Connell, J. F., and Blurton Jones, N. G. (2001). Hadza meat sharing. *Evolution and Human Behavior*, 22, 113–42.

Hechter, M. (2000). *Containing nationalism.* Oxford University Press USA, New York.
Hedges, S. B., Kumar, S., Tamura, K., and Stoneking, M. (1991). Human origins and analysis of mitochondrial DNA sequences. *Science*, 255(7 Feb.), 737–9.
Helgason, A., *et al.* (2000). Estimating Scandinavian and Gaelic ancestry in the male settlers of Iceland. *American Journal of Human Genetics*, 67(3), 697–717.
Heller, M. (1988). *Cogs in the Soviet Wheel. The formation of Soviet Man.* Collins Harvill, London.
Herman, A. (2002). *Great Scots! How the Scots invented the modern world.* Crown, New York.
Hero, R. E., and Tolbert, C. J. (1996). A racial/ethnic diversity interpretation of politics and policy in the states of the U.S. *American Journal of Political Science*, 40(3), 851–71.
Herrnstein, R., and Murray, C. (1994). *The bell curve. Intelligence and class structure in American life.* Free Press, New York.
Hewlett, S. A. (2002). *Creating a life: Professional women and the quest for children.* Talk Miramax, New York.
Hinde, R. A. (1989). Patriotism: Is kin selection both necessary and sufficient? *Politics and the Life Sciences*, 8(1), 58–61.
Hill, E. (1993). Ibn-Khaldun. In *The Oxford companion to the politics of the world*, (ed. J. Krieger), Oxford University Press, New York.
Hirschfeld, L. A. (1996). *Race in the making. Cognition, culture, and the child's construction of human kinds.* MIT Press, Cambridge, MA.
Holden, C. (1998). The tuna within. *Science*, 282(1811–Dec.4), 528.
Holper, J. J. (1996). Kin term usage. In *The Federalist*: Evolutionary foundations of Publius's rhetoric. *Politics and the Life Sciences*, 15(2), 265–72.
Hoppe, H.-H. (2001). *Democracy–the god that failed: The economics and politics of monarchy, democracy, and natural order.* Transaction, New Brunswick, NJ.
Howell, D. (2002). Who listens to the people of Europe? *International Herald Tribune*, Frankfurt am Main, 2 May, p. 9.
Hrdy, S. B. (1981). *The woman that never evolved.* Harvard University Press, Cambridge, MA.
Hrdy, S. B. (1999). *Mother nature: A history of mothers, infants, and natural selection.* Pantheon, New York.
Hroch, M. (1985). *Social preconditions of the national revival in Europe.* Cambridge University Press, Cambridge.
Huntington, S. P. (1996). *The clash of civilizations and the remaking of world order.* Simon & Schuster, New York.

Huntington, S. P. (1997). The west and the rest. *Prospect*, (February), 34–9.

Hutchinson, J., and Smith, A. D. (1996). Introduction. In *Ethnicity*, (ed. J. Hutchinson and A. D. Smith), pp. 3–14. Oxford University Press, Oxford & New York.

Huxley, J. S., Haddon, A. C., and Carr-Saunders, A. M. (1939/1935). *We Europeans. A survey of "racial" problems*. Penguin, London.

Huxley, T. H. (1894). *Evolution and ethics and other essays*. Appleton, New York.

Hyatt, M. (1990). *Franz Boas, social activist: The dynamics of ethnicity*. Greenwood Press, New York.

Irons, W. (2001). Religion as a hard-to-fake sign of commitment. In *Evolution and the capacity for commitment*, (ed. R. M. Nesse), pp. 240–61. Russell Sage Foundation, New York.

Jacobson, D. (1996). *Rights across borders: Immigration and the decline of citizenship*. Johns Hopkins University Press, Baltimore.

Jankowiak, W. and Diderich, M. (2000). Sibling solidarity in a polygamous community in the USA: Unpacking inclusive fitness. *Evolution and Human Behavior*, 21(2), 125–39.

Jensen, A. R. (1969). How much can we boost IQ and scholastic achievement? *Harvard Educational Review*, 39(Winter), 1–123.

Johnson, G. R. (1986). Kin selection, socialization, and patriotism: An integrated theory [with commentaries]. *Politics and the Life Sciences*, 4, 127–54.

Johnson, G. R. (1987). In the name of the fatherland: An analysis of kin terms usage in patriotic speech and literature. *International Political Science Review*, 8, 165–74.

Johnson, J. (2000). Review of J. Habermas, 1998, *The inclusion of the other: Studies in political theory*, MIT Press. *American Political Science Review*, 94(2), 448–9.

Johnson, P. (1987/1998). *A history of the Jews*. Perennial Library, New York.

Jones, S. (1992). The evolutionary future of humankind. In *The Cambridge encyclopedia of human evolution*, (ed. S. Jones, S. Bunney, and R. Dawkins), pp. 439–45. Cambridge University Press, Cambridge.

Kallen, H. M. (1924). *Culture and democracy in the United States*. Boni and Liveright, New York.

Kallen, H. M. (1956/1916). *Cultural pluralism and the American idea: An essay in social philosophy*. University of Pennsylvania Press, Philadelphia.

Kaplan, H. S., and Lancaster, J. B. (1999). The evolutionary economics and psychology of the demographic transition to low fertility. In *Adaptation and human behavior: An anthropological perspective*, (ed. L. Cronk, N. Chagnon and W. Irons), pp. 283–322. Aldine de Gruyter, New York.

Karow, J. (2000). The "other" genomes. *Scientific American*, 283(1), 43.

Keegan, J. (1993). *A history of warfare*. Alfred A. Knopf, New York.

Keeley, L. (1996). *War before civilization: The myth of the peaceful savage*. Oxford University Press, New York.

Keith, A. (1931). *The place of prejudice in modern civilization*. Williams and Norgate, London.

Keith, A. (1968/1947). *A new theory of human evolution*. Philosophical Library, New York.

King, P. (1976). *Toleration*. George Allen & Unwin, London.

Knack, S. and Keefer, P. (1997). Does social capital have an economic payoff? A cross-country investigation. *Quarterly Journal of Economics*, 112(4), 1251–88.

Koch, K.-F. (1974). *War and peace in Jalémó. The management of conflict in Highland New Guinea*. Harvard University Press, Cambridge, MA.

Konner, M. and Shostak, M. (1986). Ethnographic romanticism and the idea of human nature: Parallels between Samoa and !Kung San. In *The past and future of! Kung ethnography*, (ed. M. Biesele, R. Gordon and R. Lee), pp. 69–76. Buske, Hamburg.

Kotkin, J. (1992). *Tribes: How race, religion and identity determine success in the new global economy*. Random House, New York.

Kurzweil, R. (1999). *The age of spiritual machines: When computers exceed human intelligence*. Viking, New York.

Landa, J. T. (1981). A theory of the ethnically homogeneous middleman group: An institutional alternative to contract law. *Journal of Legal Studies*, 10 (June), 349–62.

Lewin, T. (2001). Whose child is this? DNA testing is unsettling paternity law. *International Herald Tribune*, 12 March, 1, 3.

Lewontin, R. C. (1972). The apportionment of human diversity. In *Evolutionary biology*, Vol. 6. (ed. T. Dobzhansky, M. K. Hecht and W. C. Steere), pp. 381–98. Appleton-Century-Crofts, New York.

Lichter, S. R., Lichter, L. S., and Rothman, S. (1994). *Prime time: How TV portrays American culture*. Regnery, Washington, DC.

Light, I., and Karageorgis, S. (1994). The ethnic economy. In *The handbook of economic sociology*, (ed. N. J. Smelser and R. Swedberg), pp. 647–71. Princeton University Press, Princeton.

Lopreato, J., and Yu, M. (1988). Human fertility and fitness optimization. *Ethology and Sociobiology*, 9, 269–89.

Low, B. S. (2000). Sex, wealth, and fertility: Old rules, new environments. In *Adaptation and human behavior: An anthropological perspective*, (ed. L.

Cronk, N. Chagnon and W. Irons), pp. 323–44. Aldine de Gruyter, New York.

Lowery, B. S., Hardin, C. D. and Sinclair, S. (2001). Social influence effects on automatic racial prejudice. *Journal of Personality and Social Psychology*, 81(5), 842–55.

Lumsden, C. J. and Wilson, E. O. (1981). *Genes, mind, and culture*. Harvard University Press, Cambridge, MA.

Lynch, A. (1996). *Thought contagion. How belief spreads through society: The new science of memes*. Basic Books, New York.

Lynn, R. (1977). The intelligence of the Japanese. *Bulletin of the British Psychological Society*, 30, 69–72.

Lynn, R. (1987). The intelligence of the Mongoloids: A psychometric, evolutionary and neurological theory. *Personality and Individual Differences*, 8, 813–44.

Lynn, R. (1994). Some reinterpretation of the Minnesota transracial adoption study. *Intelligence*, 19, 21–8.

Lynn, R. (1996). *Dysgenics. Genetic deterioration in modern populations*. Praeger Publishers, Westport, CT.

Lynn, R. (2001). *Eugenics: A reassessment*. Praeger, Westport, CT.

Lynn, R. and Vanhanen, T. (2002). *IQ and the wealth of nations*. Praeger, Westport, CT.

MacDonald, K. (1991). A perspective on Darwinian psychology: The importance of domain-general mechanisms, plasticity, and individual differences. *Ethology and Sociobiology*, 12, 449–80.

MacDonald, K. (1994). *A people that shall dwell alone: Judaism as a group evolutionary strategy*. Praeger, Westport, CT.

MacDonald, K. (1995). The establishment and maintenance of socially imposed monogamy in Western Europe [with peer commentary]. *Politics and the Life Sciences*, 14(1), 3–46.

MacDonald, K. (1998). *The culture of critique: An evolutionary analysis of Jewish involvement in twentieth-century intellectual and political movements*. Praeger, Westport, Conn.

Mann, M. (2001). Prodi urges fundamental debate on future of EU. *Financial Times*, London, 14 February, 1.

Marano, L. (2002). Researcher says delaying childbirth risks DNA 'extinction'. United Press International, Washington, DC, 25 May. http://nandotimes.com/healthscience/v-text/story/413841p-3296673c.html).

Marshall, G., ed. (1994). *The concise Oxford dictionary of sociology*. Oxford University Press. Oxford.

Maser, W., ed. (1976/1973). *Hitler's letters and notes*. (Translated by Arnold Pomerans edn). Bantam. New York.

Maslow, A. (1954). *Motivation and personality*. Harper and Row, New York.

Masters, R. D. (1975). Politics as a biological phenomenon. *Social Science Information*, 14, 7–63.

Masters, R. D. (1989). *The nature of politics*. Yale University Press, New Haven.

Masters, R. D. (1998/2002). On the evolution of political communities: The paradox of Eastern and Western Europe in the 1980s. In *Ethnic conflict and indoctrination: Altruism and identity in evolutionary perspective*, (ed. I. Eibl-Eibesfeldt and F. K. Salter), pp. 453–78. Berghahn, Oxford and New York.

Masters, R. D. (in press 2003). Why welfare states rise–and fall: Ethnicity, belief systems, and environmental influences on the support for public goods. In *Welfare, ethnicity, and altruism. New data and evolutionary theory*, (ed. F. K. Salter), Frank Cass, London.

Masters, W., and McMillan, M. (in press 2003). Ethnolinguistic diversity, government expenditure, and economic growth across countries. In *Welfare, ethnicity, and altruism. New data and evolutionary theory*, (ed. F. K. Salter), Frank Cass, London.

Mattei, R. D. (2001). The EU Charter of Fundamental Rights: A new totalitarianism. *Chronicles*, 25(10), 24–25.

Mauro, P. (1995). Corruption and growth. *Quarterly Journal of Economics*, 110(3), 681–712.

Mauss, M. (1968). *Die Gabe. Form und Funktion des Austausches in archaischen Gesellschaften [The gift. Form and function in archaic societies]*. Surhkamp, Frankfurt/M.

Maynard Smith, J. (1964). Group selection and kin selection: A rejoinder. *Nature*, 201, 1145–7.

Mayr, E., Shermer, M., and Sulloway, F. J. (2000). The grand old man of evolution. An interview with evolutionary biologist Ernst Mayr. *Skeptic*, 8(1), 76–82.

McGowan, W. (2001). *Coloring the news: How crusading for diversity has corrupted American journalism*. Encounter Books, San Francisco.

McNeill, W. H. (1979). *The human condition: An ecological and historical view*. Princeton University Press, Princeton, NJ.

McNeill, W. H. (1984). Human migration in historical perspective. *Population and Development Review*, 10, 1–18.

Menzel, P. and D'Aluisio, F. (2000). *Robo sapiens. Evolution of a new species*. MIT Press, Cambridge, MA.

Michod, R. E., and Hamilton, W. D. (2001/1980). Coefficients of relatedness in sociobiology (orig. in *Nature* 288, pp. 694–7). In *Narrow roads of gene land*.

*Volume 2: Evolution of sex*, (ed. W. D. Hamilton), pp. 108–15. Oxford University Press, New York.

Miele, F. (2002). *The battlegrounds of bio-science*. 1st Books Library, Bloomington, IN. (http.//www.1stbooks.com).

Mill, J. S. (1863). *Utilitarianism*. Parker and Bourn, London.

Mill, J. S. (1951). Representative government. In *Utilitarianism, liberty, and representative government*, (ed. J. S. Mill), Dutton, New York.

Mill, J. S. (1960). Chapter XVI: On nationality. In *Representative government. Three essays by John Stuart Mill*, (ed. J. S. Mill), pp. 380–88. Oxford University Press, London.

Millbank, A. (2001). Australia and the 1951 refugee convention. *People and Place*, 9(2), 1–13.

Miller, D. (1993). In defence of nationality. *Journal of Applied Philosophy*, 10(1), 3–16.

Miller, D. (1994). The nation-state: A modest defence. In *Political restructuring in Europe: Ethical perspectives*, (ed. C. Brown), pp. 137–62. Routledge, London.

Miller, D. (1995). *On nationality*. Oxford University Press, Oxford.

Miller, V., and Ware, R. (1999). The resignation of the European Commission. House of Commons Library. Research Paper 99/32.

Millett, K. (1971). *Sexual politics*. Rupert Hart-Davis, London.

Montagu, A. (1997/1942). *Man's most dangerous myth: The fallacy of race*. (6th edition edn). Sage, London.

Moore, G. E. (1903). *Principia Ethica*. Cambridge University Press, Cambridge.

Morton, N. E., Kenett, R., Yee, S., and Lew, R. (1982). Bioassay of kinship in populations of Middle Eastern origin and controls. *Current Anthropology*, 23(2), 157–67.

Motulsky, A. G. (1995). Jewish diseases and origins. *Nature Genetics*, 9(February), 99–101.

Mugny, G. and Pérez, J. A. (1991). *The social psychology of minority influence*. (trans. V. W. Lamongie). Cambridge University Press, New York.

Nairn, T. (1977). *The break-up of Britain*, New Left Books, London.

Neumann, J. v., and Morgenstern, O. (1955). *The theory of games and economic behavior*. Princeton University Press, Princeton, NJ.

North, D. (1990). *Institutions, institutional change, and economic performance*. Cambridge University Press, New York.

Nutini, H. G. (1997). Class and ethnicity in Mexico: Somatic and racial considerations. *Ethnology*, 36(3), 227–38.

Olson, M. (1965). *The logic of collective action: Public goods and the theory of groups*. Harvard University Press, Cambridge, MA.

Ostrom, E. (1990). *Governing the commons: the evolution of institutions for collective action.* Cambridge University Press, Cambridge.

Parsons, J. (1998). *Human population competition: The pursuit of power through numbers.* Edwin Mellen Press, Lewiston, NY.

Paul, M. (1998). Jewish-Polish relations in Soviet-occupied Eastern Poland, 1939–1941. In *The story of two Shtetls Bransk and Ejszyszki: An overview of Polish-Jewish relations in Northeastern Poland during World War II*, Vol. 2. pp. 173–230, 243–44. The Polish Educational Foundation in North America, Toronto and Chicago.

Pepper, J. W. (2000). Relatedness in trait group models of social evolution. *Journal of Theoretical Biology*, 206, 355–68.

Pérez, J. A., and Mugny, G. (1990). Minority influence, manifest discrimination and latent influence. In *Social identity theory. Constructive and critical advances*, (ed. D. Abrams and M. A. Hogg), pp. 152–68. Springer-Verlag, New York.

Phillips, K. (1999). *The cousins' wars: Religion, politics, and the triumph of Anglo-America.* Basic Books, New York.

Phillips, K. (2002). *Wealth and democracy: A political history of the American rich.* Broadway Books, New York.

Pinker, S. (1997). *How the mind works.* W.W Norton & Company Ltd., New York.

Pinker, S. (1998). *In conversation with Steven Pinker.* 2001. *The evolutionist.* Internet web site: http://www.lse.ac.uk/Depts/cpnss/darwin/evo/pinker.htm. Accessed 14.8.2001.

Plomin, R., and Petrill, S. (1997). From genetics and intelligence: What's new? *Intelligence*, 24(1), 53–77.

Price, C. A. (2000). Australians all: Who on earth are we? 19 pp. Roneo.

Proctor, R. (1999). *The Nazi war on cancer.* Princeton University Press, Princeton, NJ.

Rabbie, J. M. (1992). The effects of intragroup cooperation and intergroup compentititon on in-group cohesion and out-group hostility. In *Coalitions and alliances in humans and other animals*, (ed. A. H. Harcourt and F. B. M. d. Waal), pp. 175–205. Oxford University Press, Oxford.

Rajagopal, B. (2002). The UN casts a blind eye to justice. *International Herald Tribune*, Frankfurt, 30–31 March, 6.

Ramet, S. (1992). *Nationalism and federalism in Yugoslavia.* (2nd edn). University of Indiana Press, Bloomington, IN.

Raphael, D. D. (1981/1994). *Moral philosophy.* (2nd edn). Oxford University Press, Oxford.

Rawls, J. (1971). *A theory of justice.* Harvard University Press, Cambridge, MA.

Rawls, J. (1999). *The law of peoples*. Harvard University Press, Cambridge MA.

Raywid, M. A. (1980). The discovery and rejection of indoctrination. *Educational Theory*, 30(1), 1–10.

Reeve, H. K. (2000). Multi-level selection and human cooperation: Review of *Unto Others: The Evolution and Psychology of Unselfish Behavior*, by Elliott Sober and David Sloan Wilson, Harvard University Press, 1998, 394 pp. ISBN: 0674930460. *Evolution and Social Behavior*, 21(1), 65–72.

Richards, M., *et al.* (1996). Paleolithic and Neolithic lineages in the European mitochondrial gene pool. *American Journal of Human Genetics*, 59, 185–203.

Richerson, P. J., and Boyd, R. (1998/2002). The evolution of human ultrasociality. In *Ethnic conflict and indoctrination: Altruism and identity in evolutionary perspective*, (ed. I. Eibl-Eibesfeldt and F. K. Salter), pp. 71–95. Berghahn, Oxford.

Richerson, P. J., and Boyd, R. (2002). The biology of commitment to groups: A tribal instincts hypothesis. In *The biology of commitment*, (ed. R. M. Nesse), pp. 186–220. Russell Sage Foundation, New York.

Richerson, P. J., Borgerhoff Mulder, M., and Vila, B. J. (1996). *Principles of human ecology*. Simon & Schuster Custom Publishing, Needham Heights, MA.

Rikowski, A., and Grammer, K. (1999). Human body odour, symmetry and attractiveness. *Proceedings of the Royal Society of London B*, 266, 869–74.

Rimor, M., and Tobin, G. A. (1990). Jewish giving patterns to Jewish and non-Jewish philanthropy. In *Faith and philanthropy in America: Exploring the role of religion in America's voluntary sector*, (ed. R. Wuthnow, V. A. Hodgkinson and associates), pp. 134–64. Jossey-Bass, San Francisco.

Rodrik, D. (1999). Where did all the growth go? External shocks, social conflict, and growth collapses. *Journal of Economic Growth*, 4(4), 385–412.

Ross, E. A. (1914). *The old world in the new: The significance of past and present immigration to the American people*. The Century Company, New York.

Rubin, P. H. (2000). Does ethnic conflict pay? *Politics and the Life Sciences*, 19(1), 59–68.

Rudman, L. A., Ashmore, R. D., and Gary, M. L. (2001). "Unlearning" automatic biases: the malleability of implicit prejudice and stereotypes. *Journal of Personality and Social Psychology*, 81(5), 856–68.

Rummel, R. J. (1996). *Lethal politics. Soviet genocide and mass murder since 1917*. Transaction, New Brunswick, NJ.

Rummel, R. J. (1997). *Death by government*. Transaction, New Brunswick, NJ.

Rushton, J. P. (1989a). Genetic similarity in male friends. *Ethology and Sociobiology*, 10, 361–73.

Rushton, J. P. (1989b). Genetic similarity, human altruism, and group selection. *Behavioral and Brain Sciences*, 12, 503–559.

Rushton, J. P. (1994). The equalitarian dogma revisited. *Intelligence*, 19, 263–80.

Rushton, J. P. (1995). *Race, evolution, and behavior*. Transaction Publishers, New Brunswick, NJ.

Rushton, J. P. (1997). Race, intelligence, and the brain: The errors and omissions of the 'revised' edition of S. J. Gould's *The Mismeasure of Man* (1996). *Personality and Individual Differences*, 23(1), 169–80.

Rushton, J. P., Russell, R. J. H., and Wells, H. G. (1984). Genetic similarity theory: Beyond kin selection. *Behavior Genetics*, 14, 179–93.

Russell, R. J. H., and Rushton, J. P. (1985). Evidence for genetic similarity detection in human marriage. *Ethology and Sociobiology*, 6, 183–7.

Sachdev, I., and Bourhis, R. Y. (1990). Language and social identification. In *Social identity theory. Constructive and critical advances*, (ed. D. Abrams and M. A. Hogg), pp. 211–29. Springer-Verlag, New York.

Sack, R. D. (1986). *Human territoriality: Its theory and history*. Cambridge University Press, New York.

Sahlins, M. D. (1965). On the sociology of primitive exchange. In *The relevance of models for social anthropology*, (ed. M. Banton), pp. 139–236. Tavistock, London.

Sahlins, M. (1976/1977). *The use and abuse of biology. An anthropological critique of sociobiology*. Tavistock, London.

Salter, F. K. (1995). *Emotions in command. A naturalistic study of institutional dominance*. Oxford University Press Science Publications, Oxford.

Salter, F. K. (1998/2002). Indoctrination as institutionalised persuasion: Its limited variability and cross-cultural evolution. In *Ethnic conflict and indoctrination: Altruism and identity in evolutionary perspective*, (ed. I. Eibl-Eibesfeldt and F. K. Salter), pp. 421–52. Berghahn, Oxford and New York.

Salter, F. K. (2001). A defense and an extension of Pierre van den Berghe's theory of ethnic nepotism. In *Evolutionary theory and ethnic conflict*, (ed. P. James and D. Goetze), pp. 39–70. Praeger, Westport, CT.

Salter, F. K. (2002). Ethnic nepotism as a two-edged sword: The risk-mitigating role of ethnicity among mafiosi, nationalist fighters, middlemen, and dissidents. In *Risky transactions. Kinship, ethnicity, and trust*, (ed. F. K. Salter), pp. 243–89. Berghahn, Oxford and New York.

Salter, F. K. (in press-a 2003). Charity releasers, moral emotions, and the welfare ethic: An application of Westermarck's *secundam naturam*. In *Evolution and the moral emotions–appreciating Edward Westermarck*, (ed. J.-P. Takala and A. Wolf).

Salter, F. K. ed. (in press-b 2003). *Welfare, ethnicity, and altruism. New data and evolutionary theory*. Frank Cass, London.

Salter, F. K. (in press-c 2003). Ethnic diversity, foreign aid, economic growth, social stability, and population policy: A perspective on W. Masters and M. McMillan's findings. In *Welfare, ethnicity, and altruism. New data and evolutionary theory*, (ed. F. K. Salter), Frank Cass, London.

Salter, F. K. (in press-d 2003). The evolutionary deficit in mainstream political theory of welfare and ethnicity. In *Welfare, ethnicity, and altruism. New data and evolutionary theory*, (ed. F. K. Salter), Frank Cass, London.

Salter, F. K., Schiefenhövel, W. and Burenhult, G. (1994). The future of humankind. In *American Museum of Natural History's illustrated history of humankind*, (ed. G. Burenhult), pp. 213–226. HarperCollins/Smithsonian Museum of Natural History, San Francisco.

Sanderson, S. K. (1999). *Social transformations: A general theory of historical development*. (Expanded edn). Rowman and Littlefield, Lanham.

Sanderson, S. K. (2001). *The evolution of human sociality: A Darwinian conflict perspective*. Rowman and Littlefield, Lanham.

Sanderson, S. K. and Dubrow, J. (2000). Fertility decline in the modern world and in the original demographic transition: Testing three theories with cross-cultural data. *Population and Environment: A Journal of Interdisciplinary Studies*, 21(6), 511–37.

Sanderson, S. K. (in press 2003). Welfare spending and ethnic heterogeneity in cross-national perspective. In *Welfare, ethnicity, and altruism. New data and evolutionary theory*, (ed. F. K. Salter), Frank Cass, London.

Sanderson, S. K., and Vanhanen, T. (in press 2003). Reconciling the differences between Sanderson's and Vanhanen's results. In *Welfare, ethnicity, and altruism. New data and evolutionary theory*, (ed. F. K. Salter), Frank Cass, London.

Sassen, S. (1996). *Losing control? Sovereignty in an age of globalization*. Columbia University Press, New York.

Satel, S. (2002). I am a racially profiling doctor. *New York Times Magazine*. 5 May.

Scarr, S. (1998). On Arthur Jensen's integrity. *Intelligence*, 26(3), 227–32.

Scarr, S. and Weinberg, R. A. (1976). IQ test performance of black schoolchildren adopted by white families. *American Psychologist*, 31, 726–39.

Schelling, T. (1978). *Micromotives and macrobehavior*. W. W. Norton, New York.

Schiefenhövel, W. and Salter, F. K. (1994). The evolution of races, populations, and cultures. In *American Museum of Natural History's illustrated history of humankind*, Vol. 5. (ed. G. Burenhult), pp. 17–27. HarperCollins/Smithsonian Museum of Natural History, San Francisco.

Schiefenhövel, W. and Schiefenhövel-Barthel, S. (1999). Geburt. In *Brockhaus Mensch, Natur, Technik. Phänomen Mensch*, (ed. Brockhaus-Redaktion), pp. 41–51. R. A. Brockhaus, Leipzig u. Mannheim. [Birth]

Science for the People Editorial Collective (1977). *Biology as a social weapon.* Burgess, Minneapolis.

Schubert, G., and Masters, R. D., ed. (1991). *Primate politics*. Southern Illinois University Press. Carbondale and Edwardsville.

Schubert, J. (1983). Ethological methods for observing small group political decision making. *Politics and the Life Sciences*, 2(1), 3–41.

Schubert, J. and Tweed, M. (in press 2003). Ethnic diversity, population size and charitable giving at the local level in the United States. In *Welfare, ethnicity, and altruism. New data and evolutionary theory*, (ed. F. K. Salter), Frank Cass, London.

Scruton, R. (1990). In defence of the nation. In *The philosopher on Dover Beach*, (ed. R. Scruton), Carcanet, Manchester.

Searle, J. R. (1999). I married a computer. *New York Review of Books*, 46(6), 34–8.

Segal, N. L. (1999). *Entwined lives: Twins and what they tell us about human behavior*. Dutton, New York.

Segal, N. L., and Hershberger, S. L. (1999). Cooperation and competition between twins: Findings from a prisoner's dilemma game. *Evolution and Human Behavior*, 20, 29–51.

Segerstråle, U. (2000). *Defenders of truth: The battle for science in the sociobiology debate and beyond.* Oxford University Press, New York.

Semino, O., *et al.* (2000). The genetic legacy of Paleolithic Homo sapiens in extant Europeans: A Y-chromosome perspective. *Science*, 290, 1155–9.

Shari, M. (2002). Mahathir's change of heart? *Business Week*. 29 July 2002, 26–7.

Shaw, R. P., and Wong, Y. (1989). *Genetic seeds of warfare: evolution, nationalism, and patriotism*. Unwin Hyman, London.

Sheehan, P. (1998). *Among the barbarians: The dividing of Australia*. Random House Australia, Sydney.

Shepher, J. (1969). Familism and social structure: The case of the Kibbutz. *Journal of Marriage and the Family*, 31, 567–73.

Sherif, M. (1966). *In common predicament: The social psychology of intergroup conflict*. Houghton-Mifflin, Boston.

Sherman, P. W., Jarvis, J. U. M., and Alexander, R. D., ed. (1991). *The biology of the naked mole-rat*. Princeton University Press. Princeton, NJ.

Shipman, P. (1994). *The evolution of racism–human differences and the use and abuse of science.* Simon and Schuster, New York.

Silk, J. (1980). Adoption and kinship in Oceania. *American Journal of Sociology*, 80, 799–820.

Silverman, I., and Case, D. (1998/2002). Ethnocentrism vs. pragmatism in the conduct of human affairs. In *Ethnic conflict and indoctrination: Altruism and identity in evolutionary perspective*, (ed. I. Eibl-Eibesfeldt and F. K. Salter), pp. 389–406. Berghahn, Oxford.

Simon, H. A. (1965). *Administrative behavior: A study of decision-making processes in administrative organizations*. Free Press, New York.

Singer, P. (1975/1991). *Animal liberation: A new ethics for our treatment of animals*. New York Review/Random House, New York.

Singer, P. (1976). A utilitarian population principle. In *Ethics and population*, (ed. M. D. Bayles), pp. 81–99. Schenkman, Cambridge.

Singer, P. (1981). *The expanding circle*. Farrar, Straus & Giroux, New York.

Singer, P. (1998). Darwin for the left. *Prospect*, June, 26–30.

Singer, P. (2000). *A Darwinian left: Politics, evolution and cooperation*. Yale University Press, New Haven.

Skinner, B. F. (1948). *Walden two*. Macmillan, New York.

Skinner, B. F. (1971). *Beyond freedom and dignity*. Knopf, New York.

Skinner, B. F. (1984). The phylogeny and ontogeny of behavior [with peer commentary]. *Behavioral and Brain Sciences*, 7, 669–711.

Skinner, Q. (1998). *Liberty before liberalism*. Cambridge University Press, Cambridge.

Smart, J. J. C. (1967). Utilitarianism. In *The encyclopedia of philosophy*, Vol. 7. (ed. P. Edwards), pp. 206–212. Macmillan and Free Press, New York.

Smith, A. D. (1986). *The ethnic origins of nations*. Basil Blackwell, Oxford.

Smith, J. P., and Edmonston, B., ed. (1997). *The new Americans: Economic, demographic, and fiscal effects of immigration*. National Academy Press. Washington, DC.

Smuts, B. (1999). Multilevel selection, cooperation, and altruism. Reflections on *Unto others: The evolution and psychology of unselfish behavior. Human Nature: An Interdisciplinary Biosocial Perspective*, 10(3), 311–27.

Snyderman, M. and Rothman, S. (1988). *The IQ controversy: The media and public policy*. Transaction Books, New York.

Sober, E., and Wilson, D. S. (1998). *Unto others: The evolution and psychology of unselfish behavior*. Harvard University Press, Cambridge, MA.

Sociobiology Study Group of Science for the People (1978/1976). In *The sociobiology debate: Readings on the ethical and scientific issues concerning sociobiology*, (ed. A. L. Caplan), pp. 280–90. Harper and Row, New York.

Soltis, J., Boyd, R., and Richerson, P. J. (1995). Can group-functional behaviors evolve by cultural group selection? An empirical test [with peer commentary]. *Current Anthropology*, 36(3), 473–94.

Somit, A., ed. (1976). *Biology and politics*. Mouton. The Hague.

Spicer, E. H. (1971). Persistent cultural systems. *Science*, 174(November), 795–800.

Spiro, M. E. (1995). *Gender and culture: Kibbutz women revisited.* (2nd edn). Transaction, New Brunswick, NJ.

Stevenson, C. L. (1944). *Ethics and language*. Yale University Press, New Haven.

Stove, D. C. (1982). *Popper and after: Four modern irrationalists*. Pergamon, Oxford.

Svonkin, S. (1997). *Jews against prejudice: American Jews and the fight for civil liberties*. Columbia University Press, New York.

Sykes, B. (2001). *The seven daughters of Eve: The science that reveals our genetic ancestry*. W. W. Norton & Company, New York.

Tajfel, H. (1981). *Human groups and social categories: Studies in social psychology*. Cambridge University Press, Cambridge.

Tajfel, H., and Turner, J. C. (1986). The social identity theory of intergroup behavior. In *Psychology of intergroup relations*, (ed. S. Worchel and W. A. Austin), pp. 7–24. Nelson-Hall, Chicago.

Taylor, A. J. P. (1967/1955). *Bismarck: The man and the statesman*. Vintage, New York.

Teitelbaum, M., and Winter, J. (1998). *A question of numbers. High migration, low fertility, and the politics of national identity*. Hill & Wang, New York.

Thiessen, D. (1997). Meme over matter: Descartes *Redux*. *Mankind Quarterly*, 37(4), 415–35.

Thiessen, D., and Gregg, B. (1980). Human assortative mating and genetic equilibrium: An evolutionary perspective. *Ethology and Sociobiology*, 1111–40.

Thomas, M. G., *et al.* (2002). Founding mothers of Jewish communities: Geographically separated Jewish groups were independently founded by very few female ancestors. *American Journal of Human Genetics*, 70, 1411–20.

Thorne, A. G., and Wolpoff, M. H. (1992). The multiregional evolution of humans. *Scientific American*, 266(4), 28–33.

Thornhill, R., and Gangestad, S. W. (1993). Human facial beauty. Averageness, symmetry, and parasite resistance. *Human Nature*, 4(3), 237–69.

Tibi, B. (2002). *Europa ohne Identität? Die Krise der multikulturellen Gesellschaft*. Siedler, Frankfurt am Main.

Tiger, L. (1999). *The decline of males: The first look at an unexpected new world for men and women*. Golden Books, New York.

Tishkov, V. (1997). *Ethnicity, nationalism and conflict in and after the Soviet Union. The mind aflame*. Sage, London.
Tooby, J., and Cosmides, L. (1989). Kin selection, genic selection, and information-dependent strategies. *Behavioral and Brain Sciences*, 12(3), 542–4.
Tooby, J. and Cosmides, L. (1990). The past explains the present. Emotional adaptations and the structure of ancestral environments. *Ethology and Sociobiology*, 11, 375–424.
Triandis, H. C. (1990). Cross-cultural studies of individualism and collectivism. In *Nebraska Symposium on Motivation 1989: Cross Cultural Perspectives*, University of Nebraska Press, Lincoln.
Trivers, R. L. (1971). The evolution of reciprocal altruism. *Quarterly Review of Biology*, 14, 35–57.
Trivers, R. L. (1974). Parent-offspring conflict. *American Zoologist*, 14, 249–64.
Tullberg, J., and Tullberg, B. S. (1997). Separation or unity? A model for solving ethnic conflicts. *Politics and the Life Sciences*, 16(2), 237–48.
Turing, A. (1959). Computing machinery and intelligence. *Mind*, 59, 434–69.
van Creveld, M. (1999). *The rise and decline of the state*. Cambridge University Press, Cambridge.
van den Berghe, P. L. (1979). *Human family systems. An evolutionary view*. Elsevier, New York.
van den Berghe, P. L. (1981). *The ethnic phenomenon*. Elsevier, New York.
van den Berghe, P. L. (in press 2003). Affirmative action: Toward a sociobiologically informed social policy. In *Welfare, ethnicity, and altruism. New data and evolutionary theory*, (ed. F. K. Salter), Frank Cass, London.
van der Dennen, J. M. G. (1995). *The origin of war. The evolution of a male-coalitional reproductive strategy*. Vol. 1. Origin, Groningen, Netherlands.
Vanhanen, T. (1991). *Politics of ethnic nepotism: India as an example*. Sterling, New Delhi.
Vanhanen, T. (2000). The wealth and poverty of nations related to IQ. Paper delivered at the XVIIIth World Congress of the International Political Science Association, Quebec, 1–5 August.
Vanhanen, T. (in press 2003). An exploratory comparative study of the relationship between ethnic heterogeneity and welfare politics. In *Welfare, ethnicity, and altruism. New data and evolutionary theory*, (ed. F. K. Salter), Frank Cass, London.
Vasquez, J. A. (1993). *The war puzzle*. Cambridge University Press, Cambridge.
Vining, D. R. (1986). Social versus reproductive success: The central theoretical problem of human sociobiology. *The Behavioral and Brain Sciences*, 9, 167–216.

Voland, E. (1990). Differential reproductive success within the Krummhorn populations (Germany, 18th and 19th Centuries). *Behavioral Ecology and Sociobiology*, 26, 65–72.

Waldman, B. (1989). Sociobiology, sociology, and pseudoevolutionary reasoning. *Behavioral and Brain Sciences*, 12(3), 547–8.

Wallerstein, I. (1979). *The capitalist world economy*. Cambridge University Press, New York.

Walzer, M. (1974/1992). Civility and civic virtue in contemporary America. In *What it means to be an American*, (ed. M. Walzer), pp. 81–101. Marsilio, New York.

Walzer, M. (1983). *Spheres of justice*. Basic Books, New York.

Walzer, M. (1990/1992). What does it means to be an 'American'? In *What it means to be an American*, (ed. M. Walzer), pp. 23–49. Marsilio, New York.

Walzer, M. (1994a). Notes on the new tribalism. In *Political restructuring in Europe*, (ed. C. Brown), pp. 187–200. Routledge, London.

Walzer, M. (1994b). Toward a new realization of Jewishness. *Congress Monthly*, 61(4), 3–6.

Walzer, M., ed. (1992). *What it means to be an 'American'*. Marsilio. New York.

Walzer, M., ed. (1995). *Toward a global civil society*. International political currents. Berghahn. Oxford.

Washburn, S. L. (1976). Sociobiology. *Anthropology Newsletter*, 18(3), 3.

Watson, J. D. (2000). *A passion for DNA: Genes, genomes, and society*. Cold Spring Harbor Laboratory, Cold Spring Harbor, NY.

Weber, M. (1958). *The Protestant ethic and the spirit of capitalism*. Scribner, New York.

Weinberg, R. A., Scarr, S., and Waldman, I. D. (1992). The Minnesota transracial adoption study: A follow-up of IQ test performance at adolescence. *Intelligence*, 16, 117–35.

Weindling, P. (1989). *Health, race and German politics between national unification and Nazism, 1870–1945*. Cambridge University Press, Cambridge.

Westermarck, E. A. (1912/1971). *The origin and development of the moral ideas*. Johnson Reprint Corporation/Macmillan and Co., New York.

White, L. A. (1966). The social organization of ethnological theory. Monograph in cultural anthropology. *Rice University Studies*, 52(4), 1–66.

Wiederman, M. W., and Allgeier, E. R. (1993). Gender differences in sexual jealousy: Adaptationist or social learning explanation? *Ethology and Sociobiology*, 14, 115–40.

Wiessner, P. (1984). Reconsidering the behavioral basis for style: A case study among the Kalahari San. *Journal of Anthropological Archaeology*, 3, 190–234.

Wiessner, P. (1998/2002). Indoctrinability and the evolution of socially defined kinship. In *Ethnic conflict and indoctrination: Altruism and identity in evolutionary perspective*, (ed. I. Eibl-Eibesfeldt and F. K. Salter), pp. 133–50. Berghahn, Oxford.

Wiessner, P. (2002a). The vines of complexity: Egalitarian structures and the institutionalization of inequality among the Enga. *Current Anthropology*, 43(2), 233–69.

Wiessner, P. (2002b). Taking the risk out of risky transactions: A forager's dilemma. In *Risky transactions. Kinship, ethnicity, and trust*, (ed. F. K. Salter), pp. 21–43. Berghahn, Oxford/New York.

Wiessner, P. and Schiefenhövel, W., ed. (1996). *Food and the status quest: An interdisciplinary perspective*. Berghahn Books. Providence & Oxford.

Wiessner, P. and Tumu, A. (1998). *Historical vines: Enga networks of exchange, ritual, and warfare in Papua New Guinea*. (Translations by Nitze Pupu). Smithsonian Institute Press, Washington, DC.

Williams, G. C. (1966). *Adaptation and natural selection: A critique of some current evolutionary thought*. Princeton University Press, Princeton.

Williams, G. C. (1988). Huxley's evolution and ethics in sociobiological perspective. *Zygon: Journal of Religion and Science*, 23, 383–407.

Williams, G. C. (1997). *The pony fish's glow: And other clues to plan and purpose in nature*. Basic Books, New York.

Williamson, O. E., ed. (1990). *Organization theory. From Chester Barnard to the present and beyond*. Oxford University Press USA, New York.

Williamson, O. E. (1996). *The mechanisms of governance*. Oxford University Press USA, New York.

Wilson, A. C., and Cann, R. C. (1992). The recent African genesis of humans. *Scientific American*, 266(4), 22–27.

Wilson, D. S. (1992). On the relationship between evolutionary and psychological definitions of altruism and selfishness. *Biology and Philosophy*, 7, 61–8.

Wilson, D. S. (2002). *Darwin's cathedral: The organismic nature of religion*. University of Chicago Press, Chicago.

Wilson, D. S., and Sober, E. (1989). Reviving the superorganism. *Journal of Theoretical Biology*, 136, 337–56.

Wilson, E. O. (1971). Competitive and aggressive behavior. In *Man and beast: Comparative social behavior*, (ed. J. F. Eisenberg and W. S. Dillon), pp. 183–217. Smithsonian Institution Press, Washington DC.

Wilson, E. O. (1975). *Sociobiology: The new synthesis*. Harvard University Press, Cambridge, MA.

Wilson, E. O. (1978a). *On human nature*. Harvard University Press, Cambridge.

Wilson, E. O. (1978b). Academic vigilantism and the political significance of sociobiology. In *The sociobiology debate: Readings on the ethical and scientific issues concerning sociobiology*, (ed. A. L. Caplan), pp. 291–303. Harper and Row, New York.

Wilson, E. O. (1984). *Biophilia*. Harvard University Press, Cambridge, MA.

Wilson, E. O. (1998). *Consilience: The unity of knowledge*. Alfred A. Knopf, New York.

Wilson, E. O. (1999). *The diversity of life*. W. W. Norton, New York.

Wilson, E. O. (2002). The bottleneck [excerpted from *The future of life*, Knopf, New York]. *Scientific American*, 286(2), 70–79.

Wilson, J. Q. (1993). *The moral sense*. Free Press, New York.

Wolf, J. B., Brodie, E. D., and Wade, M. J., eds. (2000). *Epistatis and the evolutionary process*. Oxford University Press. New York.

Wolpoff, M. H. and Caspari, R. (1997). *Race and human evolution*. Simon and Schuster, New York.

Wolpoff, M. H., Hawks, J., Frayer, D. W., and Hunley, K. (2001). Modern human ancestry at the peripheries: A test of the replacement theory. *Science*, 291.

Wood, G. S. (2001). Tocqueville's lesson. *New York Review of Books*, 48(8), 46–52.

Wrangham, R., and Peterson, D. (1996). *Demonic males: Apes and the origins of human violence*. Houghton Mifflin, Boston.

Wright, S. (1943). Isolation by distance. *Genetics*, 28, 114–38.

Wright, S. (1951). The genetic structure of populations. *Eugenics Review*, 15, 323–54.

Wynne-Edwards, V. C. (1962). *Animal dispersion in relation to social behaviour*. Oliver and Boyd, Edinburgh.

Zenner, W. (1991/1996). Middlemen minorities [excerpted from *Minorities in the middle*, pp. 7–9, 11–17]. In *Ethnicity*, (ed. J. Hutchinson and A. D. Smith), pp. 179–86. Oxford University Press, Oxford & New York.

Zenner, W. P. (1982). Arabic-speaking immigrants in North America as middleman minorities. *Ethnic and Racial Studies*, 5(4), 457–77.

# Index

References to illustrations are in italic print, while note references are denoted by page number, then 'n', followed by note number.